ANNUAL REVIEW OF
COLD ATOMS
AND MOLECULES

Volume 3

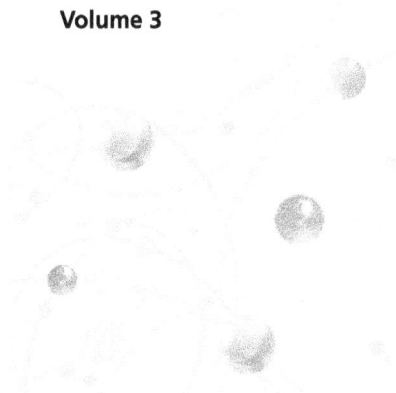

Annual Review of Cold Atoms and Molecules

ISSN: 2315-4926

Series Editors: Kirk W. Madison (*University of British Columbia, Canada*)
Kai Bongs (*University of Birmingham, UK*)
Lincoln D. Carr (*Colorado School of Mines, USA*)
Ana Maria Rey (*JILA, University of Colorado, USA*)
Hui Zhai (*Tsinghua University, China*)

Honorary Advisors: Claude Cohen-Tannoudji
(*Collège de France & Laboratoire Kastler Brossel, France*)
Yiqiu Wang (*Peking University, China*)

The aim of this book is to present review articles describing the latest theoretical and experimental developments in the field of cold atoms and molecules. Our hope is that this series will promote research by both highlighting recent breakthroughs and by outlining some of the most promising research directions in the field.

ANNUAL REVIEW OF COLD ATOMS AND MOLECULES

Volume 3

Honorary Advisors

Claude Cohen-Tannoudji
(Collège de France & Laboratoire Kastler Brossel, France)

Yiqiu Wang *(Peking University, China)*

Editors

Kirk W. Madison *(University of British Columbia, Canada)*
Kai Bongs *(University of Birmingham, UK)*
Lincoln D. Carr *(Colorado School of Mines, USA)*
Ana Maria Rey *(JILA, University of Colorado, USA)*
Hui Zhai *(Tsinghua University, China)*

World Scientific

NEW JERSEY · LONDON · SINGAPORE · BEIJING · SHANGHAI · HONG KONG · TAIPEI · CHENNAI

Published by

World Scientific Publishing Co. Pte. Ltd.

5 Toh Tuck Link, Singapore 596224

USA office: 27 Warren Street, Suite 401-402, Hackensack, NJ 07601

UK office: 57 Shelton Street, Covent Garden, London WC2H 9HE

British Library Cataloguing-in-Publication Data

A catalogue record for this book is available from the British Library.

Annual Review of Cold Atoms and Molecules — Vol. 3
ANNUAL REVIEW OF COLD ATOMS AND MOLECULES
Volume 3

ISBN 978-981-4667-73-9

Typeset by Stallion Press
Email: enquiries@stallionpress.com

CONTENTS

2. Few-Body Physics of Ultracold Atoms and Molecules
with Long-Range Interactions   77

Yujun Wang, Paul Julienne and Chris H Greene

CHAPTER 1

STRONGLY INTERACTING TWO-DIMENSIONAL FERMI GASES

Jesper Levinsen* and Meera M. Parish[†]

*Aarhus Institute of Advanced Studies, Aarhus University,
DK-8000 Aarhus C, Denmark
jfle@aias.au.dk

[†]London Centre for Nanotechnology, University College London,
Gordon Street, London, WC1H 0AH, United Kingdom
meera.parish@ucl.ac.uk

We review the current understanding of the uniform two-dimensional (2D) Fermi gas with short-range interactions. We first outline the basics of two-body scattering in 2D, including a discussion of how such a 2D system may be realized in practice using an anisotropic confining potential. We then discuss the thermodynamic and dynamical properties of 2D Fermi gases, which cold-atom experiments have only just begun to explore. Of particular interest are the different pairing regimes as the interparticle attraction is varied; the superfluid transition and associated finite-temperature phenomenology; few-body properties and their impact on the many-body system; the "Fermi polaron" problem; and the symmetries underlying the collective modes. Where possible, we include the contributions from 2D experiment. An underlying theme throughout is the effect of the quasi-2D geometry, which we view as an added richness to the problem rather than an unwanted complication.

1. Introduction

Following the successful realisation of strongly interacting atomic Fermi gases in three dimensions (3D), attention has now turned to Fermi systems that have, in principle, even stronger correlations, such as low-dimensional gases and fermions with long-range dipolar interactions. Model two-dimensional (2D) systems are of particular interest, since they may provide insight into technologically important, but complex, solid-state systems such as the high-temperature superconductors,[1] semiconductor interfaces,[2] and layered organic superconductors.[3] Moreover, 2D gases pose fundamental questions in their own right, being in the so-called marginal dimension where particle scattering can be strongly energy dependent, and quantum fluctuations are large enough to destroy long-range order at any finite temperature.[4,5]

In this review, we focus on the uniform 2D Fermi gas with short-range interactions, since this has already been successfully realised experimentally.[6–17] Here, two species of alkali atom are confined to one or more layers using a 1D optical lattice or a highly anisotropic trap. The interspecies interactions may then be tuned using a Feshbach resonance, making cold atomic gases ideal for studying the behavior of fermions in low dimensions. While the cold-atom system is clearly much simpler than solid-state systems, where the long-range Coulomb interactions are difficult to treat and there are often complex crystal structures, the usual toy models for such systems neglect the long-range interactions and consider simple contact interactions like the ones described here in this review. In particular, the attractive 2D Fermi gas provides a basic model for understanding pairing and superconductivity in 2D.[18–20] Here, by varying the attraction, one can investigate the crossover from BCS-type pairing to the Bose regime of tightly bound dimers. In the interests of space, we do not consider further extensions such as dipolar interactions, spin-orbit coupling, or any lattice within the plane. Indeed, we note that a degenerate 2D dipolar Fermi gas has yet to be achieved experimentally, while the pursuit of the 2D Hubbard model is still ongoing.

The investigation of strongly interacting 2D Fermi gases, as described in the following, may be encompassed within several broad themes. Firstly, there is the interplay between Bose and Fermi behavior as the attraction is varied. This is particularly apparent at finite temperature where the normal state evolves from a Fermi to a Bose liquid, and one has the possibility of the so-called pseudogap regime. Potentially even richer behavior may be derived from Fermi-Fermi mixtures with unequal masses and/or imbalanced "spin" populations. While attempts to confine mass-imbalanced mixtures to

2D are still underway, experiments with equal masses have already realized the regime of extreme spin imbalance,[14] corresponding to a single impurity problem. Here, it has emerged that even the strongly interacting impurity can be well described by wave functions that only contain two- and three-body correlations. A related theme is the importance of few-body phenomena in the many-body system. As well as being relevant to high temperatures, where the thermodynamic properties are well described by the behavior of few-body clusters (i.e., the virial expansion), few-body properties are also required to properly describe the Bose regime of the pairing crossover. Turning to themes unique to the 2D system, we have the existence of classical scale invariance and its impact on the collective modes in a harmonic trap. Finally, there is the question of how 2D experiments really are, since in practice there is always a finite transverse "size" of the quasi-2D geometry. To be in the 2D limit, we require the length scales associated with the quasi-2D gas (i.e., the dimer size, the thermal wavelength, and the interparticle spacing) to be much larger than the confinement length. Ultimately, it would be interesting to understand how the gas evolves from 2D to 3D.

The review is organized as follows: Section 2 surveys the basic properties of two-body scattering in a two-dimensional geometry — since the literature offers multiple different definitions in the 2D scattering problem, this may be thought of as a reference section for the remainder of the review. We also present here an alternative formulation of the scattering problem in a quasi-2D geometry, and discuss the issue of confinement induced resonances. Section 3 focuses on recent advances in the understanding of few-body physics. We discuss elastic scattering properties, as well as the bound trimer and tetramer states that are predicted to occur in the heteronuclear Fermi gas, for a sufficiently large mass imbalance. Turning to the many-body physics in a 2D Fermi gas, Sec. 4 reviews the properties of the BCS-BEC crossover, including the mean-field approach and the equation of state at zero temperature. Section 5 considers the behavior of the gas at finite temperature, which includes an outline of the high-temperature virial expansion, a sketch of the phase diagram for superfluidity, and a discussion of the existence of the pseudogap. Section 6 discusses the recent experimental and theoretical advances in the 2D Fermi polaron problem, with both metastable states and the nature of the ground state being considered. In Sec. 7, dynamical quantities such as collective modes and spin diffusion are reviewed, as well as the breakdown of classical scale invariance in the interacting quantum system — the so-called quantum anomaly. Finally, Sec. 8 provides an outlook into future investigations of strongly interacting 2D Fermi gases.

2. Basics of the Two-Dimensional System

2.1. *General properties of scattering in two dimensions*

We now summarize several properties of two-body scattering in two dimensions that are relevant to the results presented in this review. In the following discussion, we mostly follow Refs. 21 and 22. The starting point is the 2D Schrödinger equation for two particles interacting via a short-range local potential $V(\mathbf{r})$ at energy E in the center-of-mass frame:

$$-\frac{\hbar^2 \nabla^2}{2m_r} \psi(\mathbf{r}) + V(\mathbf{r})\psi(\mathbf{r}) = E\psi(\mathbf{r}). \tag{1}$$

Here, the reduced mass is defined in terms of the masses of particle 1 and 2 as $m_r = m_1 m_2/(m_1 + m_2)$, $\mathbf{r} = (m_1 \mathbf{r}_1 - m_2 \mathbf{r}_2)/m_r$ is the relative coordinate, and ∇ is the 2D gradient. We further assume that the potential only depends on $r \equiv |\mathbf{r}|$; then the Schrödinger equation is separable, the wavefunction may be written as $\psi(\mathbf{r}) = R(r)T(\theta)$, and the equation for the radial part takes the form

$$-\frac{\hbar^2}{2m_r}\frac{1}{r}\frac{d}{dr}\left(r\frac{dR}{dr}\right) + \frac{\hbar^2 \ell^2}{2m_r r^2}R + V(r)R = ER. \tag{2}$$

The quantum number ℓ is determined from the azimuthal equation $d^2 T/d\theta^2 = -\ell^2 T$ and corresponds to the angular momentum in the plane. In order for the wavefunction to be single valued we must have $T_\ell(\theta) \propto e^{i\ell\theta}$ with ℓ integer. Thus we have one s-wave component ($\ell = 0$) but two of all higher partial wave components (p, d, etc. corresponding to $\ell = \pm 1, \pm 2$, etc.). This may be thought of as clockwise and anti-clockwise rotation and should be compared with the degeneracy factor $2\ell + 1$ in 3D.[21]

In the asymptotic limit, we write the wavefunction as a sum of an incident plane wave along the $\hat{\mathbf{x}}$ direction and an outgoing circular wave

$$\psi(\mathbf{r}) \underset{r\to\infty}{\to} e^{ikx} - \sqrt{\frac{i}{8\pi kr}} f(\mathbf{k})e^{ikr}, \tag{3}$$

with the incident relative wavenumber k defined by $E = \hbar^2 k^2 / 2m_r$. The vector $\mathbf{k} \equiv k\hat{\mathbf{r}}$ is defined in the direction of the scattered wave at an angle θ with respect to the incident wave. The dimensionless scattering amplitude $f(\mathbf{k})$ may then be expanded in the partial waves as

$$f(\mathbf{k}) = \sum_{\ell=0}^{\infty} (2 - \delta_{\ell 0}) \cos(\ell\theta) f_\ell(k), \qquad (4)$$

where the Kronecker delta takes account of the degeneracy within the partial wave.

The scattering amplitude gives access to the differential elastic cross section $d\sigma/d\theta = |f(\mathbf{k})|^2 / 8\pi k$, and to both the total and elastic cross sections:

$$\sigma_\ell^{\text{tot}}(E) = -\frac{1}{k} \text{Im}[f_\ell(k)](2 - \delta_{\ell 0}), \qquad (5)$$

$$\sigma_\ell^{\text{el}}(E) = \frac{|f_\ell(k)|^2}{4k}(2 - \delta_{\ell 0}), \qquad (6)$$

where the first equation corresponds to the well-known optical theorem. For both cross sections we use the partial wave expansion $\sigma(E) = \sum_{\ell=0}^{\infty} \sigma_\ell(E)$, noting that the partial waves decouple in the cross section. The inelastic cross section simply follows as $\sigma^{\text{inel}}(E) = \sigma^{\text{tot}}(E) - \sigma^{\text{el}}(E)$. Note that in 2D the cross section has dimensions of length.

The scattering amplitude may be related to the phase shift experienced by the scatterers at distances outside the range of the potential:

$$f_\ell(k) = \frac{-4}{\cot \delta_\ell(k) - i}. \qquad (7)$$

The phase shifts are real for elastic scattering and have the low energy behavior (see, e.g., Ref. 19)

$$\cot \delta_s(k) = -\frac{2}{\pi} \ln(1/ka) + \mathcal{O}(k^2), \qquad (8)$$

$$k^2 \cot \delta_p(k) = -s^{-1} + \mathcal{O}(k^2 \ln k), \qquad (9)$$

where we denote the phase shifts $\delta_s \equiv \delta_0$, $\delta_p \equiv \delta_1$, etc. Here, $a > 0$ is a 2D scattering length, while s is a 2D scattering area (of unit length squared).

Interestingly, we see that $\cot \delta_s$ diverges logarithmically at low energies, and thus the definition of the scattering length is ambiguous (indeed several conventions are used in the literature). The logarithmic divergence means that the scattering amplitude goes to zero at zero collision energy; this is manifestly different from the 3D behavior, where the scattering amplitude at zero energy equals minus the scattering length. While the p-wave amplitude also goes to zero in this limit, we see that it does so much faster than f_s. Indeed, while the s-wave cross section is seen to diverge at zero energy, the p-wave cross section $\sigma_p \to 0$ in this limit. The low-energy behavior has important consequences in both few- and many-body physics of the 2D gas with short-range interactions.

2.2. Scattering with a short-range potential

We now specialize to the typical interactions occuring in the two-component Fermi gas in 2D. We use a spin notation for the two components, $\sigma = \uparrow, \downarrow$; the spin indices may denote different hyperfine states of the same atom or, in the case of a heteronuclear mixture, single hyperfine states of two different atomic species. The atomic interaction is characterized by a van der Waals range R_e much shorter than both the average interparticle spacing and the thermal wavelength. Thus we may consider the interaction to be effectively a contact, s-wave interaction, and model the two-body problem with the following Hamiltonian

$$\mathcal{H} = \sum_{\mathbf{k}} \frac{\hbar^2 k^2}{2m_r} |\mathbf{k}\rangle\langle\mathbf{k}| + \frac{1}{A} \sum_{\mathbf{k},\mathbf{k}'} g(\mathbf{k}, \mathbf{k}') |\mathbf{k}\rangle\langle\mathbf{k}'| . \tag{10}$$

Here, A is the system area and in the following we set $A = \hbar = 1$. The attractive contact interaction $g(\mathbf{k}, \mathbf{k}') \equiv \langle\mathbf{k}| \hat{g} |\mathbf{k}'\rangle$ has strength $g < 0$ and is taken constant up to a large ultraviolet cutoff $\Lambda \sim 1/R_e$. The reduced mass in this two-component system is $m_r = m_\uparrow m_\downarrow / (m_\uparrow + m_\downarrow)$. As we are considering low-energy s-wave scattering, interactions between the same species of fermion are suppressed by Pauli exclusion.

The interaction between two atoms is conveniently described in terms of a T matrix, illustrated in Fig. 1, which describes the sum of repeated scattering processes between two atoms. In the center of mass frame, with incoming (outgoing) momenta of $\pm\mathbf{k}_i$ ($\pm\mathbf{k}_f$), the T matrix takes

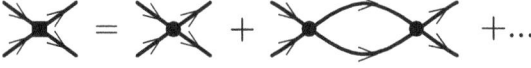

Fig. 1. The sum of all possible repeated scattering processes of two atoms, resulting in the T matrix (black square). The circles represent the interaction $\hat{g}$.

the form

$$\langle \mathbf{k}_f | \hat{T}(E + i0) | \mathbf{k}_i \rangle = \langle \mathbf{k}_f | \hat{g} + \hat{g} \frac{1}{E - \hat{H}_0 + i0} \hat{g} + \dots | \mathbf{k}_i \rangle$$

$$= \frac{1}{g^{-1} - \Pi(E)}, \qquad (11)$$

where the notation $+i0$ indicates an infinitesimal positive imaginary part. Here $\hat{H}_0$ is the non-interacting part of the Hamiltonian. The one loop polarization bubble takes the form

$$\Pi(E) = \sum_{\mathbf{q}}^{\Lambda} \langle \mathbf{q} | \frac{1}{E - \hat{H}_0 + i0} | \mathbf{q} \rangle = \sum_{\mathbf{q}}^{\Lambda} \frac{1}{E - q^2/2m_r + i0}. \qquad (12)$$

Considering scattering at negative energies, it is immediately clear that the attractive contact interaction in 2D always admits a bound diatomic molecule (dimer) state in contrast to the 3D case. The energy of the bound state, $-\varepsilon_b$ (we define ε_b positive), is determined through the pole of the T matrix, i.e.

$$\frac{1}{g} = \Pi(-\varepsilon_b). \qquad (13)$$

This relation acts to renormalize the interaction: the integral logarithmically diverges at fixed ε_b if we take $\Lambda \to \infty$, however, the physics beyond the two-body problem becomes independent of Λ once Eq. (13) is used to replace g with the binding energy. Thus we arrive at the renormalized T matrix

$$T(E) \equiv \langle \mathbf{k}_f | \hat{T}(E + i0) | \mathbf{k}_i \rangle = \frac{1}{\Pi(-\varepsilon_b) - \Pi(E)} = \frac{2\pi}{m_r} \frac{1}{\ln(\varepsilon_b/E) + i\pi}. \qquad (14)$$

As the T matrix does not depend on incoming momenta in the center of mass frame, we will simply denote it $T(E)$.

The on-shell scattering of two atoms at momenta $\pm \mathbf{k}_i$ into momenta $\pm \mathbf{k}_f$ with $k = |\mathbf{k}_i| = |\mathbf{k}_f|$ yields the scattering amplitude through the relation $f(\mathbf{k}) = 2m_r \langle \mathbf{k}_i | \hat{T}(k^2/2m_r) | \mathbf{k}_f \rangle$. Then, using Eq. (7), we find that the two-body phase shift with this contact interaction takes the form $\cot \delta_s(k) = -\frac{2}{\pi} \ln(1/k a_{2D})$, which defines the 2D atom atom scattering length a_{2D}.[a] The relation between the binding energy and the 2D scattering length is then simply

$$\varepsilon_b = \frac{1}{2m_r a_{2D}^2}. \tag{15}$$

2.3. Quasi-two-dimensional Fermi gases

Under realistic experimental conditions, the extent of the gas perpendicular to the plane is necessarily finite. The quasi-two-dimensional (quasi-2D) regime occurs when the confinement width is much smaller than both the interparticle spacing and the thermal wavelength, such that transverse degrees of freedom are frozen out. However, the length scale associated with the confinement to the quasi-2D geometry is necessarily much larger than the range of the van der Waals type interactions, and thus at short distances the two-body interactions are unaffected by the confinement. The relationship between the 2D scattering theory detailed above, and the realistic interatomic potential was considered in detail in Ref. 23. Here we present an alternative derivation of the quasi-2D scattering amplitude, and arrive at a form which is closer to that in Ref. 24.

We thus consider the experimentally relevant harmonic confinement $V_\sigma(z) = \frac{1}{2} m_\sigma \omega_z^2 z^2$ acting in the direction perpendicular to the 2D plane. While for the heteronuclear gas the confining frequency ω_z is not necessarily the same for both species, this choice in general allows a separation of the center of mass from the relative motion and provides a major simplification of the formalism. In relative coordinates, the non-interacting two-body problem in the z direction reduces to the harmonic oscillator equation

$$\left(-\frac{1}{2m_r} \frac{d^2}{dz^2} + \frac{1}{2} m_r \omega_z^2 z^2 \right) \phi_n(z) = \left(n + \frac{1}{2} \right) \omega_z \phi_n(z). \tag{16}$$

Here, the motion along the z direction is clearly quantized, with a constant spacing ω_z between energy levels. The non-interacting part of the quasi-2D

---

[a]In the literature, the alternative definition $2e^{-\gamma} a_{2D}$ of the 2D scattering length is often employed, with γ the Euler gamma constant. This definition arises naturally when considering scattering from a hard disc of radius a_c, in which case $a_{2D} = (e^\gamma/2)a_c$.

Hamiltonian is thus

$$\hat{H}_0 = \sum_{kn} \left[\frac{k^2}{2m_r} + \left(n + \frac{1}{2} \right) \omega_z \right] |kn\rangle\langle kn| , \tag{17}$$

where n is the harmonic oscillator quantum number for the z direction. The gas is considered to be kinematically 2D if motion is restricted to the $n = 0$ level.

To investigate two-body scattering in the quasi-2D geometry, we need to consider the bare interaction in three-dimensional space. For convenience, *in this section only*, we consider a separable 3D interaction of the form

$$g(\mathbf{k}_{3D}, \mathbf{k}'_{3D}) = \langle \mathbf{k}_{3D} | \hat{g} | \mathbf{k}'_{3D} \rangle \equiv g e^{-(k^2 + k'^2 + k_z^2 + k_z'^2)/\Lambda^2}. \tag{18}$$

where k_z is the z-component of the 3D momentum and k is the magnitude of the inplane momentum $\mathbf{k}$ as above. Letting the incoming (outgoing) atoms have momenta $\pm\mathbf{k}_i$ ($\pm\mathbf{k}_f$) in the plane and relative motion in the harmonic potential described by the index n_i (n_f), the matrix elements of the 3D interaction in the quasi-2D basis are

$$\begin{aligned} \langle \mathbf{k}_f n_f | \hat{g} | \mathbf{k}_i n_i \rangle &= \sum_{\mathbf{q}_{3D}\mathbf{q}'_{3D}} \langle \mathbf{k}_f n_f | \mathbf{q}_{3D} \rangle \langle \mathbf{q}_{3D} | \hat{g} | \mathbf{q}'_{3D} \rangle \langle \mathbf{q}'_{3D} | \mathbf{k}_i n_i \rangle \\ &= g f_{n_f} f_{n_i} e^{-(k_i^2 + k_f^2)/\Lambda^2} , \end{aligned} \tag{19}$$

where $f_n \equiv \sum_{q_z} \tilde{\phi}_n(q_z) e^{-q_z^2/\Lambda^2}$ and $\tilde{\phi}_n(q_z)$ is the Fourier transform[b] of the harmonic oscillator wave function. For the f coefficients, we then find

$$f_{2n} = (-1)^n \frac{1}{(2\pi l_z^2)^{1/4}} \frac{\sqrt{(2n)!}}{2^n n!} \frac{1}{\sqrt{1+\lambda}} \left(\frac{1-\lambda}{1+\lambda} \right)^n , \tag{20}$$

---

[b] The harmonic oscillator wave function is

$$\phi_n(z) = \sqrt{\frac{1}{2^n n!}} \left(\frac{m_r \omega_z}{\pi} \right)^{\frac{1}{4}} \exp\left(-\frac{m_r \omega_z z^2}{2} \right) H_n\left(\sqrt{m_r \omega_z}\, z \right),$$

where $H_n(x)$ are the Hermite polynomials. $\phi_n(z)$ also happens to be an eigenfunction of the Fourier transform, so in momentum space it is simply

$$\tilde{\phi}_n(k_z) = (-i)^n \sqrt{\frac{2}{2^n n!}} \left(\frac{\pi}{m_r \omega_z} \right)^{\frac{1}{4}} \exp\left(-\frac{k_z^2}{2m_r \omega_z} \right) H_n\left(\sqrt{\frac{1}{m_r \omega_z}} k_z \right).$$

and $f_{2n+1} = 0$. Here $l_z \equiv 1/\sqrt{2m_r \omega_z}$ is the harmonic oscillator length.[c] $\lambda \equiv 1/(\Lambda l_z)^2$ is the (squared) ratio between the length scale of the short distance physics and the harmonic oscillator length, and is very small in typical experiments. Indeed, our approach of using a 3D interaction would be invalid if this were not the case.

We then evaluate the T matrix in a manner similar to the 2D case above:

$$
\langle \mathbf{k}_f n_f | \hat{T}(E + i0) | \mathbf{k}_i n_i \rangle = \langle \mathbf{k}_f n_f | \hat{g} + \hat{g} \frac{1}{E - \hat{H}_0 + i0} \hat{g} + \ldots | \mathbf{k}_i n_i \rangle
$$

$$
= e^{-(k_i^2 + k_f^2)/\Lambda^2} f_{n_i} f_{n_f} \frac{1}{g^{-1} - \Pi_{\mathrm{Q2D}}(E)}. \tag{21}
$$

The quasi-2D polarization bubble takes the form

$$
\Pi_{\mathrm{Q2D}}(E) = \sum_{\mathbf{q}, n} |f_n|^2 \frac{e^{-2q^2/\Lambda^2}}{E - (n + 1/2)\omega_z - q^2/2m_r + i0}. \tag{22}
$$

The sum over n may be evaluated by changing variables to $u = -2\lambda \frac{q^2/2m_r}{E - (n+1/2)\omega_z}$ and using[d]

$$
\sum_{n=0}^{\infty} \frac{(2n)!}{(n!)^2} x^n = \frac{1}{\sqrt{1 - 4x}}. \tag{23}
$$

We then find

$$
\Pi_{\mathrm{Q2D}}(E) = -\frac{m_r}{(2\pi)^{3/2} l_z} \int_0^{\infty} \frac{du}{u + 2\lambda} \frac{e^{-(-E/\omega_z + 1/2)u}}{\sqrt{(1 + \lambda)^2 - (1 - \lambda)^2 e^{-2u}}}. \tag{24}
$$

Finally, we relate this result back to the 3D physics: The interaction (18) is renormalized using the relationship[e] between the T matrix at vanishing

---

[c]For equal masses, l_z reduces to the usual harmonic oscillator length for the motion of the individual atoms. For a general mass ratio it differs by a factor $\sqrt{2}$ from the usual definition of the harmonic oscillator length of the relative motion.

[d]While formally this approach is valid only for $-1/4 \leq x < 1/4$, by the analytic continuation $E \to E + i0$ the result can be extended to all energies.

[e]The 3D scattering length is related to the T matrix at vanishing energy by $a_s = (m_r/2\pi)\langle 0|T(0)|0\rangle = (m_r/2\pi)/(g^{-1} + m_r \Lambda/(2\pi)^{3/2})$.

energy and the 3D scattering length, a_s. Thus we arrive at the T matrix

$$\langle \mathbf{k}_f n_f | \hat{T}(E+i0) | \mathbf{k}_i n_i \rangle$$
$$= e^{-(k_i^2+k_f^2)/\Lambda^2} f_{n_i} f_{n_f} \frac{2\pi l_z}{m_r} \frac{1}{\frac{l_z}{a_s} - \mathcal{F}_\lambda(-E/\omega_z + 1/2)}, \quad (25)$$

with

$$\mathcal{F}_\lambda(x) = \int_0^\infty \frac{du}{\sqrt{4\pi(u+2\lambda)^3}}$$
$$\times \left[1 - \frac{e^{-xu}}{\sqrt{[(1+\lambda)^2 - (1-\lambda)^2 e^{-2u}]/(2u+4\lambda)}} \right]. \quad (26)$$

This expression reduces to that of Ref. 24 in the limit $\lambda \to 0$. In this case, the T matrix only depends on E and the quantum numbers in the harmonic potential, and we write

$$\langle \mathbf{k}_f n_f | \hat{T}(E+i0) | \mathbf{k}_i n_i \rangle \equiv \sqrt{2\pi} l_z f_{n_i} f_{n_f} \mathcal{T}(E), \quad (27)$$

where $\mathcal{T}(E) \equiv \frac{\sqrt{2\pi}}{m_r} \frac{1}{\frac{l_z}{a_s} - \mathcal{F}_0(-E/\omega_z + 1/2)}$ contains the entire energy dependence.

2.3.1. *Low-energy scattering*

At energies close to the scattering threshold, the function $\mathcal{F}$ may be expanded. Specializing to the case $\lambda = 0$, this results in

$$\mathcal{F}_0(x) \approx \frac{1}{\sqrt{2\pi}} \ln(\pi x/B) + \frac{\ln 2}{\sqrt{2\pi}} x - \frac{\pi^2 - 12\ln^2 2}{48\sqrt{2\pi}} x^2 + \mathcal{O}(x^3), \quad |x| \ll 1, \quad (28)$$

with $B \approx 0.905$; see Refs. 23, 24. This in turn yields the 2D scattering length

$$a_{2D} = l_z \sqrt{\frac{\pi}{B}} e^{-\sqrt{\frac{\pi}{2}} l_z/a_s}. \quad (29)$$

Note that the strictly 2D limit corresponds to weak interactions $l_z/a_s \ll -1$, where all bound states have binding energies much smaller than ω_z (see Sec. 3 and below). However, we emphasize that Eq. (29) is valid for all a_s as it only requires the scattering energy to be negligible compared with the strength of the confinement. In particular, if $|l_z/a_s| \gg 1$, then for a large range of energies in the continuum close to the threshold, i.e., for $|E - \omega_z/2| \ll \omega_z$, the scattering amplitude may simply be approximated by

$$f(\mathbf{k}) \approx 2\sqrt{2\pi}\, a_s/l_z. \qquad (30)$$

Thus, in this regime, the two-body interaction is approximately independent of energy, and the system may be considered scale invariant. This can have important consequences for the many-body system.

2.3.2. Bound state

The binding energy of the dimer is the solution of

$$\frac{l_z}{a_s} = \mathcal{F}_\lambda(\varepsilon_b/\omega_z), \qquad (31)$$

where we measure the binding energy from the threshold of free relative motion of the two atoms. In contrast to the situation in 3D, where a bound state only exists for $a_s > 0$, under a harmonic confinement a bound state exists for a zero-range interaction of arbitrary strength. In this sense, for negative 3D scattering length, the dimer in the quasi-2D geometry is confinement induced. This may be viewed as resulting from the increase of the continuum by $\frac{1}{2}\omega_z$, as illustrated in Fig. 2. In the following, we focus on the case $\lambda = 0$, but similar behavior should hold for $\lambda \ll 1$.

For small positive scattering length, $l_z/a_s \gg 1$, the 3D dimer with size $\sim a_s$ fits well within the confining potential and is only weakly perturbed by the harmonic confinement, as illustrated in Fig. 2. As the scattering length is increased, eventually the dimer energy becomes strongly modified; for instance at the 3D resonance the binding energy takes the universal value[23] $\varepsilon_b = 0.244\omega_z$. On the other hand, in the limit of a small negative 3D scattering length, $l_z/a_s \ll -1$, the dimer spreads out in the 2D plane and the binding energy follows from the expansion Eq. (28). Taking only the first term, the result is seen to match the 2D expression $\varepsilon_b = 1/(2m_r a_{2D}^2)$, i.e.,

$$\varepsilon_b \approx \omega_z \frac{B}{\pi} \exp(\sqrt{2\pi}\, l_z/a_s), \qquad l_z/a_s \ll -1 \qquad (32)$$

as is also seen in Fig. 2. However, this expression breaks down when $l_z/a_s > -1$. The fact that in general $\varepsilon_b \neq \frac{\hbar^2}{2m_r a_{2D}^2}$, should not come as a surprise. It

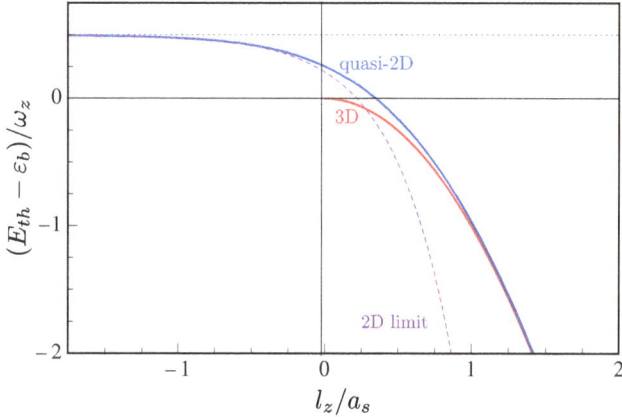

Fig. 2. The binding energy of the quasi-2D dimer with $\lambda = 0$ (blue, solid). Also shown is the 3D dimer (red, solid), and the 2D expression $\varepsilon_b = 1/(2m_r a_{2D}^2)$ (dashed). The threshold energy E_{th} is 0 in the 3D case, and $\omega_z/2$ in quasi-2D.

simply follows from the introduction of an extra length scale, l_z, into the problem, and is analogous to the problem of a narrow Feshbach resonance in the 3D gas.[25]

The dimer binding energy has been measured using radio-frequency (RF) spectroscopy in experiments on ultracold ^{6}Li (Ref. 11) and ^{40}K (Ref. 13) atoms subjected to a tight optical confinement. The results are shown in Fig. 3, and both experiments agree well with theory[23,24] across the Feshbach resonance.

2.4. *Confinement induced resonances*

One consequence of confining the gas to lower dimensions is the appearance of so-called *confinement induced resonances*. At the simplest level, these refer to any region of resonantly enhanced two-body scattering resulting from the confinement. However, the situation in 2D is slightly more subtle, given that the purely 2D system already exhibits an enhancement of the scattering amplitude for energy $E \sim \varepsilon_b$, as can be seen from the T matrix $T(E) = \frac{2\pi}{m_r} \frac{1}{\ln(\varepsilon_b/E)+i\pi}$. Thus, it is important to make a distinction between enhanced scattering that can arise from 2D kinematics, and resonances that only result from the finite extent of the gas in the confined direction.

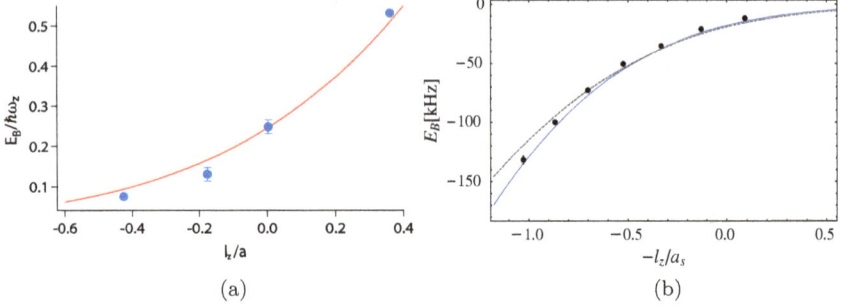

(a)   (b)

Fig. 3. (a) Experimental (filled circles) and theoretical (solid line) binding energy of a fermion pair in a gas of ^{6}Li atoms.[11] (b) Binding energy in a ^{40}K gas[13] at a confinement of $\omega_z = 2\pi \times 75$kHz. The experimental result (filled circles) is compared with the theoretical prediction at zero effective range (dashed line), and finite effective range (solid line), according to the equation for the binding energy modified by the 3D effective range r_{eff}, $\frac{l_z}{a_s} + \frac{r_{\text{eff}}}{2l_z}(\varepsilon_b/\omega_z - 1/2) = \mathcal{F}_0(\varepsilon_b/\omega_z)$. Reprinted figure in (a) with permission from: A. T. Sommer, L. W. Cheuk, M. J. H. Ku, W. S. Bakr, and M. W. Zwierlein, *Phys. Rev. Lett.* **108**, 045302 (2012). Copyright 2012 by the American Physical Society.

An example of the latter case is the confinement induced resonance associated with quasi-1D systems.[26,27] Here, a resonance occurs when a virtual bound state (arising from the excited levels of the transverse confinement) crosses the 1D atom-atom scattering threshold. This process can be captured with a simplified two-channel model[27]

$$\mathcal{H}_{\text{Q1D}} = \sum_k \frac{k^2}{2m_r} |k\rangle\langle k| + \nu |b\rangle\langle b| + \alpha \sum_k (|k\rangle\langle b| + |b\rangle\langle k|), \quad (33)$$

where $|b\rangle$ corresponds to the virtual "closed channel" bound state associated with the excited states in the harmonic confinement, ν is the energy of this state with respect to the continuum threshold, and α is the coupling between $|b\rangle$ and the scattering states $|k\rangle$ in 1D. Here, we neglect the interactions between the 1D scattering states and we take α and ν to be independent parameters.^{f} In general, this model leads to energy-dependent interactions, but for zero-energy scattering we have an effective 1D contact potential $g_{1D}\delta(x)$ with interaction strength $g_{1D} = -\alpha^2/\nu$. Thus, we obtain a scattering resonance where $g_{1D} \to \pm\infty$ when $\nu \to 0^{\mp}$. Note that this

---

^{f}In the real quasi-1D system, α and ν are not independent, as is apparent from the two-channel model in Ref. 27.

does not signal the appearance of a two-body bound state like in the 3D case where the scattering length diverges ($1/a_s = 0$). Instead, Eq. (33) always yields a two-body bound state with binding energy ε_b satisfying the condition $\sqrt{2}(\nu + \varepsilon_b)/\alpha^2 = \sqrt{m_r/\varepsilon_b}$. Moreover, we see that ε_b is finite at the resonance $\nu = 0$, thus illustrating how this resonance is a feature of confinement that goes beyond the behaviour in a purely 1D system.

On the other hand, such a confinement induced resonance does not exist in the quasi-2D system. If we consider the two-channel model (33) in 2D, where we have $|\mathbf{k}\rangle$ instead of $|k\rangle$, then we obtain the modified T matrix

$$T(E) = \frac{2\pi}{m_r} \left[\ln \left(\frac{1}{2m_r a_{2D}^2 E} \right) + \frac{2\pi E}{m_r \alpha^2} + i\pi \right]^{-1}, \qquad (34)$$

with $a_{2D} = \Lambda^{-1} \exp(\pi \nu/m_r \alpha^2)$. This is essentially the quasi-2D T matrix $T(E)$ expanded up to linear order in E/ω_z. Comparing with the terms in the expansion (28) yields $\omega_z = m_r \alpha^2 \ln(2)/2\pi$. However, this modification to the T matrix only shifts the scattering enhancement away from $E \sim \varepsilon_b$ (where $T(-\varepsilon_b)^{-1} = 0$). The resonance still remains strongly energy dependent like in the purely 2D case. However, it can still be characterized experimentally: For a Boltzmann gas in the 2D limit, the scattering is enhanced for temperature $T \sim \varepsilon_b$. The requirement $T \ll \omega_z$ then implies that $\varepsilon_b \ll \omega_z$ and thus the resonance occurs on the attractive side of the 3D Feshbach resonance, $a_s < 0$, according to Fig. 2.

Additional resonances will appear when there are anharmonicities in the confining potential, as is usually the case in experiments employing an optical lattice. Any anharmonicity inevitably leads to coupling between two-body states with different center-of-mass harmonic quantum numbers N. In particular, there will (for instance) be a coupling between scattering states $|\mathbf{k}\rangle$ with $N = 0$ and two-body bound states with $N = 2$, due to the selection rules. This leads to increased molecule formation when ε_b is close to $2\hbar\omega_z$, which in turn leads to enhanced losses due to subsequent collisions with other particles. Such an "inelastic" confinement induced resonance was recently observed experimentally in low-dimensional geometries.[28–30] In the absence of effective range corrections, the inelastic resonance in the quasi-2D geometry arising from the above mentioned coupling occurs when $l_z/a_s \simeq 1.2$, in contrast to the resonance derived from 2D kinematics.

3. Universal Few-Body Physics in a 2D Fermi Gas

Recent years have brought a wealth of experiments exploring few-
and many-body physics in ultracold atomic gases — see, for instance,
Refs. 24 and 25 and references therein. In particular, it has enabled the
study of *universal* few-body physics, since the low-energy scattering is
insensitive to the details of the short-range interactions, and this in turn has
major consequences for the many-body system. For instance, in the regime
where a dimer exists, a knowledge of the dimer–dimer[31] and atom–dimer[32]
scattering lengths is necessary for a complete description of the balanced[33]
and polarized[34] Fermi gases. The properties of few-body inelastic processes
furthermore explains the exceptional stability of 3D Fermi gases close
to the unitary limit.[31] The experimental exploration of 2D Fermi gases
is still ongoing, but few-body physics has already played an important
role in understanding many-body phenomena, as can be seen in Secs. 4
and 5.

While experiments have thus far focussed on the equal-mass case,
heteronuclear Fermi-Fermi mixtures with mass imbalance promise to
provide even richer few-body phenomena. For instance, in 1D it has been
shown that when the mass ratio exceeds one, trimers (bound states of 1
light atom, 2 heavy atoms) can form.[35] This can lead to a Luttinger liquid of
trimers in the polarized gas, while more exotic bound states such as tetramers
(1,3), pentamers (2,3), etc., can also exist at higher mass ratios.[36] Similarly,
in 2D[37] and 3D,[38] trimers are predicted to exist above a mass ratio of 3.3 and
8.2, respectively, which can lead to a trimer phase in the highly polarized
Fermi gas.[39] Additionally, tetramers have been predicted in 1D,[36,40] 2D,[41]
and 3D.[42] These bound states all share the property of being universal, in
the sense that their binding energy is a multiple of the dimer binding energy
without the need for additional parameters. For a recent review of bound
few-body states, we refer the reader to Ref. 43.

Thus far, the only heteronuclear Fermi-Fermi mixture where tunable
short-range interactions have been experimentally demonstrated[44–46] is ^{40}K-
^{6}Li with a mass ratio of 6.64. However, the periodic table offers ample
opportunity for exploring additional mass ratios, while the possibility to
tune the effective mass ratio using optical lattices also exists. Thus, for the
theoretical approach employed in this section, we may consider the mass
ratio to be a free parameter.

Finally, we note that there is a special class of states — the well-known Efimov states — where the 3D physics is manifestly different from the situation in confined geometries.[47] Efimov states have been observed experimentally as sharp peaks in the loss rate in both bosonic[48] and (three-component) fermionic[49] systems. In the heteronuclear fermionic system in 3D, the Efimov effect occurs for (2,1) trimers above a mass ratio of 13.6 and for (3,1) tetramers[50] for mass ratios exceeding 13.4. On the other hand, it is known that Efimov's scenario does not occur in the 1D and 2D geometries.[51] Very recently it was shown that under realistic experimental conditions, if Efimov trimers exist in 3D, these will impact the few-body physics in a strongly confined geometry.[52] However, in the following we ignore this effect and focus either on the idealized 2D scenario and/or on mass ratios for which the Efimov effect is absent.

3.1. *Equal mass fermions*

We now discuss the three- and four-body problem in a homonuclear gas. The first of these plays an important role in accurately determining the energy of an impurity atom immersed in a Fermi sea — see Sec. 6. It is also of practical importance when considering inelastic processes such as three-body recombination,[53] the process whereby three atoms collide to produce an atom and a dimer. The dimer–dimer scattering length, on the other hand, is important in describing the many-body system in the limit of tightly bound pairs, as shown in Sec. 4. In the present discussion we confine ourselves to on-shell scattering properties.

The interaction between a spin-↑ atom with an ↑↓ dimer may be investigated with the Skorniakov–Ter-Martirosian (STM) equation introduced in the context of neutron-deuteron scattering:[32] The atom–dimer scattering arises from the repeated exchange between identical spin-↑ atoms of the spin-↓ atom, and the STM integral equation yields the sum of diagrams with any number of such exchanges as illustrated in Fig. 4. We are interested in

Fig. 4. Illustration of the Skorniakov–Ter-Martirosian equation which governs the interaction of an atom (straight line) with a dimer (wavy line). $\tilde{f}$ is the atom-dimer scattering amplitude.

the on-shell scattering amplitude, and thus we let the incoming [outgoing] atom and dimer have four-momentum $(\mathbf{k}, \epsilon_{\mathbf{k}\uparrow})$ and $(-\mathbf{k}, E - \epsilon_{\mathbf{k}\uparrow})$ $[(\mathbf{p}, \epsilon_{\mathbf{k}\uparrow})$ and $(-\mathbf{p}, E - \epsilon_{\mathbf{k}\uparrow})]$, respectively, with $E = k^2/2m_{ad} - \varepsilon_b$ such that the incoming dimer is on shell. Here the single particle kinetic energy is $\epsilon_{\mathbf{k},\sigma} = k^2/2m_\sigma$, where m_σ is the mass. m_{ad} is the atom-dimer reduced mass $m_{ad}^{-1} = m_\uparrow^{-1} + M^{-1}$, with $M = m_\uparrow + m_\downarrow$. The on-shell condition $|\mathbf{k}| = |\mathbf{p}|$ is taken at the end of the calculation. With these definitions, the STM equation takes the form of an integral equation[53]

$$\tilde{f}_\ell(k, p) = -h(k, p) \left[g_\ell(k, p) - \int \frac{q\, dq}{2\pi} \frac{g_\ell(p, q)\tilde{f}_\ell(k, q)}{q^2 - k^2 - i0} \right], \quad (35)$$

where we note that the atom-dimer scattering preserves angular momentum, allowing a decoupling of $\tilde{f}(\mathbf{k}, \mathbf{p})$ into its partial wave components. The scattering amplitude then follows from taking the on-shell condition $f_\ell(k) = \tilde{f}_\ell(k, k)$. g_l is the partial wave projection of the spin-$\downarrow$ propagator and h is proportional to the two-body T matrix.[g]

From the resulting scattering amplitude, the low-energy scattering properties such as the s-wave scattering length and the p-wave scattering area (see Sec. 2.1) may be extracted. For equal masses,[53]

$$a_{ad} \approx 1.26a_{2D}, \qquad s_{ad} \approx -2.92a_{2D}^2. \quad (37)$$

Thus, the s-wave scattering is repulsive at low energies while the p-wave scattering is attractive in this case. The scattering amplitude furthermore gives access to the partial wave cross sections, which are shown for s- and p-waves in Fig. 5(a). While the s-wave cross section is apparently completely described in terms of the scattering length, interestingly we note that the s- and p-wave cross sections are comparable for collision energies of the order of the binding energy. This is unlike the 3D case,[54] and is due

---

[g]The projection of the propagator of the spin-$\downarrow$ fermion onto the ℓ'th partial wave is

$$g_\ell(p, q) = \int_0^{2\pi} \frac{d\phi}{2\pi} \frac{\cos(\ell\phi)}{E - \epsilon_{\mathbf{k}\uparrow} - \epsilon_{\mathbf{q}\uparrow} - \epsilon_{\mathbf{p}+\mathbf{q}\downarrow} + i0}, \quad (36)$$

with ϕ the angle between $\mathbf{p}$ and $\mathbf{q}$. We also define $h(k, p) \equiv (k^2 - p^2)T(E - p^2/2m_{ad})$ [see Eq. (14)] in order to separate out the simple pole of the two-particle propagator occuring at $|\mathbf{k}| = |\mathbf{p}|$.

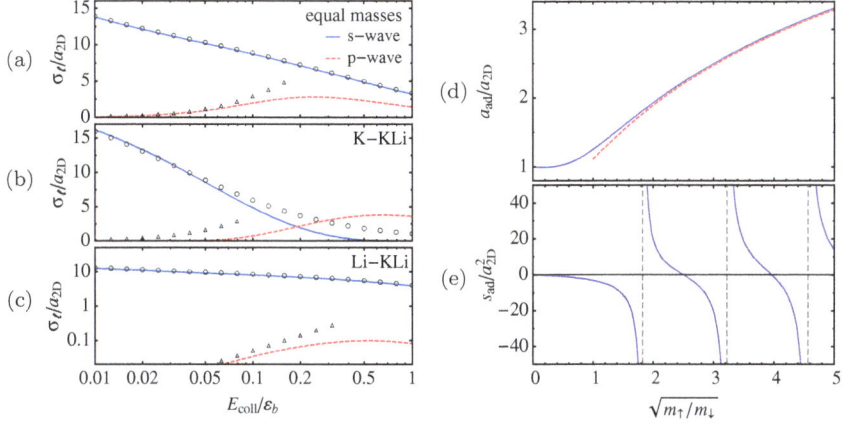

Fig. 5. (a,b,c) s- and p-wave atom–dimer cross sections as a function of collision energy $E_{coll} \equiv k^2/2M + k^2/2m_\uparrow$ for equal masses and for the K-Li mixture. The circles [triangles] correspond to the low-energy expansion, see Sec. 2.1, using the atom–dimer scattering length [area]. Note the log-scale used in (c). (d) Atom–dimer scattering length and (e) area as a function of mass ratio[53]. In (d) the dashed line is the asymptotic behavior at large mass ratio[53] and in (e) the vertical dashed lines indicate the appearance of trimers. The figure is taken from Ref. 53.

to the weaker centrifugal barrier between identical fermions in 2D. Thus three-body correlations are likely to be more important in 2D than in 3D.

Both the atom–dimer and the dimer–dimer scattering lengths have been extracted from a quantum Monte Carlo (QMC) calculation of the excitation gap in the BEC regime.[55] The result was $a_{ad} = 1.7(1)a_{2D}$ and $a_{dd} = 0.55(4)a_{2D}$. Note that these relations do not depend on the definition of the scattering length. The discrepancy between the exact result in Eq. (37) and the QMC result is likely to be due to the fact that the extraction of a_{ad} from the data relied on an equation of state that was only logarithmically accurate. On the other hand, the dimer–dimer scattering length is in perfect agreement with the exact few-body calculation of Petrov *et al.*[56]: $a_{dd} \approx 0.56a_{2D}$.

3.2. *Heteronuclear Fermi gas*

The three-body problem in a heteronuclear 2D Fermi gas with short-range interparticle interactions was first studied by Pricoupenko and Pedri.[37] Remarkably, even in the absence of Efimov physics, they still found that two heavy fermionic atoms and a light atom can form an ever increasing number of trimers as the mass ratio is increased. However, at any given mass ratio, the number of trimers in the spectrum was found to be finite. In Refs. 53 and 57 it was argued that the appearance of trimers was due to an effective $1/R$ potential in odd partial waves between heavy atoms at a separation R, mediated by the light atom.[h] Consequently, at large mass ratios, the spectrum of bound states is hydrogen-like.

Signatures of trimer formation are clearly seen in the atom–dimer scattering properties, Fig. 5. Here the p-wave scattering area diverges at the mass ratios[37] $m_\uparrow/m_\downarrow = 3.33$, 10.41, etc., when a trimer state crosses the atom–dimer scattering threshold. Thus, the p-wave interaction becomes resonant at the crossing, and while the K-Li mass ratio of 6.64 is in-between the appearance of bound states in 2D, we still observe that the p-wave cross section in K-KLi scattering dominates at a collision energy comparable to ε_b. On the other hand, the scattering length increases monotonically with mass ratio.

The strong atom–dimer scattering in higher partial waves due to the proximity of trimers may be investigated using a mixture of heavy atoms and heavy-light dimers, as in a recent experiment using a K-Li mixture in 3D.[58] In such a mixture, the energy shift of an atom due to the interaction with dimers is proportional to the real part of the atom–dimer scattering amplitude, and may be directly accessed by radiofrequency spectroscopy.

Turning now to trimers, in Fig. 6 we show their spectrum obtained from Eq. (35).[37,53] To shed light on the appearance of trimers at large mass ratio, it is instructive to turn to the Born-Oppenheimer approximation (in the following discussion we follow Ref. 53, but see also Ref. 57): Assume that the wavefunction of the light atom at position $\mathbf{r}$ adiabatically adjusts itself to the positions of the heavy atoms, $\pm\mathbf{R}/2$. Then the wavefunction of the light atom is

$$\psi_{\mathbf{R}}(\mathbf{r}) \propto K_0(\kappa_\mp(R)|\mathbf{r} - \mathbf{R}/2|) \mp K_0(\kappa_\mp(R)|\mathbf{r} + \mathbf{R}/2|), \qquad (38)$$

---

[h] As the heavy atoms are identical fermions, the wavefunction of the light atom is necessarily antisymmetric (symmetric) for scattering in even (odd) partial waves. The antisymmetric state suppresses tunneling of the light atom between heavy atoms, and as a result the effective potential between the heavy atoms is repulsive (attractive) in even (odd) partial waves. Consequently, trimers only form in odd partial waves.

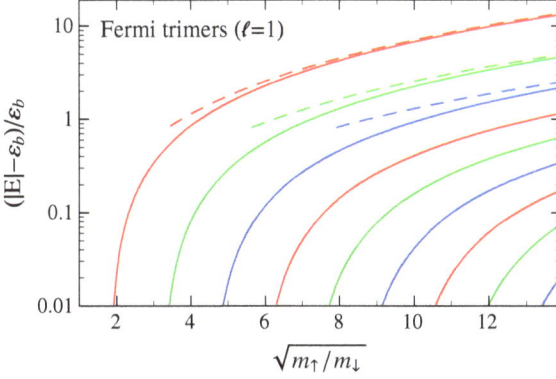

Fig. 6. The spectrum of p-wave trimers[37] as a function of mass ratio, taken from Ref. 53. The dashed lines are the large mass ratio asymptotic hydrogen-like energy levels (40).

where the upper (lower) sign describes even (odd) partial wave scattering. The modified Bessel function of the second kind $K_0(\kappa_{\mp}(R)r)$ is the decaying solution of the free single-particle Schrödinger equation with energy $\epsilon_{\mp}(R) = -\kappa_{\mp}(R)^2/2m_{\downarrow}$. The energy of the light atom as a function of separation of heavy atoms is determined from the Bethe-Peierls boundary condition in 2D: $[\tilde{r}\psi'(\tilde{r})/\psi]_{\tilde{r}\to 0} = 1/\ln(\tilde{r}/(2e^{-\gamma}a_{2D}))$ where $\tilde{r} = r \pm R/2$. This leads to the implicit equation

$$\ln\left(-\frac{\epsilon_{\mp}(R)}{\varepsilon_b}\right) = \mp 2K_0\left(\sqrt{-\frac{\epsilon_{\mp}(R)}{\varepsilon_b}}\frac{R}{a_{2D}}\right). \qquad (39)$$

The energy levels of the light atom act as potential surfaces for the motion of the heavy atoms. In the case of p-wave scattering, the effective potential $V_p(R) = \epsilon_+(R) - \epsilon(\infty) + 1/(m_{\uparrow}R^2)$ including the centrifugal barrier is shown in Fig. 7. The potential is measured from the limiting value of the potential at large separation, $\epsilon(\infty)$, which reduces to $-\varepsilon_b$ at large mass ratios. We see that when the mass ratio is small, the effective potential is always repulsive. On the other hand, the potential develops an attractive well as the mass ratio is increased and it is in this well that trimers can form. For a large mass ratio, the odd partial wave potential becomes $\epsilon_+(R) \approx -\frac{2\varepsilon_b}{e^{\gamma}}\frac{a_{2D}}{R}$ at short distances, $R \ll a_{2D}$. This potential is hydrogen-like and thus the spectrum of the deepest bound trimers is[59] (appropriately

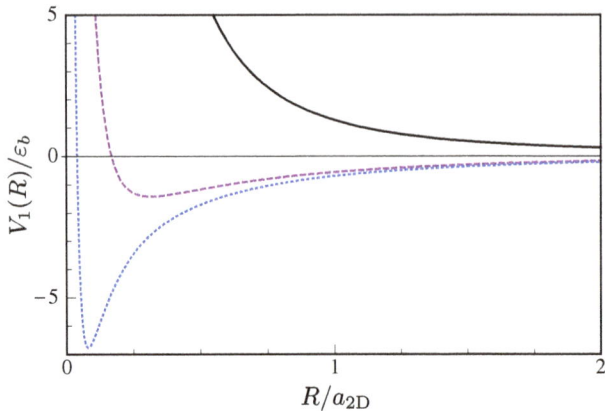

Fig. 7. Born-Oppenheimer effective p-wave potential between two heavy fermions mediated by a light atom. Top to bottom are the potentials for mass ratios $m_\uparrow/m_\downarrow = 1, 3.33$ (critical mass ratio for the appearance of trimers), and 6.64 (the ^{40}K-^{6}Li mass ratio).

shifted by the dimer binding energy)

$$E_n = -\frac{m_\uparrow}{e^{2\gamma} m_\downarrow} \frac{\varepsilon_b}{2(n+1/2)^2} - \varepsilon_b, \tag{40}$$

with integer quantum number $n \geq \ell$. In Fig. 6 this asymptotic expression is shown to agree well with the exact results for the deepest bound trimers. The expression (40) makes it clear that deeply bound trimers in different partial wave channels are quasi-degenerate[i], as observed in Ref. 37. Remarkably, at very short distances, the $1/R^2$ behavior of the centrifugal barrier always dominates. This has the important consequence that the trimers may be expected to be quite long-lived, as they are large (of size a_{2D}) and the constituent atoms do not approach each other easily. This is completely unlike the 3D case, where the mediated potential also goes as $1/R^2$ and the Efimov effect occurs at large mass ratio.

The number of bound states at any given mass ratio may be approximated[53] by noting that it is proportional to the number of nodes of the zero-energy wavefunction. The trimers exist in the hydrogen-like part of the effective potential, $R \lesssim a_{2D}$, and in this regime the relative

---

[i]In fact, as the spectrum of even partial wave trimers consisting of two identical (non-interacting) heavy bosons and a light atom are determined from the same effective potential, these are also quasi-degenerate with the fermionic trimers arising in odd partial waves.

wavefunction of the heavy particles is proportional to the Bessel function $J_{2\ell}(2\sqrt{e^{-\gamma}(m_\uparrow/m_\downarrow)}R/a_{2D})$. The wavefunction acquires an additional node each time the argument increases by π, and consequently the number of trimers is proportional to $\sqrt{m_\uparrow/m_\downarrow}$. This feature is clearly observed in Figs. 5(b) and 6. Using a semi-classical approximation, Ref. 57 found that in the limit of a large mass ratio the number of bound states is approximately $0.73\sqrt{m_\uparrow/m_\downarrow}$.

3.2.1. *Bound states of three identical fermions and a light atom*

In fact, it is also possible for three identical fermionic atoms to bind together owing to the attractive interaction mediated by a light atom.[41] The critical mass ratio for the binding of this tetramer is $m_\uparrow/m_\downarrow \approx 5.0$, and like the trimer, the tetramer binds in the p-wave state. It remains to be seen whether more tetramers bind with increasing mass ratio, as in the case of trimers, and whether they also form in higher partial waves. These questions may presumably be answered within the Born-Oppenheimer approximation. Interestingly, the tetramer is very close in energy to a trimer plus a free atom, and this should lead to strong atom–trimer interactions in the 2D heteronuclear Fermi gas for species close to the critical mass ratio. A similar bound state has been predicted in 3D above a mass ratio of 9.5 (Ref. 42) and in 1D above the mass ratio 2.0 (Ref. 40). It is likely that these states are continuously connected as the system is tuned between the different geometries.

3.3. *Universal bound states in realistic experiments: Going beyond the 2D limit*

As discussed previously, realistic experiments on 2D Fermi gases involve the presence of a tight confinement and the length scale corresponding to this confinement always greatly exceeds the range of the interatomic interactions. Thus, it is important to relate the universal 2D few-body physics presented above to realistic experiments, taking into account the 3D nature of the interactions.

In Sec. 2.3, the effects of confinement on the two-body interaction were described. As should be clear from that discussion, the 2D limit of the universal bound states described thus far constitutes the regime where the 3D scattering length is negative and much smaller than the confinement length, i.e., $l_z/a \ll -1$; in this limit we have a dimer whose binding energy ε_b is much smaller than the level spacing in the harmonic trap, ω_z, and as the energies of the universal bound states scale with ε_b we may be in a regime where these are also negligible compared with ω_z.

It is natural to ask what happens to the bound trimers and tetramers described above, once the confinement is relaxed. This question was investigated in Ref. 41 under the assumption that both species of atoms are confined by a harmonic trap of the same frequency. It was shown that the minimum energy E in the problem of N spin-↑ atoms and a single spin-↓ atom in the center of mass frame corresponds to a non-trivial solution of

$$\chi_{\mathbf{k}_2...\mathbf{k}_N}^{n_0...n_N} = -\sum_{\mathbf{k}_1', n_0' n_1'} \frac{T_{n_0 n_1}^{n_0' n_1'}(\mathbf{k}_0 + \mathbf{k}_1, E_0 + \epsilon_{\mathbf{k}_1 n_1 \uparrow})}{E_0 + \epsilon_{\mathbf{k}_1 n_1 \uparrow} - \epsilon_{\mathbf{k}_0 + \mathbf{k}_1 - \mathbf{k}_1' n_0' \downarrow} - \epsilon_{\mathbf{k}_1' n_1' \uparrow}}$$

$$\times \left\{ \chi_{\mathbf{k}_1' \mathbf{k}_3...\mathbf{k}_N}^{n_0' n_2 n_1' n_3...n_N} + \cdots + \chi_{\mathbf{k}_2...\mathbf{k}_{N-1} \mathbf{k}_1'}^{n_0' n_N n_2...n_{N-1} n_1'} \right\}. \qquad (41)$$

Here the single particle energies are $\epsilon_{\mathbf{k} n \sigma} = k^2/2m_\sigma + n\omega_z$, $\mathbf{k}_1, \ldots, \mathbf{k}_N$ are the initial momenta of the spin-↑ atoms, while $\mathbf{k}_0$ and $E_0 \equiv E - \sum_{i=1}^{N} \epsilon_{\mathbf{k}_i n_i \uparrow}$ are the initial momentum and energy of the spin-↓ atom. Since we consider scattering in the center of mass frame of the 2D motion, we have $\mathbf{k}_0 = -\sum_i^N \mathbf{k}_i$. The energy is measured from the $N + 1$ atom threshold ($N + 1)\omega_z/2$. $T_{n_0 n_1}^{n_0' n_1'}$ is related to the quasi-two-dimensional T matrix[j] Eq. (25) via a change of basis to the relative and center of mass motion in the two-atom problem. The minus sign on the *r.h.s.* arises from the antisymmetry of the vertex χ under exchange of identical fermions.

---

[j]Specifically,

$$T_{n_0 n_1}^{n_0' n_1'}(\mathbf{q}, \epsilon) = \sum_{n n_r n_r'} C_{n n_r}^{n_0 n_1}(m_\downarrow, m_\uparrow) C_{n n_r'}^{n_0' n_1'}(m_\downarrow, m_\uparrow)$$

$$\times \sqrt{2\pi} l_z f_{n_r} f_{n_r'} T\left(\epsilon - n\omega_z + \frac{1}{2}\omega_z - \frac{q^2}{2M}\right).$$

The Clebsch-Gordan coefficients $C_{n n_r}^{n_0 n_1}(m_\downarrow, m_\uparrow) \equiv \langle n_0 n_1 | n n_r \rangle$ were obtained in Ref. 60 and vanish unless $n_0 + n_1 = n + n_r$.

While Eq. (41) is quite compact and in principle allows one to capture the crossover from 2D to 3D physics, its numerical solution quickly becomes prohibitive with increasing number of atoms. Instead, the departure from the 2D limit of the few-body bound states described above may be considered in a perturbative expansion[41] in the parameter ε_b/ω_z. This amounts to expanding the function $\mathcal{F}$ in Eq. (26) to linear order in $|E|/\omega_z$ while setting all harmonic oscillator quantum numbers to zero in Eq. (41), and follows from the antisymmetry of the vertex χ under exchange of any of the N spin-$\uparrow$ atoms. For definiteness we write down here the resulting equation which determines the trimer binding energy in the three-body problem in the limit of strong confinement:

$$
\frac{m_r}{2\pi} \left(\ln\left[\frac{-E + k_2^2/2m_{\mathrm{ad}}}{\varepsilon_b} \right] - \ln(2) \frac{\varepsilon_b + E - k_2^2/2m_{\mathrm{ad}}}{\omega_z} \right) \chi_{\mathbf{k}_2}
$$
$$
= \sum_{\mathbf{k}_1} \frac{\chi_{\mathbf{k}_1}}{E - \epsilon_{\mathbf{k}_1\uparrow} - \epsilon_{\mathbf{k}_2\uparrow} - \epsilon_{\mathbf{k}_1+\mathbf{k}_2\downarrow}}. \tag{42}
$$

Figure 8 shows the behavior of the critical mass ratio of the trimer and tetramer as the system is tuned away from the strict 2D limit. We observe that the critical mass ratio increases, consistent with the corresponding results in

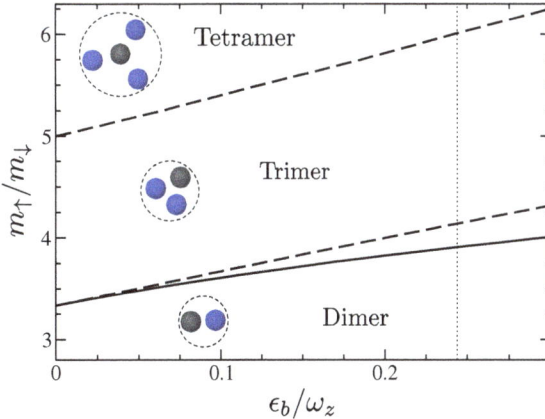

Fig. 8. The critical mass ratio for the formation of trimers and tetramers away from the 2D limit of $\varepsilon_b/\omega_z = 0$, assuming the two species are confined by a harmonic potential characterized by the same frequency ω_z. The solid line is the result of exact calculations, while the dashed lines employ a two-channel model for the confinement. The vertical dotted line corresponds to the position of the 3D resonance. The figure is taken from Ref. 41.

3D.[38,42] This behavior is likely due to the increased centrifugal barrier for
p-wave pairing in 3D. Interestingly, at unitarity the critical mass ratio for
tetramer formation in this quasi-2D geometry is below the K-Li mass ratio.

From the point of view of experiment, it is useful to determine the precise conditions under which K and Li atoms will form such stable few-body
bound states. In the above we assumed that the confinement frequencies of
the two atomic species were identical, and while this can be engineered using
species dependent optical lattices this need not be the case. However, even in
the case where the frequencies are species dependent, few-body properties
may only be weakly affected by this dependence; for instance, once the
light atom oscillator length greatly exceeds the two-body bound state, the
light atom is essentially confined by its interaction with the heavy atoms.
This is clearly illustrated in Fig. 9 which shows the critical confinement
length needed to confine two K atoms and a Li atom into a trimer, and
thus to make the atom-dimer interaction resonant in the p-wave channel.[54]
This result additionally takes into account the fact that Li-K interspecies
Feshbach resonances are narrow in magnetic field width, associated with
a weak coupling to the closed channel molecular state. This latter effect

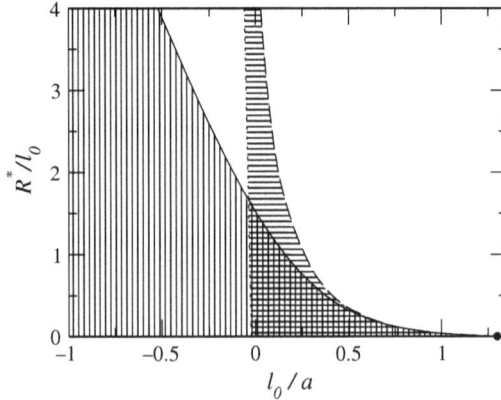

Fig. 9. Confinement induced trimers for the K-K-Li system: The solid line corresponds
to the trimer formation threshold when both species are confined by a harmonic potential
of the same frequency, while the dashed line assumes confinement of the heavy atom only.
Three atoms bind to a trimer in the shaded regions. $l_0 = 1/\sqrt{m_\uparrow \omega_\uparrow}$ is the confinement
length of the heavy atoms, while $R^* = -r_{\mathrm{eff}}/2 > 0$ is a length scale parameterizing the
width of the Feshbach resonance. The figure is taken from Ref. 54.

tends to suppress the attraction mediated by the light atom, and introduces an additional complication in the quest to obtain these few-body states.

3.4. *Identical fermions with* p-*wave interactions — Super Efimov states*

Finally, we mention that the three-body problem of identical fermions in 2D interacting via a p-wave Feshbach resonance has also been investigated.[61,62] Reference 61 studied the atom-dimer scattering problem and found that the T-matrix at a momentum $\mathbf{p}$ is $T(p) \propto \cos\left[\frac{4}{3}\ln \Lambda/p + \phi\right]$ where ϕ is a phase and Λ a momentum cutoff. Remarkably, this T-matrix displays a discrete scaling symmetry reminiscent of the Efimov effect, with a three-body parameter set by the momentum cutoff. However, unlike the standard Efimov effect, the scaling is here characterized by a double exponential factor, which quickly becomes enormous. More recently, Ref. 62 studied the bound state problem and found the existence of trimers with a doubly exponential scaling of the binding energy, naming these super Efimov states.

4. Ground State of the Many-Body System

4.1. *The BCS-BEC crossover*

We now turn to the behavior of the many-body system, where one has a finite density of spin-up and spin-down fermions, denoted $n_\uparrow = N_\uparrow/A$ and $n_\downarrow = N_\downarrow/A$, respectively. At zero temperature, in the absence of interactions, each type of fermion forms a filled Fermi sea with radius in momentum space given by the Fermi wave vector $k_{F\sigma} = \sqrt{4\pi n_\sigma}$ in 2D. For an attractive interspecies interaction, such as the short-range potential described in Eq. (10), a variety of pairing phenomena is expected to occur in the Fermi system. In particular, for the case of equal densities, $k_{F\uparrow} = k_{F\downarrow} \equiv k_F$, the ground state can smoothly evolve from the BCS regime of Cooper pairing to a Bose-Einstein condensate (BEC) of tightly bound dimers with increasing interaction strength. This constitutes the celebrated BCS-BEC crossover, which was theoretically predicted several decades ago,[63,64] and first successfully realised in 3D cold-atom experiments in 2004 (Refs. 65 and 66).

The different regimes of pairing are generally parameterized by the dimensionless quantity $k_F l$, where l is a typical length scale that defines the strength of the interaction. For short-range s-wave interactions in 3D, l is simply the s-wave scattering length a_s, while in 2D it corresponds to a_{2D}, which is related to the size of the two-body bound state and is defined in Sec. 2.2. Thus, in 2D, the BCS and BEC regimes correspond, respectively, to $k_F a_{2D} \gg 1$ and $k_F a_{2D} \ll 1$, where the pair size is much greater than the inter-particle spacing in the former case, and much smaller in the latter. The dimensionless parameter $k_F a_{2D}$ automatically implies that there are two ways of achieving the BCS-BEC crossover: by varying the interactions or by varying the density. Note, however, that this is not the case for 3D contact interactions, since a two-body bound state does not exist for arbitrarily weak attraction, and the scattering length can change sign. Thus, in 3D, one cannot traverse the entire crossover by varying density alone.

This section will focus on the situation where the masses are equal, i.e., $m_\uparrow = m_\downarrow \equiv m$. In any case, we do not expect the qualitative picture of the BCS-BEC crossover to change for a small mass imbalance. However, we can see from Fig. 6 in Sec. 3 that once $m_\uparrow/m_\downarrow \gtrsim 25$ (or equivalently when $m_\uparrow/m_\downarrow \lesssim 1/25$), the trimer state has a lower energy than two dimers and therefore, in the BEC regime, the system will prefer to form a mixed gas of trimers and light atoms. It remains an open question what happens deep in the BCS regime, where the size of the trimer becomes larger than the inter-particle distance.

4.2. Mean-field description

To gain insight into the BCS-BEC crossover, it is instructive to employ a mean-field approach to the problem. We start by considering the full quasi-2D problem as exists in experiment, and then we specialize to the 2D limit in later sections. The quasi-2D Fermi gas was first considered within the mean-field approximation in Ref. 67, but only a few harmonic oscillator levels were included. Here, we follow the approach in Ref. 68, which can in principle account for an infinite number of levels.

Building on the formalism in Sec. 2, the many-body grand-canonical Hamiltonian in the quasi-2D geometry is (setting the system area $A = 1$):

$$\hat{H} = \sum_{\mathbf{k},n,\sigma} (\epsilon_{\mathbf{k}n} - \mu) c^\dagger_{\mathbf{k}n\sigma} c_{\mathbf{k}n\sigma}$$

$$+ \sum_{\substack{\mathbf{k},n_1,n_2 \\ \mathbf{k}',n_3,n_4 \\ \mathbf{q}}} \langle n_1 n_2 | \hat{g} | n_3 n_4 \rangle c^\dagger_{\mathbf{k}n_1\uparrow} c^\dagger_{\mathbf{q}-\mathbf{k}n_2\downarrow} c_{\mathbf{q}-\mathbf{k}'n_3\downarrow} c_{\mathbf{k}'n_4\uparrow}, \qquad (43)$$

where $\epsilon_{\mathbf{k}n} = k^2/2m + n\omega_z$ are the single particle energies relative to the zero-point energy of the $n = 0$ state. Note that since we have assumed that the masses and particle densities are equal, the chemical potential must be the same for each spin σ, i.e., $\mu_\uparrow = \mu_\downarrow \equiv \mu$.

The 3D attractive short-range interaction is set by the constant g like in Sec. 2. In the many-body system, it is convenient to work in the basis of the individual atoms rather than only considering the relative pair motion as in the two-body problem. However, since the interaction only depends on the relative motion, the interaction matrix elements $\langle n_1 n_2 | \hat{g} | n_3 n_4 \rangle$ are best determined by switching to relative and center of mass harmonic oscillator quantum numbers, ν and N respectively. This yields

$$\langle n_1 n_2 | \hat{g} | n_3 n_4 \rangle = g \sum_{N\nu\nu'} f_\nu \langle n_1 n_2 | N\nu \rangle f_{\nu'} \langle N\nu' | n_3 n_4 \rangle$$

$$\equiv g \sum_N V_N^{n_1 n_2} V_N^{n_3 n_4}, \tag{44}$$

where $f_\nu = \sum_{k_z} \tilde{\phi}_\nu(k_z)$, and $\tilde{\phi}_\nu$ is the Fourier transform of the ν-th harmonic oscillator eigenfunction. These correspond to the f coefficients defined previously in Sec. 2, but with momentum cut-off $\Lambda \to \infty$, i.e., for even ν, we obtain Eq. (20) with $\lambda = 0$. In this case, f_ν reduces to the harmonic oscillator wave function evaluated at $z = 0$. The Clebsch-Gordan coefficients in the matrix elements are given by[60,69]

$$\langle n_1 n_2 | N\nu \rangle = \delta_{N+\nu, n_1+n_2} \sqrt{\frac{N!\nu!}{2^{n_1+n_2} n_1! n_2!}} \sum_{i+j=\nu} (-1)^j \binom{n_1}{i} \binom{n_2}{j}, \tag{45}$$

with $i = 0, 1, \ldots, n_1$ and $j = 0, 1, \ldots, n_2$. Note that the scattering process conserves parity since ν, ν' must be even; namely, if $n_1 + n_2$ is even (odd), then the matrix element is only non-zero when $n_3 + n_4$ is also even (odd). The 3D contact interaction parameter g can be written in terms of the quasi-2D two-body binding energy ε_b:

$$-\frac{1}{g} = \sum_{\mathbf{k}, n_1, n_2} \frac{f_{n_1+n_2}^2 |\langle n_1 n_2 | 0 \; n_1 + n_2 \rangle|^2}{\epsilon_{\mathbf{k}n_1} + \epsilon_{\mathbf{k}n_2} + \varepsilon_b}. \tag{46}$$

Here, we simply take $N = 0$ since ε_b is independent of the two-body center of mass motion. One can also connect ε_b to the 3D scattering length a_s using Eq. (31) with $\lambda = 0$.

Following Ref. 68, we define the pairing order parameter

$$\Delta_{\mathbf{q}N} = g \sum_{\mathbf{k},n_1,n_2} V_N^{n_1 n_2} \langle c_{\mathbf{q}-\mathbf{k}n_2\downarrow} c_{\mathbf{k}n_1\uparrow} \rangle, \tag{47}$$

and assume that fluctuations around this are small, thus obtaining the mean-field Hamiltonian,

$$\hat{H}_{\mathrm{MF}} = \sum_{\mathbf{k},n,\sigma} (\epsilon_{\mathbf{k}n} - \mu) c_{\mathbf{k}n\sigma}^\dagger c_{\mathbf{k}n\sigma}$$

$$+ \sum_{\mathbf{q},N} \left(\Delta_{\mathbf{q}N} \sum_{\mathbf{k},n_1,n_2} V_N^{n_1 n_2} c_{\mathbf{k}n_1\uparrow}^\dagger c_{\mathbf{q}-\mathbf{k}n_2\downarrow}^\dagger \right.$$

$$\left. + \Delta_{\mathbf{q}N}^* \sum_{\mathbf{k}',n_3,n_4} V_N^{n_3 n_4} c_{\mathbf{q}-\mathbf{k}'n_3\downarrow} c_{\mathbf{k}'n_4\uparrow} - \frac{|\Delta_{\mathbf{q}N}|^2}{g} \right). \tag{48}$$

If we further assume that the ground state has a uniform order parameter without nodes so that $\Delta_{\mathbf{q}N} = \delta_{\mathbf{q}0} \delta_{N0} \Delta_0$, then Eq. (48) only contains a single unknown parameter Δ_0. Thus $\hat{H}_{\mathrm{MF}}$ can be diagonalized to yield

$$\hat{H}_{\mathrm{MF}} = \sum_{\mathbf{k},n} (\epsilon_{\mathbf{k}n} - \mu - E_{\mathbf{k}n}) - \frac{\Delta_0^2}{g} + \sum_{\mathbf{k},n,\sigma} E_{\mathbf{k}n} \gamma_{\mathbf{k}n\sigma}^\dagger \gamma_{\mathbf{k}n\sigma}, \tag{49}$$

where $E_{\mathbf{k}n}$ are the quasiparticle excitation energies. The quasiparticle creation and annihilation operators are respectively given by

$$\gamma_{\mathbf{k}n\uparrow}^\dagger = \sum_{n'} (u_{\mathbf{k}n'n} c_{\mathbf{k}n'\uparrow}^\dagger + v_{\mathbf{k}n'n} c_{-\mathbf{k}n'\downarrow}) \tag{50}$$

$$\gamma_{-\mathbf{k}n\downarrow} = \sum_{n'} (u_{\mathbf{k}n'n} c_{-\mathbf{k}n'\downarrow} - v_{\mathbf{k}n'n} c_{\mathbf{k}n'\uparrow}^\dagger), \tag{51}$$

where the amplitudes u, v only depend on the magnitude $k \equiv |\mathbf{k}|$ and satisfy $\sum_{n'} (|u_{\mathbf{k}n'n}|^2 + |v_{\mathbf{k}n'n}|^2) = 1$. Without loss of generality, we can choose u, v to be real. Note that while the quasiparticles have a well defined spin and momentum, they involve a superposition of different harmonic oscillator levels. Since the ground state corresponds to the vacuum state for the quasiparticles, the ground-state wave function can be written

$$|\Psi_{\text{MF}}\rangle \propto \prod_{\mathbf{k}n\sigma} \gamma_{\mathbf{k}n\sigma} |0\rangle , \tag{52}$$

where $|0\rangle$ is the vacuum state for the bare operators $c_{\mathbf{k}n\sigma}$. In the 2D limit where $\omega_z \gg \mu, \varepsilon_b$ and we only have the lowest level $n = 0$, we recover the standard BCS wave function

$$|\Psi_{\text{MF}}\rangle = \prod_{\mathbf{k}} \left(u_{\mathbf{k}00} + v_{\mathbf{k}00} c^{\dagger}_{\mathbf{k}0\uparrow} c^{\dagger}_{-\mathbf{k}0\downarrow} \right) |0\rangle. \tag{53}$$

In general, we must minimize $\langle \hat{H}_{\text{MF}} \rangle = \sum_{\mathbf{k},n} (\epsilon_{\mathbf{k}n} - \mu - E_{\mathbf{k}n}) - \frac{\Delta_0^2}{g}$ with respect to Δ_0 at fixed μ to obtain the ground state. For the 2D limit, this yields the usual form $E_{\mathbf{k}} \equiv E_{\mathbf{k}0} = \sqrt{(\epsilon_{\mathbf{k}} - \mu)^2 + \Delta^2}$, with $\epsilon_{\mathbf{k}} \equiv \epsilon_{\mathbf{k}0}$ and $\Delta \equiv \Delta_0 V_0^{00}$. Using the density $2n_{\sigma} = -\partial \langle \hat{H}_{\text{MF}} \rangle / \partial \mu$ and the 2D Fermi energy $\varepsilon_F = k_F^2/2m$, we also obtain $\mu = \varepsilon_F - \varepsilon_b/2$ and $\Delta = \sqrt{2\varepsilon_F \varepsilon_b}$, as derived previously.[18,19] For the quasi-2D system, one must use the general expression for the density $n_{\sigma} = \sum_{\mathbf{k},n',n} |v_{\mathbf{k}n'n}|^2$ and define ε_F to be the chemical potential of an ideal Fermi gas with the same density.

As the interactions are varied in the quasi-2D Fermi gas, the chemical potential always evolves from ε_F in the weakly interacting BCS limit $\varepsilon_b/\varepsilon_F \ll 1$, to $-\varepsilon_b/2$ in the BEC limit $\varepsilon_b/\varepsilon_F \gg 1$. However, the precise evolution in the BCS-BEC crossover is dependent on the quasi-2D confinement frequency ω_z once we are away from the 2D limit $\omega_z \gg \varepsilon_F, \varepsilon_b$, and this has implications for current 2D experiments, as we discuss later. If we take the limit of weak confinement $\omega_z \to 0$, there will be a crossover to 3D pairing, albeit with a modified density of states due to the trapping potential in the z-direction. For instance, in the BEC regime, we should recover "3D bosons" with $\varepsilon_b \simeq 1/ma_s^2$ in the limit $\varepsilon_b \gg \omega_z$ — see the discussion in Sec. 2.3. For strong confinement, one might naively expect to observe 2D behaviour once $\varepsilon_b, \varepsilon_F < \omega_z$. However, the chemical potential can be dramatically reduced from the 2D result even in this regime, as shown in Fig. 10 (a), and the deviation from 2D involves multiple harmonic oscillator levels.[68] Moreover, once we approach $\varepsilon_F \simeq \omega_z$, the chemical potential is strongly modified even when the interactions are weak, $\varepsilon_b \ll \omega_z$. This is analogous to the behaviour of the quasi-2D two-body T matrix (27), which only resembles the 2D expression when the collision energy $E \ll \omega_z$, and is

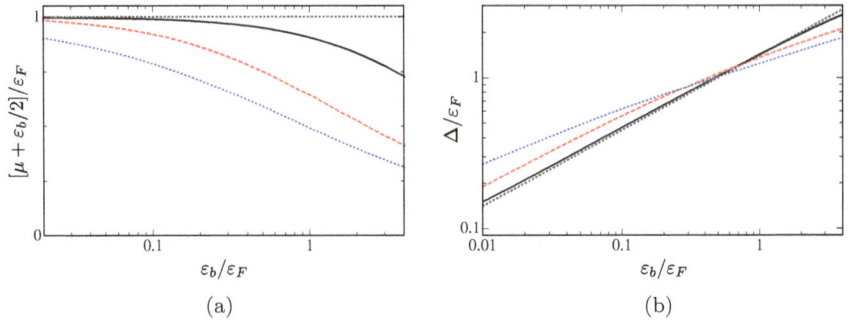

(a)   (b)

Fig. 10. Behaviour of the chemical potential (a) and the order parameter (b) as a function of the interaction parameter $\varepsilon_b/\varepsilon_F$ for different confinement strengths. The solid (black), dashed (red), and dotted (blue) lines correspond to $\varepsilon_F/\omega_z = 0.1$, 0.5 and 1, respectively. The thin dotted lines are the results in the 2D limit, $\mu = \varepsilon_F - \varepsilon_b/2$ and $\Delta = \sqrt{2\varepsilon_b\varepsilon_F}$. In (a), the term from two-body binding, $-\varepsilon_b/2$, has been subtracted from μ in order to expose the many-body corrections. The data is taken from Ref. 68.

substantially different when $E \simeq \omega_z$. The pairing order parameter Δ is also modified by the presence of ω_z, with Δ being increased in the BCS regime with increasing ε_F/ω_z, as seen in Fig. 10 (b). This suggests that pairing is enhanced by perturbing away from the 2D limit and this therefore impacts the critical temperature for superfluidity (see Sec. 5).

4.3. Perturbative regimes

In order to go beyond mean-field theory, we restrict ourselves to the 2D limit $\omega_z \gg \varepsilon_F, \varepsilon_b$, so that the different regimes are completely parameterized by $k_F a_{2D}$. In this case, the Hamiltonian in Eq. (43) reduces to

$$\hat{H} = \sum_{\mathbf{k},\sigma} (\epsilon_\mathbf{k} - \mu)\, c^\dagger_{\mathbf{k}\sigma} c_{\mathbf{k}\sigma} + g \sum_{\mathbf{k},\mathbf{k}',\mathbf{q}} c^\dagger_{\mathbf{k}\uparrow} c^\dagger_{\mathbf{q}-\mathbf{k}\downarrow} c_{\mathbf{q}-\mathbf{k}'\downarrow} c_{\mathbf{k}'\uparrow}, \qquad (54)$$

where g is now an effective 2D contact interaction like in Eq. (10), which gives rise to the required scattering length a_{2D} (or two-body binding energy ε_b). While mean-field theory provides an appealingly simple and intuitive picture of the BCS-BEC crossover in 2D, it is not expected to be quantitatively accurate and it at best provides an upper bound on the energy, being in essence a variational approach. In particular, it substantially overestimates the effective dimer-dimer interaction in the BEC regime. Even in the BCS regime where interactions are weak, it fails to capture the leading order dependence of the energy on $1/\ln(k_F a_{2D})$ since it neglects the interaction energy of the normal Fermi liquid phase. However, one can extract accurate analytic expressions for the behaviour in the limits $|\ln(k_F a_{2D})| \gg 1$ by performing a proper perturbative expansion in the interaction.

In the regime of weak attraction $\ln(k_F a_{2D}) \gg 1$, the gas behaves as a Fermi liquid in the normal state.[70-72] One can show using perturbation theory in g that the energy per particle in this limit is[71,73]

$$\frac{E}{N} = \frac{\varepsilon_F}{2}\left(1 - \frac{1}{\eta} + \frac{\mathcal{A}}{\eta^2}\right), \tag{55}$$

where $N = N_\uparrow + N_\downarrow$ and $\eta = \ln(k_F a_{2D})$. Of course, in the ground state, the gas will be a paired superfluid rather than a Fermi liquid, but the energy due to pairing scales as $\Delta^2/\varepsilon_F \sim \varepsilon_b$ in this limit, which tends to zero faster than $\varepsilon_F/\ln(k_F a_{2D})$ as $\varepsilon_b \to 0$. While the structure of the perturbative expansion is clear, there is some discrepancy in the literature regarding the constant $\mathcal{A}$. A thorough calculation for the repulsive Fermi gas using second-order perturbation theory[71,72] gives $\mathcal{A} = 3/4 - \ln(2) \simeq 0.06$. However, a recent QMC calculation[55] finds a larger value: $\mathcal{A} \simeq 0.17$.

In the opposite limit $\ln(k_F a_{2D}) \ll -1$, the system can be regarded as a weakly interacting gas of bosonic dimers. In this case, the effective dimer-dimer interaction g_d in the low-energy limit is parameterized by the scattering length $a_{dd} \simeq 0.56 a_{2D}$, as noted in Sec. 3. Moreover, the total energy of the system can be written as $E = -\varepsilon_b N_d + E_d$, where $N_d = N_\sigma = N/2$ and E_d is the energy of a repulsive gas of N_d bosons. We can likewise introduce a boson chemical potential $\mu_d = 2\mu + \varepsilon_b$. Note that in the limit $a_{dd} \to 0$ (or, equivalently, when $\varepsilon_b \to \infty$), we have $\mu_d \to 0^+$, as expected for a non-interacting BEC. However, this behavior is not captured by BCS mean-field theory, which predicts $\mu_d = 2\varepsilon_F$. This corresponds to an effective dimer-dimer interaction that only scales with the density, as one might expect from a classical theory of interacting dimers in 2D rather than an appropriately renormalized quantum one (see Sec. 7).

To extract the behaviour of the weakly repulsive Bose gas, we consider the grand potential according to Bogoliubov theory:

$$\Omega = -\frac{\mu_d^2}{2g_d} - \frac{1}{2}\sum_{k\neq 0}\left(\epsilon_{kd} + \mu_d - \sqrt{\epsilon_{kd}(\epsilon_{kd} + 2\mu_d)}\right), \tag{56}$$

where $\epsilon_{kd} = \frac{k^2}{2M}$ and $M = m_\uparrow + m_\downarrow = 2m$. After regularizing the momentum sum, we obtain (see, also, Ref. 74)

$$\Omega = \frac{M\mu_d^2}{16\pi}\left[1 - 2\ln\left(\frac{1}{Ma_{dd}^2\mu_d}\right)\right]. \tag{57}$$

To relate this back to the Fermi system, we consider the density of dimers:

$$n_d \equiv \frac{N_d}{A} = -\frac{\partial\Omega}{\partial\mu_d} = \frac{M\mu_d}{4\pi}\ln\left(\frac{1}{Ma_{dd}^2\mu_d e}\right). \tag{58}$$

Assuming that $\ln\left(\frac{1}{4\pi e n_d a_{dd}^2}\right) \gg 1$, this can then be rearranged to obtain the leading order expression for μ_d in terms of the density n_d, i.e.,

$$\mu_d \simeq \frac{4\pi n_d}{M}\frac{1}{\ln\left(\frac{1}{4\pi e n_d a_{dd}^2}\right)}\left[1 - \frac{\ln\ln\left(\frac{1}{4\pi e n_d a_{dd}^2}\right)}{\ln\left(\frac{1}{4\pi e n_d a_{dd}^2}\right)}\right]. \tag{59}$$

This finally gives us the energy density $E/N_d = -\varepsilon_b + E_d/N_d$, with

$$\frac{E_d}{N_d} \simeq \frac{2\pi n_d}{M}\frac{1}{\ln\left(\frac{1}{4\pi e n_d a_{dd}^2}\right)}\left[1 - \frac{\ln\ln\left(\frac{1}{4\pi e n_d a_{dd}^2}\right)}{\ln\left(\frac{1}{4\pi e n_d a_{dd}^2}\right)} - \frac{1}{2\ln\left(\frac{1}{4\pi e n_d a_{dd}^2}\right)}\right], \tag{60}$$

where we have used the fact that $E_d = \Omega + \mu_d N_d$ at zero temperature. This agrees with the expression used in Ref. 55.

4.4. Equation of state

In order to obtain an accurate equation of state throughout the BCS-BEC crossover, one must resort to numerical approaches such as the fixed node diffusion QMC method mentioned previously.[55] The QMC result for the energy per particle is displayed in Fig. 11(a).[k] As expected, the energy matches the known results in the BCS and BEC limits (taking the QMC value for $\mathcal{A}$ in Eq. (55)). Indeed, by interpolating between the known weak-coupling results in the BCS and BEC regimes, we can obtain a reasonable

---

[k]Note that we have used a different definition of a_{2D} compared with the original QMC paper — see the discussion in Sec. 2.2.

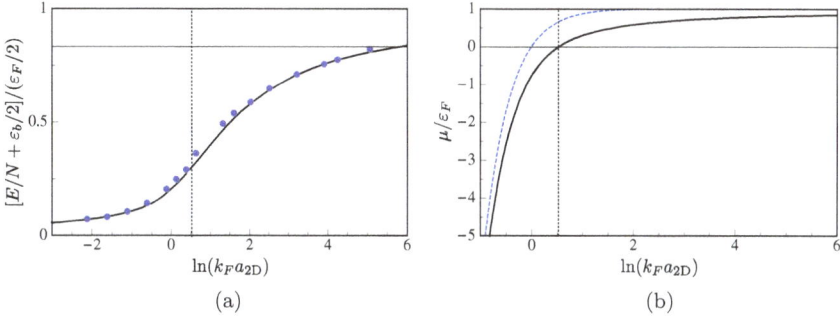

Fig. 11. (a) Energy per particle in units of $\varepsilon_F/2$ (the energy per particle in the non-interacting gas) as a function of interaction strength. The trivial two-body binding energy has been subtracted. The data points are the results of the QMC calculation,[55] while the solid line is an interpolation between the known weak coupling results in the BCS and BEC limits. (b) Chemical potential in units of the Fermi energy. The solid (black) line is an interpolation between the known limiting behaviors, while the dashed (blue) line is the mean-field result $\mu = \varepsilon_F - \varepsilon_b/2$. In both plots, the vertical dotted line indicates the point at which the chemical potential is zero.

curve for the energy throughout the crossover that matches the QMC data. This also allows us to easily extract other thermodynamic quantities such as the chemical potential and the pressure — one simply takes the appropriate derivative of the expressions in the perturbative regimes and then interpolates between the results. In particular, the chemical potential in the BEC regime can be taken from Eq. (59), while in the BCS limit it corresponds to

$$\mu = \frac{\partial E}{\partial N} = \varepsilon_F \left(1 - \frac{1}{\eta} + \frac{4\mathcal{A} + 1}{4\eta^2} \right). \tag{61}$$

Referring to Fig. 11(b), we find that the interpolated μ is lower than that from mean-field theory, but it still evolves from ε_F to $-\varepsilon_b/2$ with increasing attraction (decreasing $\ln(k_F a_{2D})$). The point $\mu = 0$ may be regarded as the "crossover point" that approximately separates Fermi and Bose regimes, as we discuss in Sec. 4.5.

The pressure as a function of interaction has recently been measured experimentally in a quasi-2D Fermi gas.[17] As shown in Fig 12, the comparison with the QMC prediction is reasonable, aside from the BCS side of the crossover (or Fermi regime), where the pressure is significantly higher. Indeed, this is in the limit where the weak-coupling result becomes

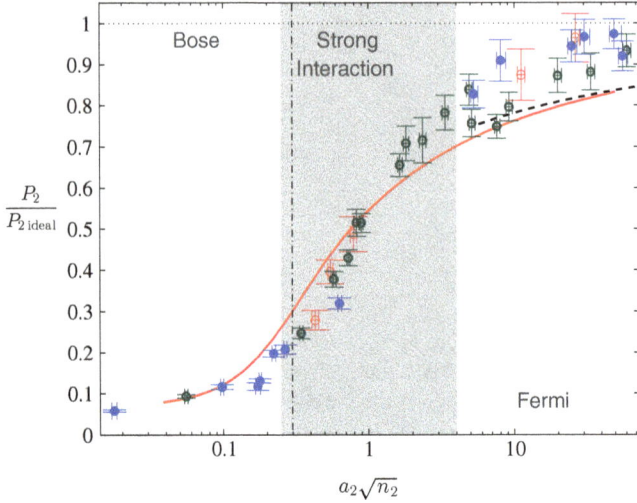

Fig. 12. The experimentally measured pressure P_2 in a quasi-2D Fermi gas.[17] The pressure is scaled with respect to that of an ideal 2D Fermi gas, $P_{2 \, \text{ideal}}$. Note that the experiment is not purely 2D since $\varepsilon_F = 2\pi n_2/m \gtrsim 0.5\omega_z$, and $\varepsilon_b > \omega_z$ in the Bose limit. Here, $n_2 = n_\uparrow = n_\downarrow$ is the planar density per spin. Therefore, the scattering length (denoted a_2) is not always simply related to ε_b, and must be defined via the quasi-2D T matrix. The solid curve is obtained from a fit to the 2D QMC data[55] while the dashed line approximately corresponds to the weak-coupling result in the Fermi regime. (Reprinted figure with permission from: V. Makhalov, K. Martiyanov, and A. Turlapov, *Phys. Rev. Lett.* **112**, 045301 (2014). Copyright 2014 by the American Physical Society.)

accurate and the pressure should simply correspond to:

$$P \simeq \frac{(2n_\uparrow)^2 \pi}{2m} \left(1 - \frac{1}{\eta} + \frac{2\mathcal{A} + 1}{2\eta^2} \right). \tag{62}$$

However, the experiment was performed at finite temperature, while the theory is for zero temperature. Indeed, a recent self-consistent T-matrix (or Luttinger-Ward) calculation[75] for the normal state finds that the deviations in this regime are consistent with a temperature of $T \simeq 0.15 T_F$, where $T_F = \varepsilon_F/k_B$. Another factor that complicates the analysis is the quasi-2D nature of the gas. The experiment is never in the 2D limit since $\varepsilon_b > \omega_z$ in the Bose regime and $\varepsilon_F \gtrsim 0.5\omega_z$ throughout. This may account for the smaller but arguably more striking deviation from the QMC prediction in the strongly interacting regime (Fig 12). Here, the pressure is consistently below the QMC curve, whereas one would generally expect the pressure to be higher at finite temperature when $k_F a_{2D}$ is fixed. However, perturbing away from the 2D limit (which is equivalent to relaxing the quasi-2D confinement) is expected to reduce the pressure in the plane since the atoms can spread out in the transverse direction.[76] This suggests that there are two competing effects in the experiment: finite temperature tends to raise the pressure, as evident in the Fermi regime, while quasi-2D effects act to reduce it, particularly for strong interactions where both ε_b/ω_z and ε_F/ω_z are sizeable.

4.4.1. *Contact*

Another important thermodynamic quantity is the contact density C, which fixes the tail of the momentum distribution: $n(\mathbf{k}) \sim C/k^4$ as $k \to \infty$. This is related to the 2D equation of state by the adiabatic theorem[77,78]

$$C = 2\pi m \frac{dE}{d \ln a_{2D}} . \qquad (63)$$

Since the contact determines the short-distance behavior of the gas (i.e., it essentially gives the probability of finding a pair of ↑ and ↓ fermions close together), it can also be related to the high-frequency and large-momentum limits of other correlation functions in 2D such as the current response function.[79] Mean-field theory simply gives $C = m^2\Delta^2$, which is consistent with the fact that C should monotonically increase with increasing attraction, and it yields the correct two-body contact in the Bose limit. However, the mean-field result is not quantitatively accurate in the Fermi regime since it does not capture the leading order dependence on the interaction, as discussed in Sec. 4.3.

The 2D contact may be experimentally determined from the high-frequency tail of the RF spectrum,[80] as well as from the momentum profile. In contrast to the pressure, the contact appears to be surprisingly insensitive to temperature in the degenerate regime $T < T_F$. Figure 13 shows that there is good agreement between the experimentally measured contact density[15] at $T/T_F = 0.27$ and the $T = 0$ QMC result.[55] The contact determined using the Luttinger-Ward approach at the same temperature ($T/T_F = 0.27$) confirms that it is relatively unchanged for low temperatures.[75]

4.5. *The 2D crossover "point"*

Finally, we turn to the crossover from Fermi to Bose behavior in the regime of strong interactions $|\ln(k_F a_{2D})| \lesssim 1$. From the point of view

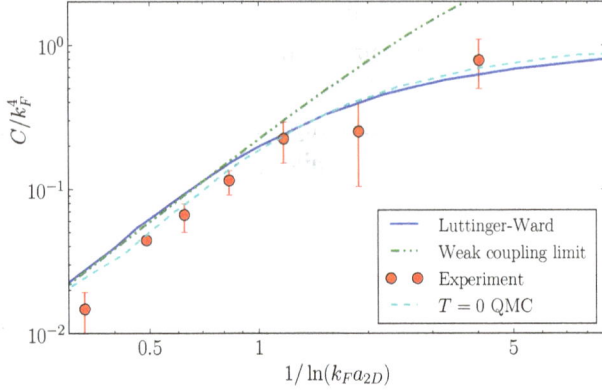

Fig. 13. Contact density C in the regime $\ln(k_F a_{2D}) > 0$. The filled circles correspond to the data in Ref. 15 at $T/T_F = 0.27$, the dashed line is a fit to the 2D QMC result,[55] and the solid line is determined from a Luttinger-Ward, or self-consistent T-matrix, approach described in Ref. 75. The $T = 0$ weak-coupling limit is extracted from Eq. (55). The figure is adapted from Ref. 75.

of mean-field theory, the natural crossover point is at $\mu = 0$, since this marks a qualitative change in the quasiparticle excitation spectrum $E_\mathbf{k} = \sqrt{(\epsilon_\mathbf{k} - \mu)^2 + \Delta^2}$. When $\mu > 0$, the minimum energy gap Δ occurs at finite momentum, $k = \sqrt{2m\mu}$, corresponding to the remnants of a Fermi surface. However, for $\mu < 0$, the minimum gap occurs at $k = 0$ and no longer corresponds to Δ. Indeed, for sufficiently strong attraction, the gap in the single-particle spectrum becomes $-\varepsilon_b/2$, as expected. Thus, the $\mu < 0$ regime resembles the behavior of a gas of bosonic dimers. According to mean-field theory, the point where $\mu = 0$ corresponds to $\ln(k_F a_{2D}) = 0$, and this is generally viewed as playing a role analogous to the unitarity point $1/a_s = 0$ in the 3D BCS-BEC crossover.[1]

However, if we go beyond mean-field theory and consider the more accurate QMC calculations,[55] then we find[81] that the point where $\mu = 0$ instead occurs at much weaker attraction, with $\ln(k_F a_{2D}) \simeq 0.5$, as shown in Fig. 11. This suggests that the Fermi side of the crossover occurs at larger $\ln(k_F a_{2D})$ than previously assumed, and is consistent with the observation

---

[1]Note that the analogy between $1/a_s = 0$ in 3D and $\ln(k_F a_{2D}) = 0$ in 2D is far from perfect, since we have $\mu > 0$ at $1/a_s = 0$ and thus the unitarity point lies on the Fermi side of the crossover.

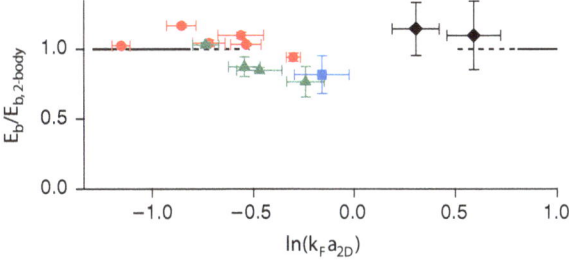

Fig. 14. Experimental measurement of the pairing gap E_b in the 2D Fermi gas using RF spectroscopy, taken from Ref. 11. The straight line corresponds to the mean-field result, which is simply the two-body binding energy (denoted here by $E_{b,2\text{-body}}$). (Reprinted figure with permission from: A. T. Sommer, L. W. Cheuk, M. J. H. Ku, W. S. Bakr, and M. W. Zwierlein, *Phys. Rev. Lett.* **108**, 045302 (2012). Copyright 2012 by the American Physical Society.)

in QMC simulations[55] that a variational wave function based on dimers outperforms the one for a Fermi liquid once $\ln(k_F a_{2D}) \lesssim 1$. Of course, it remains an open question how μ is connected to the quasiparticle dispersion beyond the mean-field approximation, but a negative μ already indicates strong deviations from fermionic behavior.

A recent experiment[11] on pairing in the 2D Fermi gas also suggests that the Bose regime extends beyond $\ln(k_F a_{2D}) = 0$. Figure 14 shows the pairing gap $E_b \equiv E_{k=0} - \mu$, corresponding to the onset of the pairing peak in the RF spectrum. At first glance, this measurement appears to validate mean-field theory, which simply predicts $E_b = \varepsilon_b$ throughout the crossover. However, this result is also expected for a gas of dimers; thus an alternative explanation is that the experiment only probes the Bose limit of the crossover, with $\ln(k_F a_{2D}) \lesssim 0.5$.

5. Finite-Temperature Phenomenology

The behaviour of the 2D Fermi gas at finite temperature is even richer than at zero temperature, since we have the possibility of superfluid phase transitions and pairing without superfluidity. Above the critical temperature T_c for superfluidity, the normal phase is also markedly different in the two perturbative limits, with a Fermi liquid for $\ln(k_F a_{2D}) \gg 1$ and a Bose liquid for $\ln(k_F a_{2D}) \ll -1$. This raises the question of whether the normal gas within the crossover can display features intermediate between Fermi and

Bose behavior. In particular, there may exist a so-called "pseudogap" regime in the normal phase, where there is a suppression of spectral weight at the Fermi surface that is reminiscent of a pairing gap.[82] Such a phenomenon has been observed in the quasi-2D cuprate superconductors, but its origin still remains a mystery.[83] By investigating its existence in attractive Fermi gases, cold-atom experiments may help settle the question of whether or not a pseudogap can be produced by pairing alone, in principle.

In this section, we will review our current understanding of the normal phase of the 2D Fermi gas, including the transition to superfluidity at low temperatures. We will also briefly discuss how the behavior is affected by the quasi-2D nature of the gas and the in-plane trapping potential present in experiment. To simplify the equations, we set $k_B = 1$ in the following.

5.1. Critical temperature for superfluid transition

Two-dimensional gases are marginal in the sense that true long-range order (i.e., condensation) only exists at $T = 0$. Instead, the superfluid phase at finite temperature exhibits quasi-long-range order where the correlations decay algebraically.[84] Increasing temperature further eventually results in a Berezinskii-Kosterlitz-Thouless (BKT) transition to the normal phase.[84–86]

In the limit $\ln(k_F a_{2D}) \ll -1$, the system corresponds to a weakly interacting Bose gas and the BKT transition temperature is:[56]

$$\frac{T_c}{T_F} = \frac{1}{2}\left[\ln\left(\frac{\mathcal{B}}{4\pi}\ln\left(\frac{4\pi}{k_F^2 a_{2D}^2}\right)\right)\right]^{-1}, \tag{64}$$

where $\mathcal{B} \simeq 380$. Note that $T_c \to 0$ in the limit $\ln(k_F a_{2D}) \to -\infty$, but since the dependence on $\ln(k_F a_{2D})$ is logarithmic, in practice we obtain $T_c/T_F \simeq 0.1$ for the interaction regime accessible in experiment (see Fig. 15).

In the BCS limit $\ln(k_F a_{2D}) \gg 1$, the critical temperature is set by the energy required to break pairs, which is the lowest energy scale in the problem. Thus, one can estimate T_c by taking the mean-field Hamiltonian (49) and determining the point at which Δ vanishes.[87] From the resulting linearized gap equation (or Thouless criterion) one obtains[88]

$$\frac{T_c}{T_F} = \frac{2e^\gamma}{\pi k_F a_{2D}}. \tag{65}$$

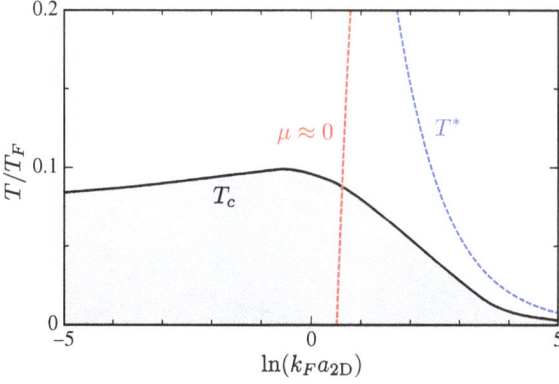

Fig. 15. Schematic phase diagram throughout the BCS-Bose crossover. The critical temperature for superfluidity is represented by the solid line, and corresponds to an interpolation between the known limits. The dashed lines correspond to $\mu \approx 0$ and the onset of pairing T^*, which approximately bound the pseudogap region above T_c. The $\mu(T) \approx 0$ line is obtained by setting $T = \mu(0)$, while T^* is estimated from the Thouless criterion (65).

A more thorough calculation that includes Gorkov–Melik-Barkhudarov corrections[56] yields the BCS result above reduced by a factor of e.

Referring to Fig. 15, we see that the results for T_c in the BCS and Bose limits can be smoothly interpolated, suggesting that T_c/T_F never exceeds 0.1. Note that T_c has a maximum in the regime $|\ln(k_F a_{2D})| < 1$. As yet, there is no experimental observation of T_c in the 2D Fermi gas.

5.1.1. *Quasi-2D case*

Given that experiments deal with quasi-2D Fermi gases, it is important to understand the effect of a finite confinement length on T_c. This is in general a challenging problem to address throughout the BCS-Bose crossover, but it is possible to estimate the dependence on ε_F/ω_z in the BCS limit. Using the mean-field approach for the quasi-2D system described in Sec. 4.2, one obtains a natural generalization of the Thouless criterion to quasi-2D:[76]

$$-\frac{1}{g} = \sum_{\mathbf{k},n_1,n_2} (V_0^{n_1 n_2})^2 \frac{\tanh\left(\beta_c \xi_{\mathbf{k}n_1}/2\right) + \tanh\left(\beta_c \xi_{\mathbf{k}n_2}/2\right)}{2(\xi_{\mathbf{k}n_1} + \xi_{\mathbf{k}n_2})}, \quad (66)$$

where g is the strength of the 3D contact interaction, $\beta_c = 1/T_c$, and $\xi_{\mathbf{k}n} = \epsilon_{\mathbf{k}n} - \mu$. Solving for T_c, we arrive at the result plotted in Fig. 16. While

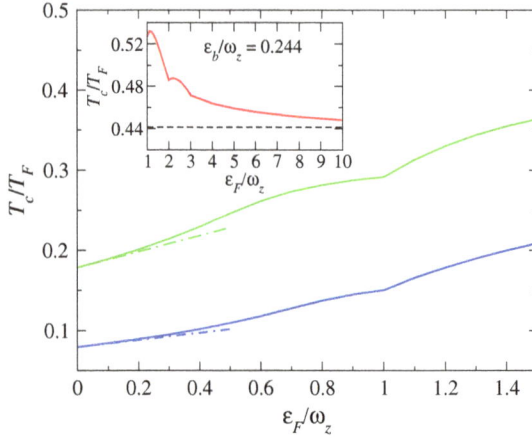

Fig. 16. Evolution of the mean-field critical temperature as the system is perturbed away from the 2D limit, taken from Ref. 76. The solid curves correspond to $\varepsilon_b/\varepsilon_F = 0.05$ and $\varepsilon_b/\varepsilon_F = 0.01$, going from top to bottom. The dash-dotted lines are the leading order behavior in ε_F/ω_z. Inset: The critical temperature at unitarity for large ε_F/ω_z. The line tends towards the 3D result (dashed line) as $\varepsilon_F/\omega_z \to \infty$.

the mean-field approach will overestimate T_c, it should be qualitatively accurate in the BCS regime and we clearly see that T_c/T_F increases as we perturb away from 2D at fixed $\varepsilon_b/\varepsilon_F$. Indeed, the leading order behavior in ε_F/ω_z is[76]

$$\frac{T_c}{T_F} = \frac{2e^\gamma}{\pi k_F a_{2D}} \left[1 + \frac{\varepsilon_F}{\omega_z} \ln\left(\frac{7 + 4\sqrt{3}}{8}\right) \right]. \tag{67}$$

This suggests that experiments will have a better chance of observing T_c and superfluidity if they are not purely 2D. For intermediate values of the confinement, we clearly see the presence of cusps at integer values of ε_F/ω_z, which correspond to discontinuities in the density of states every time the Fermi energy crosses a harmonic oscillator level. In the limit $\omega_z \to 0$, Eq. (66) yields the 3D expression for the Thouless criterion, as expected.[m] An interesting possibility is that T_c/T_F is maximized at intermediate confinement strengths, where the geometry is between two and three dimensions, but one would need to go beyond mean-field theory to assess this.

---

[m]The correct 3D expression is obtained by treating the confining potential in the z-direction within the local density approximation — see, also, Sec. 5.4.

5.2. High temperature limit

For high temperatures $T \gg T_F$, the gas is no longer quantum degenerate and the behavior tends towards that of a classical Boltzmann gas where the particle statistics are unimportant. In this limit, one may exploit the virial expansion described below, which has the advantage of being a controlled approach at high temperatures throughout the Fermi-Bose crossover. As such, the virial expansion can be used to investigate pairing phenomena at finite temperature and thus provide a benchmark for both theory and experiment.

5.2.1. Virial expansion

In the following, we outline the basic idea of the virial expansion, as applied to the uniform 2D Fermi gas. Working in the grand canonical ensemble, we define the virial coefficients b_j such that the grand potential $\Omega(T, \mu)$ is given by:

$$\Omega = -2T\lambda^{-2} \sum_{j \geq 1} b_j z^j, \tag{68}$$

where the thermal wavelength $\lambda = \sqrt{2\pi/mT}$, the fugacity $z = e^{\beta\mu}$, and $\beta \equiv 1/T$. In the high-temperature limit, the thermodynamics of the system can be accurately described by just the first few terms in the above power series. For a typical Fermi gas, z is the relevant expansion parameter, but this is not the case in the Bose limit of the crossover where $k_F a_{2D} \ll 1$. In this limit, $\mu \simeq -\varepsilon_b/2$ at low temperatures so that $z \simeq e^{-\beta\varepsilon_b/2} \to 0$ as $T \to 0$, which naively suggests that the virial expansion is valid at arbitrarily low temperatures. However, it can be shown that the coefficients b_j also contain powers of $e^{\beta\varepsilon_b/2}$ that cancel the contribution from the binding energy in z when j is even.[81] Thus, the relevant expansion parameter in the Bose limit is instead $z^{(\text{Bose})} = ze^{\beta\varepsilon_b/2}$, with corresponding coefficients $b_j^{(\text{Bose})} = e^{-j\beta\varepsilon_b/2}b_j$.

The virial expansion effectively amounts to a cluster expansion, whereby one determines the correlations between particles in a cluster of a given size, and then increases the size of the cluster at each order. For instance, b_2 only contains contributions from the one- and two-body problems, b_3 further includes three-body scattering, and so on. As such,

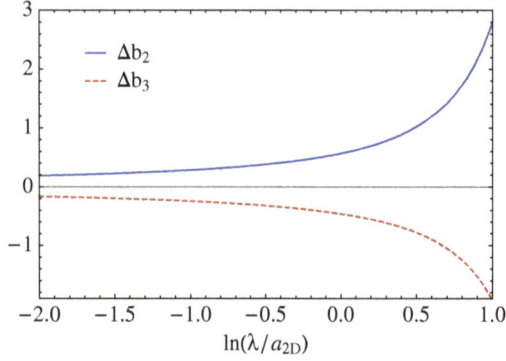

Fig. 17. The contribution from interactions to the second and third virial coefficients of the uniform 2D Fermi gas, taken from Ref. 81. The coefficients for the non-interacting gas are $b_j^{(\text{free})} = (-1)^{j-1} j^{-2}$ for $j \geq 1$, and $\Delta b_j \equiv b_j - b_j^{(\text{free})}$. Note that the virial coefficients are functions of $\ln(\lambda/a_{2D})$, or equivalently $\beta \varepsilon_b$, only. In the limit $\beta \varepsilon_b \to \infty$, both Δb_2 and Δb_3 are dominated by the two-body bound state and thus they both go like $e^{\beta \varepsilon_b}$ (but with different signs).

one can make use of the few-body results described in Sec. 3. For a recent review of the virial expansion in cold gases, see Ref. 89.

The first calculation of the virial coefficients in a 2D Fermi gas was for the trapped system.[90] Indeed, one typically determines each virial coefficient by solving the relevant few-body problem in a harmonic trap. The coefficents for the trapped gas can be straighforwardly mapped to those in the uniform case using the relation: $b_j = j b_j^{\text{trap}}$ (Ref. 81). The lowest order coefficients are plotted in Fig. 17. We see that the correction to the second virial coefficient due to interactions is attractive, as expected, since it lowers the grand potential at fixed μ and T. However, this lowest order term is expected to overestimate the attraction at lower temperatures and thus the third-order correction acts to increase the energy.

One can also determine b_j directly using a diagrammatic approach[91] where the single-particle propagator G is expanded in z, and this was first performed in 2D in Ref. 81. This approach also makes it straightforward to determine the virial expansion for the spectral function $A_\sigma(\mathbf{k}, \omega) = -2\text{Im}\,G_\sigma(\mathbf{k}, \omega)$, which is related to the probability of extracting an atom in state σ with momentum $\mathbf{k}$ and frequency ω. This allows one to investigate pairing gaps in the spectrum at high temperature, as we now discuss.

5.3. Pseudogap

The pseudogap regime is often synonymous with "pairing above T_c" in the cold-atom literature. However, such a scenario is trivially achieved in a classical gas of diatomic molecules, where the gap in the spectrum corresponds to the dimer binding energy. To reproduce the phenomenology of high-T_c superconductors, one requires the presence of a Fermi surface, since the pseudogap in these systems manifests itself as a loss of spectral weight at the Fermi surface.[83] Indeed, it is not a priori obvious that such a phenomenon can be replicated with an attractive Fermi gas: a large attraction will surely lead to a pronounced pairing gap above T_c, but it will also destroy the Fermi surface. It is therefore reasonable to assume that any pseudogap regime must have $\mu > 0$ in addition to pairing, as schematically depicted in Fig. 15.

The possibility of a pseudogap regime has been investigated in 3D Fermi gases,[92–95] but its existence is still under debate. In 2D, the pseudogap regime is expected to be much more pronounced than in 3D, since quantum fluctuations suppress superfluid long-range order, and the system more readily forms two-body bound states. Already, a recent measurement[10] of the spectral function in 2D has found indications of a pairing gap above T_c. However, a similar pairing gap is found using the lowest order virial expansion of the spectral function, which only includes two-body correlations, i.e., no Fermi surface.[81,96] In Fig. 18, the agreement between experiment and theory suggests that the observed pairing effectively arises from two-body physics only and therefore does not correspond to a pseudogap. Furthermore, most of the experimental measurments of the spectral function were apparently performed in the regime where $\mu < 0$ (see Fig. 15 and Ref. 81). Thus, it is likely that lower temperatures and lower attraction are required to observe a pseudogap. In particular, both non-self-consistent[97,98] and self-consistent[75] T-matrix approximations predict the existence of a pseudogap in the regime $T/T_F \lesssim 0.2$ for $\ln(k_F a_{2D}) \simeq 1$.

5.4. Equation of state in a trapped gas

The fact that the interaction parameter $\ln(k_F a_{2D})$ can be tuned by varying the density has important consequences for the trapped 2D gas. Specifically, it implies that $\ln(k_F a_{2D})$ decreases as we move from the

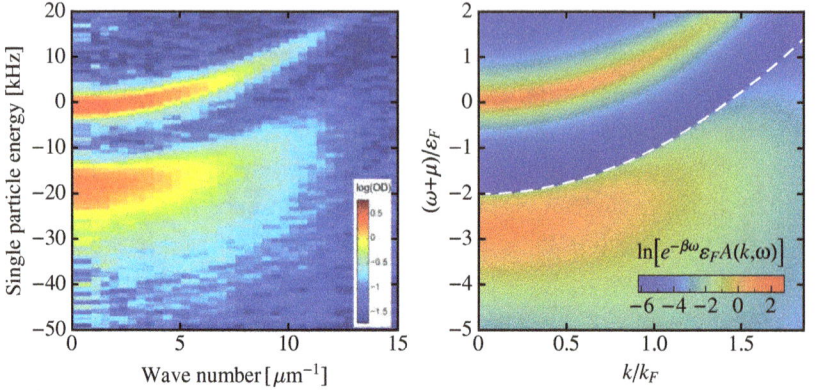

Fig. 18. The occupied part of the spectral function at $\ln(k_F a_{2D}) = 0$. (left) Measured momentum-resolved photoemission signal, taken from Ref. 10. (right) Theoretical prediction at $T/T_F = 1$ from the virial expansion up to second order.[81] The white dashed line marks the edge of the band of bound dimers (the incoherent part of the spectrum) and corresponds to the free atom dispersion shifted by the two-body binding energy, i.e., $\epsilon_{\mathbf{k}} - \varepsilon_b$. ((left) Reprinted by permission from Macmillan Publishers Ltd: Nature **480**, 75 (2011) copyright 2011.)

high-density region at the center of the trapped gas to the low-density region at the edge. Thus, we can in principle observe the entire Fermi-Bose crossover in a single experiment. This argument relies on the local density approximation (LDA), where the in-plane trapping potential can be incorporated into the chemical potential, $\mu(r) = \mu - V(r)$, and thus each point in the trap corresponds to a different $\ln(k_F(r)a_{2D})$. For a harmonic potential with frequency $\omega_\perp$, we require $\omega_\perp \ll T, \varepsilon_F$ in order for LDA to be valid.

One can make a direct connection with trapped-gas experiments by considering the density $n(\beta\mu, \beta\varepsilon_b)$ as a function of $\beta\mu$ for different values of the interaction parameter $\beta\varepsilon_b$ (see Fig. 19). Such an equation of state can be straightforwardly extracted from the measured density profile in a trap.[99] To reveal the effects of interactions, we normalize the density n by that of the ideal Fermi gas, $n_0 = 2\ln(1 + e^{\beta\mu})/\lambda^2$. In the high-temperature (low-density) limit where $\beta\mu \to -\infty$, the behaviour approaches that of an ideal Boltzmann gas, as expected. However, with decreasing temperature, n/n_0 eventually exhibits a maximum around $\beta\mu \simeq 0$, implying that interactions are strongest at intermediate rather than low temperatures. This results from

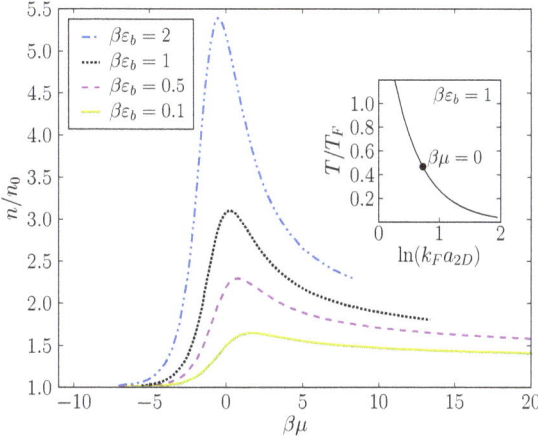

Fig. 19. The equation of state for the density at finite temperature, taken from Ref. 75. The density n is normalized by $n_0(\beta\mu)$, the density of the non-interacting Fermi gas. The curves for large $\beta\varepsilon_b$ are shown up to the critical value $\mu_c(\beta\varepsilon_b)$ where the system is expected to enter the BKT phase. The inset shows a typical trajectory corresponding to fixed $\beta\varepsilon_b$ in the phase space of T/T_F versus $\ln(k_F a_{2D})$. Along this line, $\beta\mu$ increases with decreasing T/T_F.

the fact that decreasing T/T_F at fixed $\beta\varepsilon_b$ corresponds to an increasing $\ln(k_F a_{2D})$. Thus, we likewise expect the system to approach a weakly interacting gas in the low temperature regime. This behavior is qualitatively different from that observed in 3D,[99] and is a direct consequence of the fact that one can traverse the Fermi-Bose crossover in 2D by only varying the density.

6. The 2D Polaron Problem

The properties of an impurity immersed in a quantum-mechanical medium constitutes a fundamental problem in many-body physics. A classic example in the solid state is the Fröhlich polaron, an electron moving in a crystal and interacting with the resulting bosonic lattice vibrations. Due to the interactions, the system of impurity plus lattice vibrations is better described in terms of a quasiparticle, the polaron, which has modified effective mass, chemical potential, charge, etc., compared with the free electron. The quasiparticle thus encompasses both the electron and the cloud of excitations of the medium.

In the context of two-component Fermi gases, the spin components may be imbalanced straightforwardly, leading naturally to a polaron problem in the limit of a large spin polarization, i.e., the problem of a single spin-down impurity. However, in contrast to the case of the Fröhlich polaron, the medium is now fermionic, and this can strongly modify the character of the impurity quasiparticle, as we discuss below. Furthermore, the properties of the polaron will directly impact the topology of the whole phase diagram for the spin-imbalanced Fermi gas. It is well known that BCS pairing is very sensitive to mismatched Fermi surfaces, and such a spin imbalance can thus lead to more exotic superfluid phases. For instance, the formation of Cooper pairs at finite momentum may occur, giving rise to the so-called Fulde-Ferrell-Larkin-Ovchinnikov (FFLO) state.[100, 101] For sufficiently large spin imbalance, the system encounters the Chandrasekhar-Clogston limit and ceases to display paired-fermion superfluidity. This limit has recently been experimentally investigated in the strongly interacting Fermi gas in both 3D[102–105] and 1D,[106] but it remains to be seen how the breakdown of superfluidity occurs in the 2D Fermi gas. For a further discussion of the polarized Fermi gas in 3D, we refer the reader to, e.g., the reviews of Refs. 107 and 108.

An important question concerns the nature of the ground state of a spin-down impurity atom in a spin-up Fermi sea. For weak attractive interactions, the quasiparticle has properties similar to that of the bare impurity and will be termed the "monomeron".[n] However, as the interaction strength is increased, the impurity can bind a majority particle to form a two-body bound state dressed by particle-hole fluctuations of the Fermi sea.[113–115] This is illustrated in Fig. 20, which shows the quasiparticle branches for equal masses. Interestingly, in the fermionic problem, the impurity can undergo a sharp transition in the ground state and effectively change its statistics by binding fermions from the majority fermions, an effect absent in the classic Fröhlich polaron example above. Quasiparticles in a 2D Fermi gas have been investigated in two experiments: Fermi polarons have been

---

[n]In this work we use the terminology monomeron, dimeron, trimeron, and tetrameron to denote the impurity bound to 0, 1, 2, and 3 majority atoms, respectively, in the presence of interactions with the Fermi sea. This replaces previous terminology (attractive polaron, molecule, dressed trimer, and dressed tetramer, respectively). As there is only one repulsive branch (see below) this is referred to as the repulsive polaron.

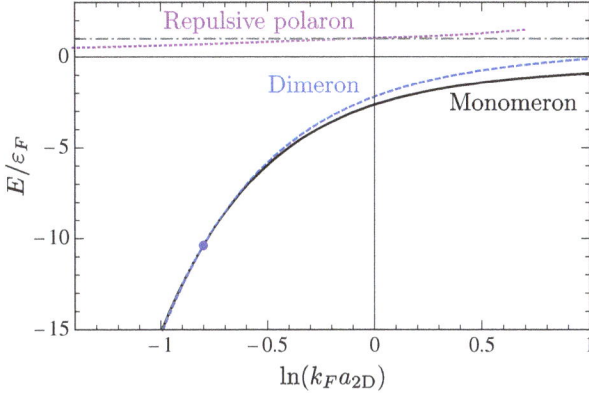

Fig. 20. The relevant quasiparticle branches in 2D for $m_\uparrow = m_\downarrow$: The monomeron (solid line),[109] the dimeron (dashed line),[110] and the repulsive polaron (dotted line).[111,112] The filled circle marks the monomeron-dimeron transition in the ground state. The dot-dashed line marks the Fermi energy. All quasiparticle energies displayed follow from variational wavefunctions limited to one particle-hole pair excitation (see text).

observed,[14] while radio-frequency spectra of the unpolarized Fermi gas have been interpreted in terms of monomerons.[12]

The polaron problem also has relevance to the phenomenon of *itinerant ferromagnetism*. Here, a two-component equal-mass Fermi gas with repulsive short-range interactions is predicted to spontaneously undergo a transition to spin-polarized domains for sufficiently strong repulsion. This classic Stoner transition received renewed interest when its observation was reported in a recent MIT experiment on the repulsive branch in 3D Fermi gases.[116] Subsequently, however, it was shown that the experiment had instead only observed a fast decay into pairs;[117] this realization led to the assertion that the Fermi gas with strong short-range repulsive interactions can never undergo a ferromagnetic transition.[118] The central issue is that strongly repulsive interactions can only be truly short-ranged if the underlying potential is attractive, and thus any magnetic phase in such a system will be metastable at best. As we describe below, the properties of the repulsive polaron (see Fig. 20) are crucial for determining whether saturated ferromagnetism may exist in the 2D Fermi gas, and it, in fact, appears that the fast decay into the attractive branch also precludes saturated ferromagnetism in 2D.[112] For a recent review on polaron physics in ultracold gases with an emphasis on the relation to itinerant ferromagnetism, we refer the reader to Ref. 119.

6.1. *Variational approach*

An intuitive way to describe the Fermi polaron theoretically is through variational wavefunctions. The simplest is Chevy's ansatz:[120]

$$|P\rangle = \alpha_0^{(p)} c_{\mathbf{p}\downarrow}^\dagger |FS\rangle + \sum_{\mathbf{kq}} \alpha_{\mathbf{kq}}^{(p)} c_{\mathbf{p+q-k}\downarrow}^\dagger c_{\mathbf{k}\uparrow}^\dagger c_{\mathbf{q}\uparrow} |FS\rangle. \tag{69}$$

Here and in the following we assume that $|\mathbf{k}| > k_F$ ($|\mathbf{q}| < k_F$) describes a particle (hole). For simplicity, we define k_F as the Fermi momentum of the spin-↑ atoms. The wavefunction describes the spin-↓ impurity as a quasiparticle at momentum $\mathbf{p}$ using two terms: the first is simply the bare impurity on top of the non-interacting majority Fermi sea, denoted by $|FS\rangle$, while the second incorporates how the impurity can distort the Fermi sea by exciting a particle out of it, leaving a hole behind.

The energy of the polaron state is obtained by minimizing the expectation value $\langle P|\mathcal{H} - E|P\rangle$ with respect to the variational parameters $\alpha_0^{(p)}$ and $\alpha_{\mathbf{kq}}^{(p)}$, where $\mathcal{H}$ is the 2D Hamiltonian (54). This yields the equation

$$E - \epsilon_{\mathbf{p}\downarrow} = \sum_{\mathbf{q}} \left[\frac{1}{g} - \sum_{\mathbf{k}} \frac{1}{E - \epsilon_{\mathbf{k}\uparrow} + \epsilon_{\mathbf{q}\uparrow} - \epsilon_{\mathbf{p+q-k}\downarrow}} \right]^{-1}. \tag{70}$$

Formally, the variational approach as introduced here only admits one solution: the monomeron,[120] which has energy less than the impurity in vacuum. However, the variational approach may be extended to include metastable states where the energy is allowed to have a finite imaginary part — see Ref. 121. In this case, one also obtains a second solution, the "repulsive polaron",[122, 123] which has an energy E_{rep} exceeding that of the impurity in vacuum and potentially even exceeding the Fermi energy for strong interactions. The wavefunction (69) may straightforwardly be extended by considering further excitations; however the present approximation of one particle-hole pair excitation gives a surprisingly good estimate of the energy and the residue $Z = \left|\alpha_0^{(p)}\right|^2$. This is due to an approximate cancellation of higher order terms in the expansion in particle-hole pairs.[124] A recent work in 3D has demonstrated an impressive agreement between the variational approach and experiment.[125]

In addition to the states described by Eq. (69), the impurity may also (depending on the ↑-↓ mass ratio) form dimeron, trimeron, and tetrameron states by binding one or several majority particles, in a natural analogy to the possible vacuum bound states such as the dimer, trimer, and tetramer described in Sec. 3. Remarkably, these states may be the ground states even when they do not bind in vacuum. The variational wavefunctions for such states can be generated in a similar fashion to Eq. (69) above, but rather than displaying them here, we instead refer the reader to the original works on the dimeron[110, 126–128] and trimeron.[39, 121]

6.2. The repulsive polaron and itinerant ferromagnetism

Following the observation that recombination processes preclude itinerant ferromagnetism in the 3D atomic Fermi gas,[118] it is pertinent to ask the question whether the Stoner transition can take place in a 2D Fermi gas.[129] The main difference between the 2D and 3D Fermi gases with short-range interactions is that in 3D the vacuum two-body bound state appears in the regime of strongest interactions, $1/k_F a_s = 0$, whereas in 2D, the bound state only approaches the continuum in the limit of weak attraction. Thus, one may speculate that the pairing mechanism that prevented the appearance of itinerant ferromagnetism in 3D could be suppressed. Indeed, the three-body recombination mechanism by which three atoms recombine into an atom and a dimer takes completely different forms in 3D[130] and in 2D.[53] However, despite this difference, the decay into the attractive branch is still strong enough to exclude fully polarized itinerant ferromagnetism, as we now discuss.

Following Ref. 112, we investigate the stability of *fully polarized* domains. To preserve $SU(2)$ symmetry and make a direct connection with ferromagnetism, we confine the discussion to equal-mass fermions. The fully polarized domains are illustrated in Fig. 21(a), and the central question is whether there is an energy cost associated with moving a spin-↓ atom from its domain to that of the spin-↑ atoms. Assuming purely repulsive interactions, this is the case if the energy of the dressed impurity exceeds the Fermi energy in the spin-↑ region. If one further assumes mechanical equilibrium, where the pressures of the domains are equal, then the Fermi

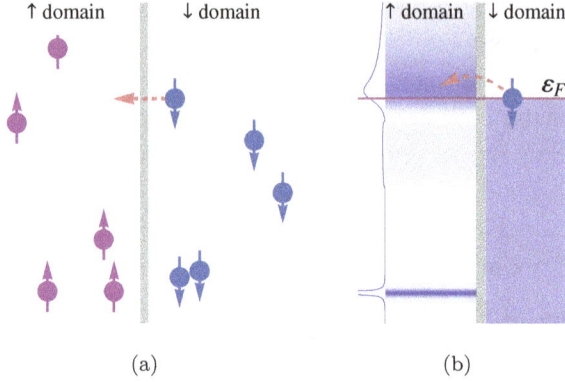

Fig. 21. Illustration of the stability condition for stable spin-polarized domains, taken from Ref. 131. (a) A spin-$\downarrow$ atom can tunnel across the interface and become an impurity in the spin-$\uparrow$ domain. (b) Density plot of the energy levels available to the fermion at $\ln(k_F a_{2D}) = 0.5$. The spectral function at $\mathbf{k} = \mathbf{0}$ of the impurity in the $\uparrow$ domain is evaluated in the one particle-hole pair dressing approximation.

energies $\varepsilon_F^\uparrow = \varepsilon_F^\downarrow$. Thus, referring to Fig. 20 and assuming that the impurity would tunnel into the repulsive polaron state, the domains appear mechanically stable if $\ln(k_F a_{2D}) > -0.15$ and one concludes that itinerant ferromagnetism is possible.

However, we must consider two other effects: The first is the finite lifetime of the repulsive polaron, as the quasiparticle decay rate is predicted to be a significant fraction of the Fermi energy in the strongly interacting regime[112] (the decay rate may also be investigated as a pairing instability — see Ref. 132). This in turn leads to a large uncertainty in the energy of the repulsive state, allowing atoms to tunnel across the interface and depolarize the domains. Eventually, in the weakly interacting regime $\ln(k_F a_{2D}) \gg 1$, the quasiparticle decay rate becomes suppressed; however, as the tunneling probability is proportional to the residue Z of the corresponding quasiparticle, and the residue of the repulsive branch is strongly suppressed in this regime,[111,112] the atoms will tunnel directly into the attractive branch. Combining the knowledge of the residue and the lifetime of the repulsive polaron allows one to conclude that even if spin polarized domains were to be artificially created, these would not be dynamically stable.[112]

The repulsive polaron has been observed in a recent experiment[14] and, in accordance with the theory, no ferromagnetic transition was observed.

In fact, the experiment[o] was limited to the regime $-2.5 < \ln(k_F a_{2D}) < -1.3$, i.e., away from the limit where the variational approach predicts $E_{rep} > \varepsilon_F$. In this regime, it may be expected that $E_{rep} \gtrsim 0.3\varepsilon_F$ (see Fig. 20), whereas the experimentally observed energies ranged from 10% to 20% of the Fermi energy. The discrepancy may in part be due to the trap averaging[111] and finite temperature effects. However, in agreement with the theory,[112] the lifetime was severely suppressed, preventing the detection of a coherent repulsive quasiparticle for stronger interactions.

6.3. *Ground state of an impurity in a 2D Fermi gas*

In the following, we initially focus on the equal-mass case. For a single impurity attractively interacting with a 3D Fermi gas of identical atoms, the existence of a sharp quasiparticle transition from the monomeron to the dimeron state has been predicted.[113–115] Such a transition was recently observed experimentally for a finite density of impurities.[134] On the other hand, in the 1D case, the exact Bethe ansatz solution[135] implies that no such transition takes place. It is therefore natural to ask whether a transition in the ground state occurs for an impurity in a 2D Fermi gas, where quantum fluctuations are expected to be stronger than in 3D. The existence or otherwise of such a transition will impact the overall phase diagram for the spin-imbalanced 2D Fermi gas.[136–139]

The first work on this subject[109] did not find any ground-state transition, the issue being that the authors did not consider the monomeron and dimeron on an equal footing in terms of particle-hole pair dressing of the variational wavefunctions. Later, one of us[110] included a particle-hole pair excitation in the dimeron variational wavefunction to show that there is indeed a ground state transition. We recently extended this analysis to argue that, under a minimal set of assumptions, the critical interaction strength for the

---

[o]In the experiment, the 2D scattering length was taken directly from the quasi-2D dimer binding energy, i.e., $a_{2D}^* = 1/\sqrt{m\varepsilon_b}$. The relation between the present definition of a_{2D} and the one used in experiment is: $a_{2D} = a_{2D}^* \sqrt{\frac{\pi}{B}\frac{\varepsilon_b}{\omega_z}} e^{-\sqrt{\frac{\pi}{2}}\mathcal{F}_0(\varepsilon_b/\omega_z)}$, where $\mathcal{F}_0$ was introduced in Eq. (26). As argued in Ref. 133, the convention used for a_{2D} in this review yields a better agreement between the results of the quasi-2D experiments and the strict 2D theory presented here.

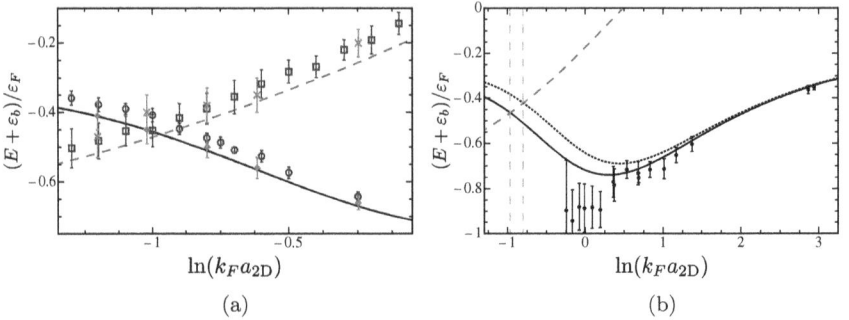

(a)   (b)

Fig. 22. Energy of the impurity measured from the two-body binding energy. The dotted line corresponds to the energy of the monomeron within the Chevy ansatz,[109] Eq. (70), while the solid line is within the two particle-hole pair approximation.[121] The dashed line is the energy of the dimeron within the one particle-hole pair approximation.[110] (a) The data points are from the two recent Monte Carlo simulations: for the monomeron (dimeron) these are marked by red open squares (circles)[140] and by green diamonds (crosses).[141] Note that the variational approach provides an upper bound on the energy. (b) The ground state transition within the two approximations for the monomeron energy are illustrated by vertical dashed lines, while the experimental data^{0} is taken from Ref. 14.

monomeron-dimeron transition must lie in the interval[121]

$$-0.97 < \ln(k_F a_{2D})_{\text{crit}} < -0.80. \tag{71}$$

The lower (upper) bound corresponds to comparing the dimeron dressed by one particle-hole pair excitation with the monomeron dressed by two (one) excitations. As seen in Fig. 22(a), our result agrees with the critical interaction found in two recent diagrammatic Monte Carlo studies: $\ln(k_F a_{2D})_{\text{crit}} = -0.95(0.15)$ [Ref. 140] and $\ln(k_F a_{2D})_{\text{crit}} = -1.1(0.2)$ [Ref. 141].

The monomeron was investigated in a recent experiment,[14] and Fig. 22(b) shows that for $\ln(k_F a_{2D}) \geq 0.3$ the comparison between theory and experiment is excellent. For stronger attraction, the agreement becomes progressively worse until at $\ln(k_F a_{2D}) \simeq -0.6$ the measured effective mass appears to diverge, which was taken to be a signature of the monomeron-dimeron transition.[14] However, if one extrapolates the measured residue to zero,[142] one instead obtains a critical interaction strength of $\ln(k_F a_{2D})_{\text{crit}} = -0.88(0.20)$, which is in good agreement with theory.[110,121] As mentioned previously, the experimental investigation of the polaron problem can be complicated by temperature effects and trap averaging. In addition, one must consider the fact that the high polarization limit typically corresponds to a *finite density* of spin-$\downarrow$ impurities.[143] Thus, we are faced with the question of whether the single-impurity transitions are thermodynamically stable, i.e., whether they are preempted by first-order transitions in the thermodynamic limit. We have recently shown[121] that a first-order superfluid-normal phase transition preempts the single-impurity transition at zero temperature, similarly to the situation in 3D.[39] However, this result requires the presence of a superfluid, and thus it is an open question whether single-impurity transitions may exist at higher temperatures.

In the present discussion, we have mapped[o] the results of the experiment onto a pure 2D theory. However, let us now discuss the validity of such an approach.[133] The transverse confinement applied in the experiment[14] was $\omega_z = 2\pi \times 78.5\text{kHz}$, while the Fermi energy of the majority component was $2\pi \times 10\text{kHz}$. This in turn means that the pure 2D theory[110,121] predicts the transition to occur when $\varepsilon_b \geq 2\pi \times 100\text{kHz}$, i.e. when the binding energy exceeds the transverse confinement strength. In this regime, the binding energy is strongly modified from the 2D prediction — see Fig. 2 and the discussion in Sec. 2. On the other hand, the ground state transition is governed by interactions that take place at the typical energy scale $\sim \varepsilon_F$. Since $\varepsilon_F \ll \omega_z$ the low-energy quasi-2D theory described by Eqs. (28) and (29) is still approximately valid, explaining our choice of using a definition of a_{2D} which derives from low-energy scattering rather than the binding energy.

The deviation from the pure 2D limit of the monomeron-dimeron transition may be further investigated[133] by including harmonic oscillator levels in the variational wavefunction and using the full quasi-2D Hamiltonian (43). The results of such an analysis are shown in Fig. 23, where we see how the transition point indeed changes very little in $\ln(k_F a_{2D})$ for $\varepsilon_F/\omega_z \lesssim 1/10$, while on the other hand the change is rather large in terms of the parameter $\varepsilon_b/\varepsilon_F$.

We finally turn to the mass-imbalanced system, where the single impurity phase diagram[121] as a function of interaction strength takes the form displayed in Fig. 24. In Sec. 3 we discussed how in vacuum the $\uparrow\uparrow\downarrow$ trimer appears[37] when $m_\uparrow/m_\downarrow = 3.33$. Remarkably it is seen that the presence of a Fermi sea favors trimer formation: within the approximation used, the

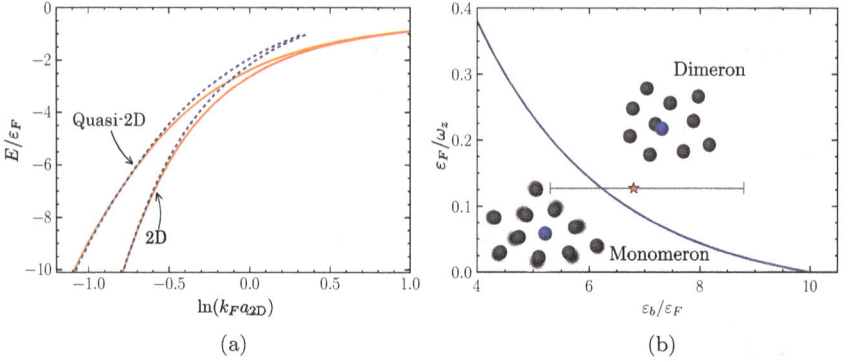

(a)   (b)

Fig. 23. The behavior of the quasi-2D polaron taken from Ref. 133. (a) Monomeron (solid lines) and dimeron (dashed) energies in 2D and in quasi-2D at $\varepsilon_F/\omega_z = 1/10$. (b) The single impurity phase diagram. The ground state of the impurity is a monomeron (dimeron) to the left (right) of the line. The theory lines are all within the one particle-hole pair approximation. The star shows the experimental transition point with error bars.[142]

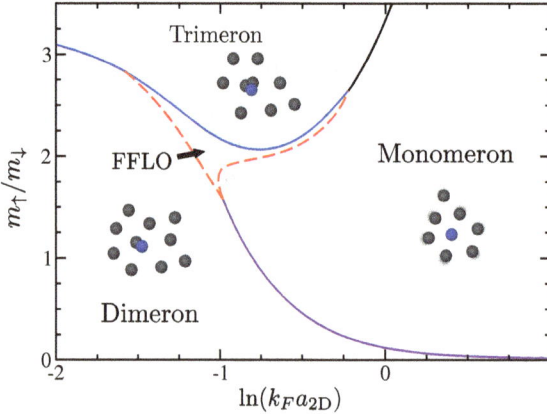

Fig. 24. Ground-state phase diagram for a single impurity atom of mass $m_\downarrow$ immersed in a gas of fermions of mass $m_\uparrow$, adapted from Ref. 121. The phase boundaries are derived within the one particle-hole pair dressing approximation. The single-impurity analog of the FFLO phase corresponds to a ground state dimeron at non-zero momentum.

trimeron is predicted to be the ground state for mass ratios $m_\uparrow/m_\downarrow \geq 2.1$. This lower critical mass ratio may be understood as a consequence of the kinetic energy cost involved in forming a dimeron at rest: in the simplest approximation, the impurity at momentum $+\mathbf{k}_F$ binds a majority atom at $-\mathbf{k}_F$ and, if the impurity is sufficiently light, it may be energetically favorable to instead form a dimeron at finite momentum or a trimeron. The same effect was predicted in 3D.[39] While the trimeron is favored by the Fermi sea, we found[121] that the tetrameron appears disfavored, i.e., the critical mass ratio for tetrameron formation in the strongly interacting regime increases from its vacuum value[41] of $m_\uparrow/m_\downarrow = 5.0$.

The possibility of a dimeron at finite momentum is of considerable interest, since it is a single-particle analog of the FFLO phase — for a small but finite density of impurities this has been shown to lead to a spatially-modulated superfluid.[144, 145] We see in Fig. 24 that the FFLO dimeron occupies a considerable part of the phase diagram, making it possible that FFLO physics may be observed in the strongly spin-imbalanced 2D Fermi gas.

7. Dynamics

Dynamical properties provide a powerful probe into the nature of interactions in strongly correlated quantum systems. For instance, it has been predicted that the harmonically trapped 2D quantum gas features an $SO(2, 1)$ dynamical scaling symmetry due to the (classical) scale invariance of the uniform gas with contact interactions. A consequence of this symmetry is the existence of an undamped monopole breathing mode with frequency exactly twice that of the trap.[146] While true in the absence of interactions, the scale invariance which exists at the classical level is broken by the procedure of renormalization,[146] the so-called *quantum anomaly*.[147] Thus, the shift of the breathing mode frequency probes the breaking of scale invariance in the interacting 2D quantum gas. Another dynamical phenomenon is that of spin diffusion, the process that evens out differences in spin polarization across the gas. Here the diffusivity in the strongly interacting and degenerate regime is naturally of order $\hbar/m$ and an interesting possibility is that there is a universal lower bound set by quantum mechanics.

As of now, there have been experiments on the collective modes in a harmonic trap[16] and on spin transport.[148] The results of the experiments have indicated several surprising features of the 2D Fermi gas: an undamped breathing mode with a frequency compatible with no shift from the classical (non-interacting) result; a quadrupole mode strongly damped even in the weakly interacting regime; and a transverse spin diffusivity three orders of magnitude smaller than in any other system. The strong damping of the quadrupole mode may be explained,[149] at least in part,[150] by the anisotropy of the trapping potential used. However at first sight the other two features appear contradictory, as the results of the breathing mode experiment indicate that the effect of interactions is much weaker than expected by theories, while the spin diffusivity experiment indicates the opposite. Ultimately, further experiments as well as possibly finite temperature QMC calculations for the equation of state will likely be needed to shed light on the discrepancy.

In this section we assume a purely 2D geometry, such that the transverse confinement frequency ω_z drops out of the problem. The experiments described here are indeed all in the regime where $T \le T_F \lesssim 0.1\omega_z$, so this approximation is reasonable.

7.1. Classical scale invariance, a hidden $SO(2, 1)$ symmetry, and the breathing mode

It has been predicted[146] that a 2D quantum gas in a harmonic transverse trapping potential features an undamped monopole breathing mode with frequency exactly twice that of the trap. The origin of this surprising result is the (classical) scale invariance of the Hamiltonian with a short-range δ-function interaction. Define the $2N$ dimensional vector $\mathbf{X} = (\rho_1, \cdots, \rho_N)$, with ρ_i the positions of the atoms $i = 1, \cdots, N$, and the hyperradius $X \equiv |\mathbf{X}|$. In real space, both terms in the Hamiltonian

$$H_0 = -\frac{\Delta_\mathbf{X}^2}{2m} + g \sum_{i<j} \delta(\rho_i - \rho_j) \tag{72}$$

scale as λ^{-2} under the scale transformation $\mathbf{X} \to \lambda\mathbf{X}$, and consequently the Hamiltonian is scale invariant. While this is true classically, the procedure of renormalization of the quantum theory introduces a scale, the 2D scattering length a_{2D}, as discussed in Section 2; the absence of a scale in the classical theory and the introduction of one through renormalization is known as a quantum anomaly. The scale invariance is still approximately valid in the limits $a_{2D} \to 0$ and $a_{2D} \to \infty$ where the following results apply.[p]

---

[p]Whereas the scale invariance in 2D is only exact in the trivial non-interacting limits, it is, in fact, quantum mechanically exact for the 3D unitary Fermi gas[151] as well as for the 1D gas in the Tonks limit, both strongly interacting systems.

The presence of a harmonic trapping potential

$$H_{\text{trap}} = \frac{1}{2} m \omega_0 X^2 \tag{73}$$

in the 2D plane obviously breaks the scale invariance as it scales as λ^2 under $X \to \lambda X$. However, it leads to a very interesting algebra:[146] using the usual commutation relations for X and $P \equiv i \partial_X$, one may easily show $[H_{\text{trap}}, H] = i \omega_0^2 Q$. Here $Q \equiv \frac{1}{2}(P \cdot X + X \cdot P)$, $e^{-\ln(\lambda)Q}$ is the generator of scale transformations,[152] and $H = H_0 + H_{\text{trap}}$ is the total Hamiltonian. Then defining the operators

$$L_1 = \frac{1}{2\omega_0}(H_0 - H_{\text{trap}}), \qquad L_2 = Q/2, \qquad L_3 = \frac{1}{2\omega_0}(H_0 + H_{\text{trap}}), \tag{74}$$

these satisfy

$$[L_1, L_2] = -i L_3, \qquad [L_2, L_3] = i L_1, \qquad [L_3, L_1] = i L_2, \tag{75}$$

which is the algebra of the Lorentz group in 2D, $SO(2, 1)$. As usual, one may then define raising and lowering operators $L_\pm = \frac{1}{\sqrt{2}}(L_1 \pm i L_2)$. From the commutation relations $[H, L_\pm] = \pm 2\omega_0 L_\pm$ it follows that if $|\Psi_g\rangle$ is the ground state with energy E_g, the state $L_+|\Psi_g\rangle$ has energy $E_g + 2\omega_0$ while $L_-|\Psi_g\rangle = 0$. Thus the repeated action of L_+ generates a tower of states, separated by $2\omega_0$ and these may be identified with the breathing modes of the system. For instance, if the system is initially in a stationary state with a constant trap frequency ω_0 at time $t < 0$, the trap frequency is slightly perturbed during the interval $0 < t < t_f$, and returns to its initial value at time $t > t_f$, one finds[151] that the final state scale oscillates around unity with frequency $2\omega_0$. That is, the lowest breathing mode has been excited.

In the above scale invariant (and non-interacting) regimes, the breathing mode is undamped and its frequency is independent of amplitude. On the other hand, in the interacting quantum system, the breathing mode becomes damped and is shifted to $\omega_B = 2\omega_0 + \delta\omega_B$ from its non-interacting value, as discussed in the Bose case in Ref. 147. In the 2D Fermi gas,[q] this shift has been modelled[154] (see also Refs. 153, 155 and 156) by assuming a hydrodynamic description of the strongly interacting regime,

---

[q]In fact, within mean-field theory, the breathing mode has frequency $2\omega_0$ in the entire BCS-BEC crossover.[153]

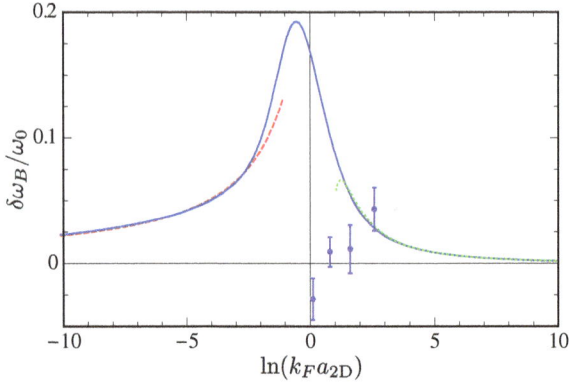

Fig. 25. Shift of the breathing mode as a function of interaction parameter. The solid line (full theory), dashed (BEC limit), and dotted (BCS limit) are the theory curves from Ref. 154 while the data points are from the experiment.[16]

and a polytrope $P \sim n^{\gamma+1}$ for the dependence of pressure on density. These assumptions allow for a solution of the linearized hydrodynamic equations, and in turn for the breathing mode frequency, $\omega_B = \omega_0\sqrt{2 + 2\gamma}$. γ itself was obtained by comparing with the zero-temperature QMC data[55] discussed in Sec. 4. The resulting frequency shift is shown in Fig. 25, and is seen to be of order 10% in the regime of strong interactions, $\ln(k_F a_{2D}) \sim 0$.

Experimentally, the breathing mode was investigated[16] using a procedure essentially as described above: The in-plane confinement was adiabatically lowered from the initial configuration and then abrubtly returned to its original configuration. After a variable wait time, the confinement was switched off and the density distribution was revealed by an absorption image after time of flight. The experiment investigated a large range of interaction strengths, $0 \lesssim \ln k_F a_{2D} \lesssim 500$. Surprisingly, the results of the experiment were consistent with the scale invariant assumption above, i.e., no significant frequency shift was observed, even in the regime of strong interactions (see Fig. 25). The results beg the question whether the zero-temperature equation of state[55] is appropriate for the comparison with the experiment at $T/T_F = 0.4$, i.e., whether the apparent scale invariance arises due to finite temperature effects.[154] Indeed, in the high temperature limit the shift of the breathing mode may be analyzed by combining the virial expansion of the equation of state with a variational method in the hydrodynamic regime.[157] The results of this analysis are consistent with the experiment and the theoretical curve in Fig. 25 in the regime of validity of the approach, $\ln(k_F a_{2D}) \gtrsim 1.75$.

The damping of the breathing mode is related to the bulk viscosity. In particular both the bulk viscosity and the damping are expected to vanish in the normal phase in the regime where the $SO(2, 1)$ symmetry is exact, as first pointed out in the context of the unitary Fermi gas.[158] Using a sum rule, the bulk viscosity has been argued[153] to vanish in the weakly interacting limits $|\ln(k_F a_{2D})| \gg 1$. However, in the intermediate strongly interacting regime one expects a non-vanishing bulk viscosity and related damping of the breathing mode. Therefore it is surprising that the experiment[16] measures a damping consistent with vanishing bulk viscosity across the entire interaction range.

7.2. *Quadrupole mode*

In addition to the breathing mode, the experiment[16] considered the quadrupole mode, corresponding to an excitation with velocity field $\mathbf{x} - \mathbf{y}$ oscillating with frequency ω_Q. The excitation procedure was similar to the monopole mode described above: the radial trap was adiabatically made elliptical, followed by an abrupt return to the original configuration, a short free oscillation, and an absorption image after time of flight. The results of the experiment are shown in Fig. 26(a). Two regimes are immediately identifiable: the collisionless regime, where $\ln(k_F a_{2D}) \gg 1$ and $\omega_Q \approx 2\omega_0$, and the hydrodynamic regime, where $\omega_Q \approx \sqrt{2}\omega_0$. The theory curves[149] (see also Ref. 159) are calculated using kinetic theory and correctly identify the onset of the hydrodynamic regime. The theory is not expected to be valid when $\ln(k_F a_{2D}) \lesssim 0.5$, where the (zero-temperature) chemical potential is negative[81] (see the discussion in Sec. 4.4) and pairing becomes significant.

The damping of the quadrupole mode is shown in Fig. 26(b): it is seen that this is a sizeable fraction of the trap frequency, and curiously this is the case even in the weakly interacting regime $\ln(k_F a_{2D}) \gg 1$ — in fact, the experiment[16] showed that the large damping persists up to very large $\ln(k_F a_{2D}) \sim 500$, far into the collisionless regime where the damping is expected to vanish, and kinetic theory is valid. Ref. 149 argued that the large damping in the weakly interacting regime arises mainly from systematic effects, which generate an approximately constant damping across all interaction strengths (indeed a smaller constant damping was

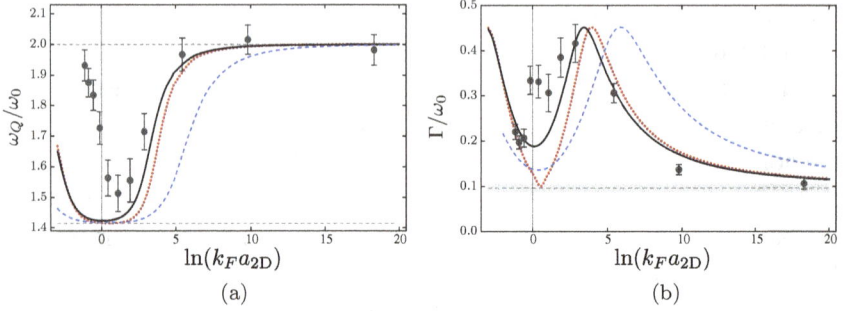

(a)   (b)

Fig. 26. (a) Quadrupole frequency and (b) damping as a function of interaction parameter. The lines are the theoretical curves[149] in the Boltzmann limit (blue, dashed), with Pauli blocking only (black, solid), and with additional medium effects (red, dotted). The black dots are the results of the experiment.[16] The trap frequency here is $\omega_0 = \sqrt{\omega_x \omega_y}$. In (b) a constant shift has been applied to the theory to account for systematic effects. (Figure adapted with permission from Ref. 149. Copyrighted by the American Physical Society.)

also observed in the breathing mode experiment described above). The anisotropy of the trap before time of flight may then account for the remaining discrepancy in this limit.[149] However, a recent analysis using a realistic trapping potential and including even the effect of gravity concluded that the damping of the quadrupole mode in the weakly interacting regime could not be explained by the specific geometry of the trap.[150] In the strongly interacting hydrodynamic regime, the damping may be related to the shear viscosity of the gas.[160–162]

7.3. Spin diffusion

An interesting application of fermionic quantum gases is the study of spin transport. These systems may provide particularly clean experimental realizations compared with, e.g., ^{3}He-^{4}He solutions, since in the quantum gases the interactions are tunable and spin states may be manipulated in a coherent manner by radio-frequency pulses. One basic transport process is that of spin diffusion, recently investigated in the context of ultracold atomic gases.[148, 163] This process acts to even out differences in polarization. Writing the magnetization as a product of the magnitude and the direction, $\mathcal{M} = \mathcal{M}\hat{\mathbf{e}}$, there are two contributions to the magnetization gradient $\nabla \mathcal{M} = (\nabla \mathcal{M})\hat{\mathbf{e}} + \mathcal{M}\nabla\hat{\mathbf{e}}$. The first of these, longitudinal diffusion, acts between regions of different magnitude of magnetization, while the second, transverse diffusion, acts between regions of different orientation.

In the light of the proposed quantum limit of the ratio of shear viscosity to entropy density,[164] it is interesting to ask the question whether quantum mechanics provides a lower bound for other transport phenomena such as spin diffusion in a strongly interacting Fermi gas. As decoherence is introduced by collisions, the resulting spin diffusivity may be expected to go as the collision speed of two atoms multiplied by the mean free path. In the degenerate regime, the former may be taken to be $\hbar k_F / m$, while the mean free path is $1/n\sigma \sim 1/k_F$; the density $n \sim k_F^2$ and the cross section in the degenerate regime takes its strongest value allowed by quantum mechanics, i.e., k_F^{-1}. Thus the diffusivity may be expected to be of order $\hbar/m$ in the degenerate regime and indeed the lowest spin diffusivity for longitudinal spin currents has been measured to be $6.3\hbar/m$ in a 3D quantum degenerate Fermi gas at unitarity.[163] In general this argument is too simple; for instance it neglects the effect of Pauli blocking which causes the longitudinal spin diffusivity in the Fermi liquid to diverge as $1/T^2$ at low temperature. Note that for the purpose of this discussion we have displayed $\hbar$ explicitly.

Surprisingly, a recent experiment[148] has found a transverse spin diffusivity in the strongly interacting regime that is orders of magnitude smaller than $\hbar/m$. Starting from a fully polarized 2D gas of ^{40}K atoms, the experiment used a spin-echo technique consisting of three consecutive radio-frequency pulses: First, a $\pi/2$ pulse was applied to rotate the spin into a coherent superposition of $\uparrow$ and $\downarrow$ states. A magnetic field gradient ensured a transverse spin wave due to the difference in gyromagnetic ratio of the two spin states, thus lifting the spin polarization and allowing the different spin states to collide and diffuse. Trivial dephasing due to the magnetic field gradient was reversed by the application of a π pulse after a time τ. This ensured that the spin state would refocus at time 2τ in the absence of decoherence, in which case the final $\pi/2$ pulse would rotate the spin state back to the original one. The experimental observable was the final magnetization $\langle M \rangle \equiv (N_\uparrow - N_\downarrow)/(N_\uparrow + N_\downarrow)$, and by measuring this for different spin evolution times 2τ, the transverse spin diffusivity was extracted.[r] The results are shown in Fig. 27, and it is seen that $\mathcal{D}_0$ has a shallow minimum around $\ln(k_F a_{2D}) = 0$, with values as low as $0.006\hbar/m$.

---

[r]$\mathcal{D}_0$ was obtained from the magnetization using the time evolution $\langle M_z \rangle \propto e^{-(2/3)\mathcal{D}_0(\delta\gamma\, B')^2 \tau^3}$ with $\delta\gamma$ the difference in gyromagnetic ratio between the two spin states and $B' = \partial B_z/\partial x$ the magnetic field gradient.

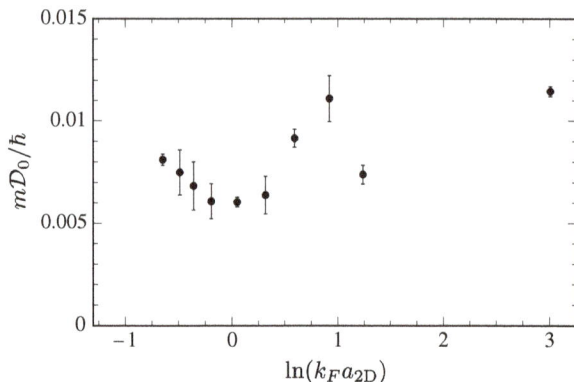

Fig. 27. Transverse spin diffusivity measured in experiment[148] across the strongly interacting regime.

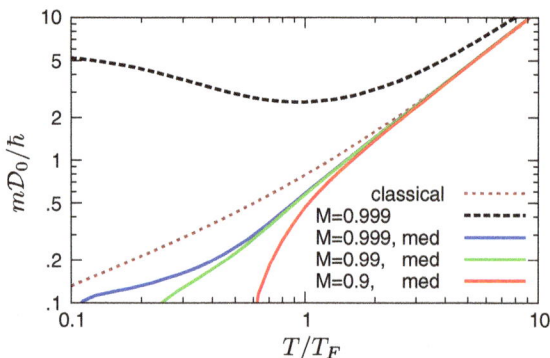

Fig. 28. $\mathcal{D}_0$ as a function of temperature at $\ln(k_F a_{2D}) = 0$ for various magnetizations.[165] The dashed (black) line includes Pauli blocking only, while the dotted line is the high-temperature Boltzmann limit. For comparison, the experiment[148] was carried out at $T/T_F = 0.24(3)$. (Reprinted figure with permission from: T. Enss, *Phys. Rev. A* **88**, 033630 (2013). Copyright 2013 by the American Physical Society.)

The transverse spin diffusivity in the 2D Fermi gas was recently investigated using a kinetic theory based on a many-body T matrix.[165] Indeed, it was found that medium effects could substantially suppress the spin diffusion below $\hbar/m$, see Fig. 28. As shown in the figure, the theory also predicts that at temperatures below T_F, the transverse spin diffusivity is quite sensitive to magnetization and gets suppressed as the magnetization decreases. The origin of the suppression lies in the enhanced cross section in the many-body system close to the Thouless pole and, as discussed in Sec. 4, the Thouless pole overestimates the critical temperature. However, it is likely that the theory captures the correct qualitative behavior; thus this feature may have implications for the interpretation of the experiment, which assumes a constant transverse spin diffusivity over the timescale 2τ.

8. Discussion and Outlook

Low-dimensional Fermi gases are expected to feature stronger correlations and larger quantum fluctuations than their 3D counterparts. Yet, some of the first experiments on the 2D Fermi gas appear to have observed the opposite. The monopole breathing mode apparently displays no shift from the predicted value in the absence of interactions, indicating a scale invariant system.[16] Likewise, the energy of the repulsive branch of the polarized Fermi gas was found to be much smaller than that predicted theoretically .[14] This could imply one of two things: either our expectation of strong correlations in the 2D Fermi gas is incorrect, or there are additional factors present in 2D experiments that need to be taken into account. For instance, the apparent scale invariance may be influenced by finite temperature, the quasi-2D nature of the gas, or even the in-plane trapping potential and trap averaging. Thus, a detailed theoretical understanding of these effects is important.

Indeed, a major challenge currently facing experiments on 2D Fermi gases is to achieve ultracold temperatures under strong confinement. As such, superfluidity in the 2D Fermi gas has not yet been realized experimentally. Given that the BKT transition has already been observed in 2D Bose gases,[86] it is likely that superfluidity in the Bose regime of the crossover in Fermi gases will soon be realized. It may prove more difficult to observe the superfluid phase in the BCS regime because of the reduced T_c in this limit. However, as one of us has recently argued,[76] the quasi-2D nature of the gas could turn out to be advantageous here, since mean-field theory predicts that T_c/ε_F is increased as the confinement is relaxed at fixed $\varepsilon_F/\varepsilon_b$ and the Fermi system is tuned away from 2D. This raises the tantalizing possibility of T_c being maximal in the regime intermediate between 2D and 3D.

Thus far, cold-atom experiments have only just begun to explore the behavior of fermions in 2D. Even above T_c, a pseudogap regime has not yet been conclusively observed: while a gap in the spectra has been nicely demonstrated,[10] it seems likely that this is due to two-body effects only, and any apparent reduction of the gap at finite temperature is due to thermal broadening.[81] Thus, the interaction strength vs temperature phase diagram requires further investigation. In the future, we expect an increasing array of tuning "knobs" to be added to the exploration of 2D Fermi gases. There is the prospect of varying the spin imbalance and achieving superfluid-normal transitions at zero temperature. Moreover, heteronuclear Fermi-Fermi mixtures promise a fascinating new playground, where novel bound states become possible as the mass ratio is increased. Ultimately, one would like to fully uncover the fundamental differences between 2D and other dimensions.

Acknowledgements

We gratefully acknowledge our collaborators on 2D Fermi gases and related subjects for many illuminating discussions. In particular, we thank Marianne Bauer, Stefan Baur, Georg Bruun, Nigel Cooper, Tilman Enss, Andrea Fischer, Peter Littlewood, Francesca Marchetti, Pietro Massignan, Vudtiwat Ngampruetikorn, Dmitry Petrov, and Gora Shlyapnikov. Michael Köhl is thanked for several very useful discussions on experiments in 2D Fermi gases, and for sharing the data of Refs. 14, 16, 148. We also wish to thank Stefan Baur, Georg Bruun, Pietro Massignan, and Vudtiwat Ngampruetikorn for helpful feedback on the manuscript. Stefan Baur and Vudtiwat Ngampruetikorn are also thanked for help with figures. We thank Johannes Hofmann for sharing the data of Ref. 154, Jonas Vlietinck for sharing the data of Ref. 140, and Peter Kroiß for sharing the data of Ref. 141. This work was supported in part by the National Science Foundation under Grant No. PHYS-1066293 and the hospitality of the Aspen Center for Physics. MMP acknowledges support from the EPSRC under Grant No. EP/H00369X/2.

References

1. M. R. Norman, The challenge of unconventional superconductivity, *Science.* **332**, 196 (2011).
2. D. L. Smith and C. Mailhiot, Theory of semiconductor superlattice electronic structure, *Rev. Mod. Phys.* **62**, 173–234 (1990).

3. J. Singleton and C. Mielke, Quasi-two-dimensional organic superconductors: A review, *Contemporary Physics*. **43**, 63–96 (2002).
4. N. D. Mermin and H. Wagner, Absence of ferromagnetism or antiferromagnetism in one- or two-dimensional isotropic Heisenberg models, *Phys. Rev. Lett.* **17**, 1133–1136 (1966).
5. P. C. Hohenberg, Existence of long-range order in one and two dimensions, *Phys. Rev.* **158**, 383–386 (1967).
6. K. Günter, T. Stöferle, H. Moritz, M. Köhl, and T. Esslinger, p-wave interactions in low-dimensional fermionic gases, *Phys. Rev. Lett.* **95**, 230401 (2005).
7. K. Martiyanov, V. Makhalov, and A. Turlapov, Observation of a two-dimensional Fermi gas of atoms, *Phys. Rev. Lett.* **105**, 030404 (2010).
8. B. Fröhlich, M. Feld, E. Vogt, M. Koschorreck, W. Zwerger, and M. Köhl, Radio-frequency spectroscopy of a strongly interacting two-dimensional Fermi gas, *Phys. Rev. Lett.* **106**, 105301 (2011).
9. P. Dyke, E. D. Kuhnle, S. Whitlock, H. Hu, M. Mark, S. Hoinka, M. Lingham, P. Hannaford, and C. J. Vale, Crossover from 2D to 3D in a weakly interacting Fermi gas, *Phys. Rev. Lett.* **106**, 105304 (2011).
10. M. Feld, B. Fröhlich, E. Vogt, M. Koschorreck, and M. Köhl, Observation of a pairing pseudogap in a two-dimensional Fermi gas, *Nature*. **480**, 75 (2011).
11. A. T. Sommer, L. W. Cheuk, M. J. H. Ku, W. S. Bakr, and M. W. Zwierlein, Evolution of fermion pairing from three to two dimensions, *Phys. Rev. Lett.* **108**, 045302 (2012).
12. Y. Zhang, W. Ong, I. Arakelyan, and J. E. Thomas, Polaron-to-polaron transitions in the radio-frequency spectrum of a quasi-two-dimensional Fermi gas, *Phys. Rev. Lett.* **108**, 235302 (2012).
13. S. K. Baur, B. Fröhlich, M. Feld, E. Vogt, D. Pertot, M. Koschorreck, and M. Köhl, Radio-frequency spectra of Feshbach molecules in quasi-two-dimensional geometries, *Phys. Rev. A*. **85**, 061604 (2012).
14. M. Koschorreck, D. Pertot, E. Vogt, B. Fröhlich, M. Feld, and M. Kohl, Attractive and repulsive Fermi polarons in two dimensions, *Nature*. **485**, 619 (2012).
15. B. Fröhlich, M. Feld, E. Vogt, M. Koschorreck, M. Köhl, C. Berthod, and T. Giamarchi, Two-dimensional Fermi liquid with attractive interactions, *Phys. Rev. Lett.* **109**, 130403 (2012).
16. E. Vogt, M. Feld, B. Fröhlich, D. Pertot, M. Koschorreck, and M. Köhl, Scale invariance and viscosity of a two-dimensional Fermi gas, *Phys. Rev. Lett.* **108**, 070404 (2012).
17. V. Makhalov, K. Martiyanov, and A. Turlapov, Ground-state pressure of quasi-2D Fermi and Bose gases, *Phys. Rev. Lett.* **112**, 045301 (2014).
18. M. Randeria, J.-M. Duan, and L.-Y. Shieh, Bound states, Cooper pairing, and Bose condensation in two dimensions, *Phys. Rev. Lett.* **62**, 981 (1989).
19. M. Randeria, J.-M. Duan, and L.-Y. Shieh, Superconductivity in a two-dimensional Fermi gas: Evolution from Cooper pairing to Bose condensation, *Phys. Rev. B*. **41**, 327 (1990).
20. S. Schmitt-Rink, C. M. Varma, and A. E. Ruckenstein, Pairing in two dimensions, *Phys. Rev. Lett.* **63**, 445–448 (1989).
21. L. D. Landau and E. M. Lifshitz, *Quantum Mechanics*. Butterworth-Heinemann, Oxford, UK (1981).

22. S. K. Adhikari, Quantum scattering in two dimensions, *American Journal of Physics.* **54**, 362 (1986).
23. D. S. Petrov and G. V. Shlyapnikov, Interatomic collisions in a tightly confined Bose gas, *Phys. Rev. A.* **64**, 012706 (2001).
24. I. Bloch, J. Dalibard, and W. Zwerger, Many-body physics with ultracold gases, *Rev. Mod. Phys.* **80**, 885 (2008).
25. D. S. Petrov. Few-atom problem. In ed. L. C. C. Salomon, G. V. Shlyapnikov, *Many-body physics with ultra-cold gases: Lecture Notes of the Les Houches Summer Schools, vol. 94.* Oxford University Press, Oxford, England (2012).
26. M. Olshanii, Atomic scattering in the presence of an external confinement and a gas of impenetrable bosons, *Phys. Rev. Lett.* **81**, 938 (1998).
27. T. Bergeman, M. G. Moore, and M. Olshanii, Atom-atom scattering under cylindrical harmonic confinement: Numerical and analytic studies of the confinement induced resonance, *Phys. Rev. Lett.* **91**, 163201 (2003).
28. E. Haller, M. J. Mark, R. Hart, J. G. Danzl, L. Reichsöllner, V. Melezhik, P. Schmelcher, and H.-C. Nägerl, Confinement-induced resonances in low-dimensional quantum systems, *Phys. Rev. Lett.* **104**, 153203 (2010).
29. S. Sala, P.-I. Schneider, and A. Saenz, Inelastic confinement-induced resonances in low-dimensional quantum systems, *Phys. Rev. Lett.* **109**, 073201 (2012).
30. S. Sala, G. Zürn, T. Lompe, A. N. Wenz, S. Murmann, F. Serwane, S. Jochim, and A. Saenz, Coherent molecule formation in anharmonic potentials near confinement-induced resonances, *Phys. Rev. Lett.* **110**, 203202 (2013).
31. D. S. Petrov, C. Salomon, and G. V. Shlyapnikov, Weakly bound dimers of fermionic atoms, *Phys. Rev. Lett.* **93**, 090404 (2004).
32. G. V. Skorniakov and K. A. Ter-Martirosian, Three body problem for short range forces. I. Scattering of low energy neutrons by deuterons, *Sov. Phys. JETP.* **4**, 648 (1957).
33. J. Levinsen and V. Gurarie, Properties of strongly paired fermionic condensates, *Phys. Rev. A.* **73**, 053607 (2006).
34. R. Combescot, S. Giraud, and X. Leyronas, Normal state of highly polarized Fermi gases: The bound state, *Laser Physics.* **20**, 678–682 (2010).
35. O. I. Kartavtsev, A. V. Malykh, and S. A. Sofianos, Bound states and scattering lengths of three two-component particles with zero-range interactions under one-dimensional confinement, *ZhETF.* **135**, 419 (2009).
36. G. Orso, E. Burovski, and T. Jolicoeur, Luttinger liquid of trimers in Fermi gases with unequal masses, *Phys. Rev. Lett.* **104**, 065301 (2010).
37. L. Pricoupenko and P. Pedri, Universal $(1 + 2)$-body bound states in planar atomic waveguides, *Phys. Rev. A.* **82**, 033625 (2010).
38. O. I. Kartavtsev and A. V. Malykh, Low-energy three-body dynamics in binary quantum gases, *J. Phys. B: At. Mol. Opt. Phys.* **40**, 1429–1441 (2007).
39. C. J. M. Mathy, M. M. Parish, and D. A. Huse, Trimers, molecules and polarons in imbalanced atomic Fermi gases, *Phys. Rev. Lett.* **106**, 166404 (2011).
40. N. P. Mehta, Born-Oppenheimer study of two-component few-particle systems under one-dimensional confinement, *Phys. Rev. A.* **89**, 052706 (May, 2014).

41. J. Levinsen and M. M. Parish, Bound states in a quasi-two-dimensional Fermi gas, *Phys. Rev. Lett.* **110**, 055304 (2013).
42. D. Blume, Universal four-body states in heavy-light mixtures with a positive scattering length, *Phys. Rev. Lett.* **109**, 230404 (2012).
43. O. I. Kartavtsev and A. V. Malykh, Recent advances in description of few two-component fermions, *Yad. Fiz.* **77**, 458 (2014).
44. E. Wille, F. M. Spiegelhalder, G. Kerner, D. Naik, A. Trenkwalder, G. Hendl, F. Schreck, R. Grimm, T. G. Tiecke, J. T. M. Walraven, S. J. J. M. F. Kokkelmans, E. Tiesinga, and P. S. Julienne, Exploring an ultracold Fermi-Fermi mixture: Inter-species Feshbach resonances and scattering properties of ^{6}Li and ^{40}K, *Phys. Rev. Lett.* **100**, 053201 (2008).
45. L. Costa, J. Brachmann, A.-C. Voigt, C. Hahn, M. Taglieber, T. W. Hänsch, and K. Dieckmann, s-wave interaction in a two-species Fermi-Fermi mixture at a narrow Feshbach resonance, *Phys. Rev. Lett.* **105**, 123201 (2010).
46. C.-H. Wu, I. Santiago, J. W. Park, P. Ahmadi, and M. W. Zwierlein, Strongly interacting isotopic Bose-Fermi mixture immersed in a Fermi sea, *Phys. Rev. A.* **84**, 011601 (2011).
47. V. N. Efimov, Energy levels of three resonantly interacting particles, *Nucl. Phys. A.* **210**, 157 (1973).
48. T. Kraemer, M. Mark, P. Waldburger, J. G. Danzl, C. Chin, B. Engeser, A. D. Lange, K. Pilch, A. Jaakkola, H.-C. Nägerl, and R. Grimm, Evidence for Efimov quantum states in an ultracold gas of caesium atoms, *Nature.* **440**, 315 (2006).
49. T. B. Ottenstein, T. Lompe, M. Kohnen, A. N. Wenz, and S. Jochim, Collisional stability of a three-component degenerate Fermi gas, *Phys. Rev. Lett.* **101**, 203202 (2008).
50. Y. Castin, C. Mora, and L. Pricoupenko, Four-body Efimov effect for three fermions and a lighter particle, *Phys. Rev. Lett.* **105**, 223201 (2010).
51. Y. Nishida and S. Tan, Liberating Efimov physics from three dimensions, *Few-Body Systems.* **51**, 191 (2011).
52. J. Levinsen, P. Massignan, and M. M. Parish, Efimov trimers under strong confinement, *Phys. Rev. X.* **4**, 031020 (2014).
53. V. Ngampruetikorn, M. M. Parish, and J. Levinsen, Three-body problem in a two-dimensional Fermi gas, *EPL*, **102**, 13001 (2013).
54. J. Levinsen, T. G. Tiecke, J. T. M. Walraven, and D. S. Petrov, Atom-dimer scattering and long-lived trimers in fermionic mixtures, *Phys. Rev. Lett.* **103**, 153202 (2009).
55. G. Bertaina and S. Giorgini, BCS-BEC crossover in a two-dimensional Fermi gas, *Phys. Rev. Lett.* **106**, 110403 (2011).
56. D. S. Petrov, M. A. Baranov, and G. V. Shlyapnikov, Superfluid transition in quasi-two-dimensional Fermi gases, *Phys. Rev. A.* **67**, 031601 (2003).
57. F. F. Bellotti, T. Frederico, M. T. Yamashita, D. V. Fedorov, A. S. Jensen, and N. T. Zinner, Mass-imbalanced three-body systems in two dimensions, *Journal of Physics B: Atomic, Molecular and Optical Physics.* **46**, 055301 (2013).
58. M. Jag, M. Zaccanti, M. Cetina, R. S. Lous, F. Schreck, R. Grimm, D. S. Petrov, and J. Levinsen, Observation of a strong atom-dimer attraction in a mass-imbalanced Fermi-Fermi mixture, *Phys. Rev. Lett.* **112**, 075302 (2014).
59. S. Flügge and H. Marschall, *Rechenmethoden der Quantentheorie*. Springer-Verlag, Berlin (1952).

60. Y. F. Smirnov, Talmi transformation for particles with different masses (ii), *Nucl. Phys.* **39**, 346–352 (1962).
61. J. Levinsen, N. R. Cooper, and V. Gurarie, Stability of fermionic gases close to a *p*-wave Feshbach resonance, *Phys. Rev. A.* **78**, 063616 (2008).
62. Y. Nishida, S. Moroz, and D. T. Son, Super Efimov effect of resonantly interacting fermions in two dimensions, *Phys. Rev. Lett.* **110**, 235301 (2013).
63. A. J. Leggett. Diatomic molecules and Cooper pairs. In eds. A. Pekalski and J. Przystawa, *Modern Trends in the Theory of Condensed Matter*, p. 14. Springer-Verlag, Berlin (1980).
64. D. M. Eagles, Possible pairing without superconductivity at low carrier concentrations in bulk and thin-film superconducting semiconductors, *Phys. Rev.* **186**, 456 (1969).
65. C. A. Regal, M. Greiner, and D. S. Jin, Observation of resonance condensation of fermionic atom pairs, *Phys. Rev. Lett.* **92**, 040403 (2004).
66. M. W. Zwierlein, C. A. Stan, C. H. Schunck, S. M. F. Raupach, A. J. Kerman, and W. Ketterle, Condensation of pairs of fermionic atoms near a Feshbach resonance, *Phys. Rev. Lett.* **92**, 120403 (2004).
67. J.-P. Martikainen and P. Törmä, Quasi-two-dimensional superfluid fermionic gases, *Phys. Rev. Lett.* **95**, 170407 (2005).
68. A. M. Fischer and M. M. Parish, BCS-BEC crossover in a quasi-two-dimensional Fermi gas, *Phys. Rev. A.* **88**, 023612 (2013).
69. R. Chasman and S. Wahlborn, Transformation scheme for harmonic-oscillator wave functions, *Nuclear Physics A.* **90**, 401 (1967).
70. J. R. Engelbrecht and M. Randeria, New collective mode and corrections to Fermi-liquid theory in two dimensions, *Phys. Rev. Lett.* **65**, 1032 (1990).
71. J. R. Engelbrecht, M. Randeria, and L. Zhang, Landau f function for the dilute Fermi gas in two dimensions, *Phys. Rev. B.* **45**, 10135 (1992).
72. J. R. Engelbrecht and M. Randeria, Low-density repulsive Fermi gas in two dimensions: Bound-pair excitations and Fermi-liquid behavior, *Phys. Rev. B.* **45**, 12419 (1992).
73. P. Bloom, Two-dimensional Fermi gas, *Phys. Rev. B.* **12**, 125 (1975).
74. C. Mora and Y. Castin, Ground state energy of the two-dimensional weakly interacting Bose gas: First correction beyond Bogoliubov theory, *Phys. Rev. Lett.* **102**, 180404 (2009).
75. M. Bauer, M. M. Parish, and T. Enss, Universal equation of state and pseudogap in the two-dimensional Fermi gas, *Phys. Rev. Lett.* **112**, 135302 (2014).
76. A. M. Fischer and M. M. Parish. Quasi-two-dimensional Fermi gases at finite temperature. *Phys. Rev. B* **90**, 214503 (2014).
77. S. Tan, Large momentum part of a strongly correlated Fermi gas, *Annals of Physics.* **323**, 2971 (2008).
78. F. Werner and Y. Castin, General relations for quantum gases in two and three dimensions: Two-component fermions, *Phys. Rev. A.* **86**, 013626 (2012).
79. J. Hofmann, Current response, structure factor and hydrodynamic quantities of a two- and three-dimensional Fermi gas from the operator-product expansion, *Phys. Rev. A.* **84**, 043603 (2011).
80. C. Langmack, M. Barth, W. Zwerger, and E. Braaten, Clock shift in a strongly interacting two-dimensional Fermi gas, *Phys. Rev. Lett.* **108**, 060402 (2012).

81. V. Ngampruetikorn, J. Levinsen, and M. M. Parish, Pair correlations in the two-dimensional Fermi gas, *Phys. Rev. Lett.* **111**, 265301 (2013).
82. N. Trivedi and M. Randeria, Deviations from Fermi-liquid behavior above T_C in 2D short coherence length superconductors, *Phys. Rev. Lett.* **75**, 312 (1995).
83. V. M. Loktev, R. M. Quick, and S. G. Sharapov, Phase fluctuations and pseudogap phenomena, *Physics Reports.* **349**, 1 (2001).
84. D. S. Fisher and P. C. Hohenberg, Dilute Bose gas in two dimensions, *Phys. Rev. B.* **37**, 4936 (1988).
85. M. Holzmann, G. Baym, J.-P. Blaizot, and F. Laloë, Superfluid transition of homogeneous and trapped two-dimensional Bose gases, *Proc. Natl. Acad. Sci.* **104**, 1476 (2007).
86. Z. Hadzibabic, P. Krüger, M. Cheneau, B. Battelier, and J. Dalibard, Berezinskii–Kosterlitz–Thouless crossover in a trapped atomic gas, *Nature.* **441**, 1118 (2006).
87. S. S. Botelho and C. A. R. Sá de Melo, Vortex-antivortex lattice in ultracold fermionic gases, *Phys. Rev. Lett.* **96**, 040404 (2006).
88. K. Miyake, Fermi liquid theory of dilute submonolayer ^{3}He on thin ^{4}He II film: dimer bound state and Cooper pairs, *Progr. Theor. Phys.* **69**, 1794 (1983).
89. X.-J. Liu, Virial expansion for a strongly correlated Fermi system and its application to ultracold atomic Fermi gases, *Physics Reports.* **524**, 37 (2013).
90. X.-J. Liu, H. Hu, and P. D. Drummond, Exact few-body results for strongly correlated quantum gases in two dimensions, *Phys. Rev. B.* **82**, 054524 (2010).
91. X. Leyronas, Virial expansion with Feynman diagrams, *Phys. Rev. A.* **84**, 053633 (2011).
92. Y. He, Q. Chen, and K. Levin, Radio-frequency spectroscopy and the pairing gap in trapped Fermi gases, *Phys. Rev. A.* **72**, 011602 (2005).
93. Q. Chen, C. A. Regal, M. Greiner, D. S. Jin, and K. Levin, Understanding the superfluid phase diagram in trapped Fermi gases, *Phys. Rev. A.* **73**, 041601 (2006).
94. J. P. Gaebler, J. T. Stewart, T. E. Drake, D. S. Jin, A. Perali, P. Pieri, and G. C. Strinati, Observation of pseudogap behaviour in a strongly interacting Fermi gas, *Nature Phys.* **6** (2010).
95. A. Perali, F. Palestini, P. Pieri, G. C. Strinati, J. T. Stewart, J. P. Gaebler, T. E. Drake, and D. S. Jin, Evolution of the normal state of a strongly interacting Fermi gas from a pseudogap phase to a molecular Bose gas, *Phys. Rev. Lett.* **106**, 060402 (2011).
96. M. Barth and J. Hofmann, Pairing effects in the nondegenerate limit of the two-dimensional Fermi gas, *Phys. Rev. A.* **89**, 013614 (2014).
97. R. Watanabe, S. Tsuchiya, and Y. Ohashi, Low-dimensional pairing fluctuations and pseudogapped photoemission spectrum in a trapped two-dimensional Fermi gas, *Phys. Rev. A.* **88**, 013637 (2013).
98. F. Marsiglio, P. Pieri, A. Perali, F. Palestini and G. C. Strinati, Pairing effects in the normal phase of a two-dimensional Fermi gas, arXiv:1406.7761.
99. M. J. H. Ku, A. T. Sommer, L. W. Cheuk, and M. W. Zwierlein, Revealing the superfluid lambda transition in the universal thermodynamics of a unitary Fermi gas, *Science.* **335**, 563 (2012).
100. P. Fulde and R. A. Ferrell, Superconductivity in a strong spin-exchange field, *Phys. Rev.* **135**, A550–A563 (1964).

101. A. I. Larkin and Y. N. Ovchinnikov, Inhomogeneous state of superconductors, *Sov. Phys. JETP.* **20**, 762–769 (1965).
102. M. W. Zwierlein, A. Schirotzek, C. H. Schunck, and W. Ketterle, Fermionic superfluidity with imbalanced spin populations, *Science.* **311**, 492 (2006).
103. G. B. Partridge, W. Li, R. I. Kamar, Y. Liao, and R. G. Hulet, Pairing and phase separation in a polarized Fermi gas, *Science.* **311**, 503 (2006).
104. Y. Shin, C. H. Schunck, A. Schirotzek, and W. Ketterle, Phase diagram of a two-component Fermi gas with resonant interactions, *Nature.* **451**, 689–693 (2008).
105. S. Nascimbène, N. Navon, K. J. Jiang, L. Tarruell, M. Teichmann, J. McKeever, F. Chevy, and C. Salomon, Collective oscillations of an imbalanced Fermi gas: Axial compression modes and polaron effective mass, *Phys. Rev. Lett.* **103**, 170402 (2009).
106. Y.-A. Liao, A. S. C. Rittner, T. Paprotta, W. Li, G. B. Partridge, R. G. Hulet, S. K. Baur, and E. J. Mueller, Spin-imbalance in a one-dimensional Fermi gas, *Nature.* **467**, 567–569 (2010).
107. F. Chevy and C. Mora, Ultra-cold polarized Fermi gases, *Reports on Progress in Physics.* **73**, 112401 (2010).
108. D. E. Sheehy and L. Radzihovsky, BEC-BCS crossover, phase transitions and phase separation in polarized resonantly-paired superfluids, *Annals of Physics.* **322**, 1790 (2007).
109. S. Zöllner, G. M. Bruun, and C. J. Pethick, Polarons and molecules in a two-dimensional Fermi gas, *Phys. Rev. A.* **83**, 021603 (2011).
110. M. M. Parish, Polaron-molecule transitions in a two-dimensional Fermi gas, *Phys. Rev. A.* **83**, 051603 (2011).
111. R. Schmidt, T. Enss, V. Pietilä, and E. Demler, Fermi polarons in two dimensions, *Phys. Rev. A.* **85**, 021602 (2012).
112. V. Ngampruetikorn, J. Levinsen, and M. M. Parish, Repulsive polarons in two-dimensional Fermi gases, *EPL.* **98**, 30005 (2012).
113. N. Prokof'ev and B. Svistunov, Fermi-polaron problem: Diagrammatic Monte Carlo method for divergent sign-alternating series, *Phys. Rev. B.* **77**, 020408 (2008).
114. N. V. Prokof'ev and B. V. Svistunov, Bold diagrammatic Monte Carlo: A generic sign-problem tolerant technique for polaron models and possibly interacting many-body problems, *Phys. Rev. B.* **77**, 125101 (2008).
115. G. M. Bruun and P. Massignan, Decay of polarons and molecules in a strongly polarized Fermi gas, *Phys. Rev. Lett.* **105**, 020403 (2010).
116. G.-B. Jo, Y.-R. Lee, J.-H. Choi, C. A. Christensen, T. H. Kim, J. H. Thywissen, D. E. Pritchard, and W. Ketterle, Itinerant ferromagnetism in a Fermi gas of ultracold atoms, *Science.* **325**, 1521 (2009).
117. D. Pekker, M. Babadi, R. Sensarma, N. Zinner, L. Pollet, M. W. Zwierlein, and E. Demler, Competition between pairing and ferromagnetic instabilities in ultracold Fermi gases near Feshbach resonances, *Phys. Rev. Lett.* **106**, 050402 (2011).
118. C. Sanner, E. J. Su, W. Huang, A. Keshet, J. Gillen, and W. Ketterle, Correlations and pair formation in a repulsively interacting Fermi gas, *Phys. Rev. Lett.* **108**, 240404 (2012).
119. P. Massignan, M. Zaccanti, and G. M. Bruun, Polarons, dressed molecules and itinerant ferromagnetism in ultracold Fermi gases, *Reports on Progress in Physics.* **77**, 034401 (2014).

120. F. Chevy, Universal phase diagram of a strongly interacting Fermi gas with unbalanced spin populations, *Phys. Rev. A.* **74**, 063628 (2006).

121. M. M. Parish and J. Levinsen, Highly polarized Fermi gases in two dimensions, *Phys. Rev. A.* **87**, 033616 (2013).

122. X. Cui and H. Zhai, Stability of a fully magnetized ferromagnetic state in repulsively interacting ultracold Fermi gases, *Phys. Rev. A.* **81**, 041602 (2010).

123. P. Massignan and G. M. Bruun, Repulsive polarons and itinerant ferromagnetism in strongly polarized Fermi gases, *Eur. Phys. J. D.* **65**, 83–89 (2011).

124. R. Combescot and S. Giraud, Normal state of highly polarized Fermi gases: Full many-body treatment, *Phys. Rev. Lett.* **101**, 050404 (2008).

125. C. Kohstall, M. Zaccanti, M. Jag, A. Trenkwalder, P. Massignan, G. M. Bruun, F. Schreck, and R. Grimm, Metastability and coherence of repulsive polarons in a strongly interacting Fermi mixture, *Nature.* **485**, 615 (2012).

126. R. Combescot, S. Giraud, and X. Leyronas, Analytical theory of the dressed bound state in highly polarized Fermi gases, *Europhys. Lett.* **88**, 60007 (2009).

127. M. Punk, P. T. Dumitrescu, and W. Zwerger, Polaron-to-molecule transition in a strongly imbalanced Fermi gas, *Phys. Rev. A.* **80**, 053605 (2009).

128. C. Mora and F. Chevy, Ground state of a tightly bound composite dimer immersed in a Fermi sea, *Phys. Rev. A.* **80**, 033607 (2009).

129. G. J. Conduit, Itinerant ferromagnetism in a two-dimensional atomic gas, *Phys. Rev. A.* **82**, 043604 (2010).

130. D. S. Petrov, Three-body problem in Fermi gases with short-range interparticle interaction, *Phys. Rev. A.* **67**, 010703 (2003).

131. V. Ngampruetikorn, Ph.D. thesis.

132. V. Pietilä, D. Pekker, Y. Nishida, and E. Demler, Pairing instabilities in quasi-two-dimensional Fermi gases, *Phys. Rev. A.* **85**, 023621 (2012).

133. J. Levinsen and S. K. Baur, High-polarization limit of the quasi-two-dimensional Fermi gas, *Phys. Rev. A.* **86**, 041602 (2012).

134. A. Schirotzek, C.-H. Wu, A. Sommer, and M. W. Zwierlein, Observation of Fermi polarons in a tunable Fermi liquid of ultracold atoms, *Phys. Rev. Lett.* 102:230402 (2009).

135. J. B. McGuire, Interacting fermions in one dimension. II. Attractive potential, *J. Math. Phys.* **7**, 123 (1966).

136. L. He and P. Zhuang, Phase diagram of a cold polarized Fermi gas in two dimensions, *Phys. Rev. A.* **78**, 033613 (2008).

137. G. J. Conduit, P. H. Conlon, and B. D. Simons, Superfluidity at the BEC-BCS crossover in two-dimensional Fermi gases with population and mass imbalance, *Phys. Rev. A.* **77**, 053617 (2008).

138. J. Tempere, S. N. Klimin, and J. T. Devreese, Effect of population imbalance on the Berezinskii-Kosterlitz-Thouless phase transition in a superfluid Fermi gas, *Phys. Rev. A.* **79**, 053637 (2009).

139. S. Yin, J.-P. Martikainen, and P. Törmä, Fulde-Ferrell states and Berezinskii-Kosterlitz-Thouless phase transition in two-dimensional imbalanced Fermi gases, *Phys. Rev. B.* **89**, 014507 (2014).

140. J. Vlietinck, J. Ryckebusch, and K. Van Houcke, Diagrammatic Monte Carlo study of the Fermi polaron in two dimensions, *Phys. Rev. B.* **89**, 085119 (2014).
141. P. Kroiss and L. Pollet, Diagrammatic Monte Carlo study of quasi-two-dimensional Fermi polarons, *Phys. Rev. B* **90**, 104510 (2014).
142. M. Köhl, talk at the 2012 APS March meeting.
143. C. Lobo, A. Recati, S. Giorgini, and S. Stringari, Normal state of a polarized Fermi gas at unitarity, *Phys. Rev. Lett.* 97:200403 (2006).
144. M. M. Parish, F. M. Marchetti, and P. B. Littlewood, Supersolidity in electron-hole bilayers with a large density imbalance, *Europhys. Lett.* **95**, 27007 (2011).
145. S. Gopalakrishnan, A. Lamacraft, and P. M. Goldbart, Universal phase structure of dilute Bose gases with Rashba spin-orbit coupling, *Phys. Rev. A.* **84**, 061604 (2011).
146. L. P. Pitaevskii and A. Rosch, Breathing modes and hidden symmetry of trapped atoms in two dimensions, *Phys. Rev. A.* **55**, R853 (1997).
147. M. Olshanii, H. Perrin, and V. Lorent, Example of a quantum anomaly in the physics of ultracold gases, *Phys. Rev. Lett.* **105**, 095302 (2010).
148. M. Koschorreck, D. Pertot, E. Vogt, and M. Köhl, Universal spin dynamics in two-dimensional Fermi gases, *Nature Physics.* **9**, 405 (2013).
149. S. K. Baur, E. Vogt, M. Köhl, and G. M. Bruun, Collective modes of a two-dimensional spin-1/2 Fermi gas in a harmonic trap, *Phys. Rev. A.* **87**, 043612 (2013).
150. S. Chiacchiera, D. Davesne, T. Enss, and M. Urban, Damping of the quadrupole mode in a two-dimensional Fermi gas, *Phys. Rev. A.* **88**, 053616 (2013).
151. F. Werner and Y. Castin, Unitary gas in an isotropic harmonic trap: Symmetry properties and applications, *Phys. Rev. A.* **74**, 053604 (2006).
152. Y. Nishida and D. T. Son, Nonrelativistic conformal field theories, *Phys. Rev. D.* **76**, 086004 (2007).
153. E. Taylor and M. Randeria, Apparent low-energy scale invariance in two-dimensional Fermi gases, *Phys. Rev. Lett.* **109**, 135301 (2012).
154. J. Hofmann, Quantum anomaly, universal relations, and breathing mode of a two-dimensional Fermi gas, *Phys. Rev. Lett.* **108**, 185303 (2012).
155. C. Gao and Z. Yu, Breathing mode of two-dimensional atomic Fermi gases in harmonic traps, *Phys. Rev. A.* **86**, 043609 (2012).
156. S. Moroz, Scale-invariant Fermi gas in a time-dependent harmonic potential, *Phys. Rev. A.* **86**, 011601 (Jul, 2012).
157. C. Chafin and T. Schäfer, Scale breaking and fluid dynamics in a dilute two-dimensional Fermi gas, *Phys. Rev. A.* **88**, 043636 (2013).
158. D. T. Son, Vanishing bulk viscosities and conformal invariance of the unitary Fermi gas, *Phys. Rev. Lett.* **98**, 020604 (2007).
159. L. Wu and Y. Zhang, Applicability of the Boltzmann equation for a two-dimensional Fermi gas, *Phys. Rev. A.* **85**, 045601 (2012).
160. G. M. Bruun, Shear viscosity and spin-diffusion coefficient of a two-dimensional Fermi gas, *Phys. Rev. A.* **85**, 013636 (2012).
161. T. Schäfer, Shear viscosity and damping of collective modes in a two-dimensional Fermi gas, *Phys. Rev. A.* **85**, 033623 (2012).
162. T. Enss, C. Küppersbusch, and L. Fritz, Shear viscosity and spin diffusion in a two-dimensional Fermi gas, *Phys. Rev. A.* **86**, 013617 (2012).

163. A. Sommer, M. Ku, G. Roati, and M. W. Zwierlein, Universal spin transport in a strongly interacting Fermi gas, *Nature.* **472**, 201 (2011).
164. G. Policastro, D. T. Son, and A. O. Starinets, Shear viscosity of strongly coupled N = 4 supersymmetric Yang-Mills plasma, *Phys. Rev. Lett.* **87**, 081601 (2001).
165. T. Enss, Transverse spin diffusion in strongly interacting Fermi gases, *Phys. Rev. A.* **88**, 033630 (2013).

CHAPTER 2

FEW-BODY PHYSICS OF ULTRACOLD ATOMS AND MOLECULES WITH LONG-RANGE INTERACTIONS

Yujun Wang[*], Paul Julienne[†], and Chris H Greene[‡]

*Department of Physics, Kansas State University,
Manhattan, Kansas, 66506, USA*

†*Department of Physics and Astronomy, Purdue University,
West Lafayette, Indiana, 47907, USA*

‡*Joint Quantum Institute, University of Maryland and NIST
College Park, Maryland 20742, USA*
*yujunw@phys.ksu.edu
†chgreene@purdue.edu
‡psj@umd.edu

1. Introduction

The quantum mechanical few-body problem at ultracold energies poses severe challenges to theoretical techniques, particularly when long-range interactions are present and decay only as a power-law potential. One familiar result is the modification of the elastic scattering Wigner threshold laws[1] in the presence of power-law interaction potentials. Analogously, the near-threshold behaviors of all two-body scattering and bound state observables require generalizations in order to correctly describe the role of long range potentials. These generalizations are often termed modified effective range theories[2–5] or generalized quantum defect theories[6–18] in the literature.

For simple and isotropic long-range interaction potentials involving one or more power-law potentials, analytical solutions are often known at least at zero energy, and allow extensive development in terms of closed-form solutions through the techniques of classical mathematics.[8,18,19]

With anisotropic long-range interactions however, such as the dipole–dipole interaction characteristic for a gas composed of polar molecules or else of strong magnetic dipoles, often the theoretical heavy lifting relies on numerical solution techniques.[20–22] For dipolar scattering, numerical methods provide critically detailed information beyond the insights gleaned from simpler analytical approaches such as the partial-wave Born approximation.[23–29]

Moving from the problem of two interacting particles to three or more involves a tremendous leap in the complexity of the theoretical description, as well as a commensurate richness in the phenomena that will occur. For three interacting particles that have no long range Coulomb interactions, the realm of universal Efimov physics has received tremendous theoretical interest[30–51] during the several decades since its original prediction[52,53] in 1970. Moreover, it has received extensive experimental attention since the first clear observation of the Efimov effect in the 2006 measurement by the Innsbruck group of Grimm and co-workers.[54] The community of theorists interested in the Efimov effect started initially in nuclear physics, but the recent studies have concentrated on ultracold atomic and molecular systems. This change in the relevant subfield occurred because the key parameter controlling Efimov states is the atom–atom scattering length, which is nowadays controllable near a Fano-Feshbach resonance,[55–59] through the application of external fields.[60–63]

One of the interesting aspects of the Efimov effect is that for sufficiently large two-body scattering lengths, the sum of pairwise particle interactions produces a net effective attraction, namely a hyperradial potential energy curve proportional to $-1/R^2$, where R is the three-body hyperradius [see Eq. (14)]. Such a potential is sometimes called a "dipole potential" because it also arises in various atomic and molecular two-body systems, in particular when a charge moves in the field of a permanent dipole, which is either a polar molecule[64–67] or else an excited, degenerate hydrogen atom.[68–70] A peculiarity of such an attractive inverse square potential is that it is too singular to produce a unique solution near the origin, and that behavior must be regularized in any given physical system. For the charge-dipole system, it occurs naturally because any dipole has finite extent and so the dipole potential does not hold once the charge moves inside the charge distribution that produces the dipole. In Efimov physics, the short-range hyperradial phase as the system moves into the effective dipole potential (Efimov) region of the hyperradius is regulated by the so-called "three-body parameter".[71,72]

We discuss the details of the three-body parameter in Sec. 3.1. The main idea is the following: in the presence of zero-range two-body interactions, the energy of the lowest Efimov resonance would drop to negative infinity were it not for the fact that the $-1/R^2$ potential energy does not hold all the way down to $R = 0$ for any real system. Zero-range interactions are a convenient model of Efimov physics, but they require the introduction of a three-body parameter to truncate the ground state energy to a finite value. One can view the three-body parameter as setting either a characteristic length or energy related to the deepest Efimov trimer. When expressed as an energy, it is the lowest Efimov energy level at unitarity, and when expressed as a length it is the value of the negative scattering length a_-^* at which the lowest Efimov state binding energy goes to zero (for $a < 0$). In some models, such as effective field theory[71] or in a model analysis of the adiabatic hyperspherical potential curves,[73,74] it is possible to derive an approximate relationship between these two different ways of defining the three-body parameter. In the first few decades of theoretical study of Efimov physics, it was taken as a matter of fact that the three-body parameter could not be predicted in general on the basis of two-body interactions alone, and that for different real systems that parameter would vary almost randomly. This view was expressed, for instance, in Ref. 72. Thus it was a major surprise when experiments on the system of three bosonic Cs atoms began to show[75] that several different Efimov resonances have almost the same value of the three body parameter expressed as a_-^*. In particular, it can be expressed in units of the van der Waals length r_{vdW}, which is defined by[60,76]

$$r_{\mathrm{vdW}} = \frac{1}{2}(2\mu_2 C_6/\hbar^2)^{1/4}. \tag{1}$$

For two atoms of reduced mass μ_2 interacting via the $-C_6/r^6$ van der Waals potential (r is the two-body distance), experiments showed that $a_-^*/r_{\mathrm{vdW}} \approx -10$. Around one year later, theory was able to explain this quasi-universality of systems having a long-range van der Waals interaction between each pair of particles, but this eventually was explained for a system of three identical bosonic atoms[48] which was largely confirmed later by Naidon *et al.*[45] and for a three-boson system with only two of the atoms identical by Wang *et al.*[77]

It remains of considerable interest to map out this recently identified quasi-universality of the three-body parameter to see how it depends on the nature of the long-range two-body interactions. Some hints that there could be a universality for interacting dipoles was in fact suggested in two studies of the three-dipole problem, first for three oriented bosonic dipoles where an Efimov effect was predicted for the first time in a 3D system where angular momentum is not conserved.[78] That study was later extended to three oriented identical fermionic dipoles where there is no predicted Efimov effect, but nevertheless an interesting universal state could be predicted.[79]

The present review summarizes recent developments that have led to an improved understanding of long-range interactions, focusing on two-body and three-body systems with ultracold atoms or molecules. While there have been tremendous accomplishments in experimental few-body physics in recent years, the present review concentrates on the theoretical understanding that has emerged from combined experimental and theoretical efforts. The theory has been developed to the point where we increasingly understand which aspects of few-body collisions, bound states, and resonances are universal and controlled only by the long-range Hamiltonian, and which aspects differ and distinguish one species from another having identical long-range forces.

2. van der Waals Physics for Two Atoms

2.1. *van der Waals universality and Feshbach resonances with atoms*

Let us consider two atoms interacting by a long range potential of the form $-C_N/r^N$, where r is the distance between the atoms. While the physics associated with various values of N has been widely studied, e.g., $N = 1$ (the Coulomb potential), $N = 3$ (two dipoles, see Sec. 5.1), or $N = 4$ (an ion and an atom, see Sec. 6), in this section we consider the specific case of $n = 6$ for the van der Waals (vdW) potential between two neutral S-state atoms. When the length $r' = r/r_{vdW}$ and potential $v(r') = V(r)/E_{vdW}$ are scaled by their respective van der Waals units of length and energy

$E_{vdW} = \hbar^2/(2\mu_2 r_{vdW}^2)$, the long range potential $v(r')$ between the two atoms including the centrifugal barrier for partial wave ℓ is[80]

$$v(r') \to -\frac{16}{r'^6} + \frac{\ell(\ell+1)}{r'^2} \qquad (2)$$

It is important to note that there are other conventions in the literature defining a length and corresponding energy associated with the van der Waals potential. Gao[8] uses $\beta_6 = 2r_{vdW}$ and Gribakin and Flambaum introduced the mean scattering length $\bar{a} = [4\pi/\Gamma(\frac{1}{4})^2]r_{vdW} \approx 0.955978\ldots r_{vdW}.$[76] The latter is especially useful for giving a simpler form to theoretical expressions.

Gao[8] has worked out the analytic solutions for the bound and scattering states of this potential, which are especially relevant to two- and few-body physics with cold atoms. This physics is to a large extent governed by the states near the $E = 0$ collision threshold, where E represents energy. These analytic solutions are parameterized in terms of the s-wave scattering length a,[81] a threshold property of the scattering phase shift $\eta(k) \to -ka$ for the $\ell = 0$ partial wave as collision momentum $\hbar k = \sqrt{2\mu_2 E} \to 0$. Specifying a specifies the bound ($E < 0$) and scattering ($E \geq 0$) solutions away from $E = 0$ for all ℓ.[7,9,82] For example, if $a = \infty$, there is not only an s-wave bound state at $E = 0$, but also an $E = 0$ bound state for partial waves $\ell = 4, 8, 12 \ldots$; similarly, an $E = 0$ p-wave bound state exists if $a = 2\bar{a}$ and a d-wave bound state if $a = \bar{a}$.

It is much more effective to use the analytic solutions to the van der Waals potential instead of the solutions to a zero-range pseudopotential to characterize near-threshold cold atom physics, since the former are accurate over a much wider range of energy near threshold. This is a consequence of the fact that r_{vdW} tends to be much larger than the range of strong chemical interactions that occur when $r \ll r_{vdW}$. Thus, the van der Waals potential spans a wide range of r between the chemical and asymptotic regions, and its solutions accurately span a range of bound state and collision energies large compared to E_{vdW} and large compared to energy scales relevant to cold atom phenomena; see Refs. 60, 80. For example, Chin et al.[60] give examples showing the near threshold spectrum of bound states for s-waves ($\ell = 0$) and other partial waves ($\ell > 0$) on a scale spanning hundreds of E_{vdW}. These bound state energies in units of E_{vdW} are universal functions of a/r_{vdW}, that is, species-specific parameters like the reduced mass and the magnitude of C_6 are scaled out by scaling by r_{vdW} and E_{vdW}, and the scaled a captures the effect of all short range physics.

When the scattering length is large and positive, the energy of the last s-wave bound state of the potential is universally related to the scattering length by the simple relation,

$$E_{-1}^{\text{U}} \approx -\frac{\hbar^2}{2\mu_2 a^2}. \tag{3}$$

It is simple to make universal corrections to this expression due to the van der Waals potential. Using reduced units of $\epsilon = E/\bar{E}$ for energy and $\alpha = a/\bar{a}$ length, Gribakin and Flambaum[76] and Gao[81] have developed corrections to the universal binding energy $\epsilon_{-1}^{\text{U}} = -1/\alpha^2$, valid when $1 \ll \alpha < \infty$:

$$\epsilon_{-1}^{\text{GF}} = -\frac{1}{(\alpha - 1)^2}, \quad \epsilon_{-1}^{\text{Gao}} = -\frac{1}{(\alpha - 1)^2}\left[1 + \frac{g_1}{\alpha - 1} + \frac{g_2}{(\alpha - 1)^2}\right], \tag{4}$$

where $g_1 = \Gamma(\frac{1}{4})^4/6\pi^2 - 2 \approx 0.9179$ and $g_2 = (5/4)g_1^2 - 2 \approx -0.9468$. Furthermore, the universal van der Waals effective range correction r_e (defined in Sec. 2.2) to the scattering phase shift for $E > 0$ is given by[8,83]:

$$\frac{r_e}{\bar{a}} = \frac{\Gamma(\frac{1}{4})^4}{6\pi^2}\left(1 - \frac{2}{\alpha} + \frac{2}{\alpha^2}\right). \tag{5}$$

Note that the effective range expansion breaks down near $\alpha \to 0$, or $|\alpha| \ll 1$.

Fano-Feshbach resonances are possible when a closed channel bound state is tuned in the vicinity of an open channel threshold. A closed spin channel is one with separated atom energy (or threshold energy) $E_c > E > 0$ that, in the case of alkali metal species with Zeeman spin structure, has a bound state with a different magnetic moment than the separated atoms of the threshold open channel, the difference being μ_{dif}. The "bare" closed channel bound state energy tunes with magnetic field B as $\mu_{\text{dif}}(B - B_c)$, whereas when the bare closed and open channels are mixed by interaction terms in the Hamiltonian, the scattering length takes on the following form near the resonance pole:

$$a = a_{bg} - a_{bg} \frac{\Delta}{B - B_0}, \tag{6}$$

where a_{bg} represents the "'background" or "bare" open channel scattering length in the absence of coupling to the closed channel bound state and Δ is the resonance "width." The field position of the pole at B_0 where $E_{-1}^U = 0$ is normally shifted from B_c, as described in the next paragraph. It is the ability to tune the scattering length to any value using resonance tuning that makes cold atoms such good probes of few-body physics.

It turns out to be very useful to characterize the properties of various Feshbach resonance in terms of universal van der Waals parameters. In fact, Chin *et al.*[60] introduced a dimensionless parameter $s_{res} = (a_{bg}/\bar{a})(\Delta \mu_{dif}/\bar{E})$ to classify the strength of a resonance. Broad, open channel dominated resonances are those with $s_{res} \gg 1$, and narrow, closed channel dominated ones are those with $s_{res} \ll 1$. The former tends to behave like single open channel states characterized by the scattering length alone. The latter tend to exhibit their mixed closed/open channel character, and can not be fully characterized by their scattering length alone. However, all isolated resonances take on the following simple threshold Breit-Wigner form for the phase shift,[84,85] here given in its near-threshold limiting form using the leading terms in an expansion in $\kappa = k\bar{a}$, valid when $\kappa a_{bg} \to 0$:

$$\eta(\epsilon, B) = \eta_{bg}(\epsilon) - \tan^{-1} \left[\frac{\kappa s_{res}}{\epsilon - m_{dif}(B - B_0)} \right], \tag{7}$$

where the reduced slope $m_{dif} = \mu_{dif}/\bar{E}$, and the pole is shifted from the "bare" crossing by $B_0 - B_c = \Delta a_{bg}(1 - \alpha_{bg})/[1 + (1 - \alpha_{bg})^2]$, where $\alpha_{bg} = a_{bg}/\bar{a}$. The universal van der Waals Eq. (7) depends on the background scattering length α_{bg}, the "pole strength" s_{res} in the numerator of the resonance term, and the slope m_{dif}. When collision energy is large enough that the condition $\kappa a_{bg} \ll 1$ is not satisfied, then the simple threshold forms for the numerator and shift term in Eq. (7) need to be replaced by the more complex energy-dependent universal van der Waals functions of α_{bg} as explained in Refs. 84, 85

Julienne and Hutson[86] give examples with Li atoms illustrating the universal binding energy formulas in Eqs. (3) and (4). The experimentally

measured and coupled channels calculated binding energies for the broad
832G resonance of ^{6}Li with $s_{res} = 59$ agree well with the Gao expression
E_{-1}^{Gao}, but clear departures are seen from the conventional universal formula
E_{-1}^U. On the other hand, the more closed channel dominant 738G resonance
of ^{7}Li with $s_{res} = 0.54$ shows clear departures of calculated and measured
binding energies from these simple formulas as the field is tuned away from
the pole position. Blackley *et al.*[84] test the effectiveness of the effective
range expansion for $\eta(E, B)$ for finite energies near the resonance pole for
several different cases. While this expansion tends to be quite good near the
poles of broad resonances, it tends to fail for narrow resonances, whereas
the universal van der Waals expression of Eq. (7), or its energy-dependent
generalization, tends to be much more accurate.

Figure 1 illustrates some basic features of near-threshold bound states
and the scattering length as B is tuned for the case of two Cs atoms in their

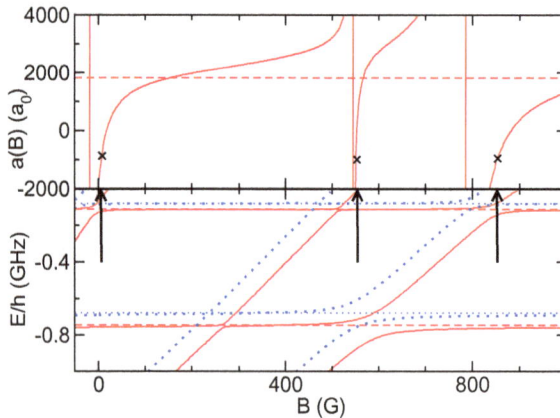

Fig. 1. Scattering length (upper panel) and bound state energies (lower panel) versus
magnetic field B for two interacting Cs atoms, each in its lowest Zeeman level with spin
projection quantum number +3. The solid lines show the s-wave results calculated using
coupled channels calculations with an s-wave basis set only;[87] the heavy dotted lines show
the corresponding d-wave levels with the same total spin projection (Berninger *et al.*[87]
also show many other d levels with other projection quantum numbers not shown here, as
well as other resonances due to bound states of higher partial waves.) The dashed lines
in the lower panel show the -2 and -3 universal van der Waals s-wave bound states
corresponding to a "background" scattering length of $\alpha_{bg} = 18.3$ (dashed line in upper
panel); the light dotted lines in the lower panel show the corresponding d-wave bound states.
The arrows and crosses indicate the regions where universal three-body Efimov states were
observed.[75]

lowest energy Zeeman sub level, based on the theoretical coupled channels model in Ref. 87. There are three prominent broad s-wave resonances present with poles near 0 G, 550 G, and 800 G. The corresponding ramping states are evident; the displacement between the apparent crossing of the linear ramping level from the nearby resonance pole position is a manifestation of the $B_0 - B_c$ shift discussed above. While the last -1 s-wave bound state is too close to threshold to be seen in Fig. 1, the flat -2 and -3 bound states for a scattering length $\alpha_{bg} = 18.3$ are quite close to the actual calculated levels. Here the integer quantum number $-n$ counts the bound states down from the threshold at $E = 0$. The agreement of calculated and universal van der Waals s- and d-wave levels illustrates how the pure van der Waals levels are good indicators of the levels of the real system, even on an energy scale large compared to $\bar{E}$ ($\bar{E}/h = 2.9$ MHz in this case).

It is also apparent that the different resonances in Fig. 1 overlap to some extent with one another. Jachymski *et al.*[88] show how to extend the isolated resonance treatment above to a series of interacting resonances. While it is still a good approximation to treat each resonance in an overlapping series as an isolated resonance near its pole, with a uniquely defined s_{res} parameter, its "local" background (in B near a pole) can be a consequence of its interactions with other nearby resonances. While it is usually adequate to consider resonances as isolated ones, we will give an example below in Sec. 3.2 where the three-body Efimov physics of Cs atoms is affected by two overlapping resonances.

2.2. *Feshbach resonances and multichannel quantum defect theory (MQDT)*

Theories that are typically grouped under the name "quantum defect theory" can be viewed as having two main elements: (1) They systematically separate energy dependences arising from long-range interactions from those associated with short-range physics; and (2) They connect and interrelate the physics of high-lying bound states (spectroscopy) with the properties of low-energy scattering states (collisions). Historically, the first study along these lines was the development of effective range theory by Schwinger, Bethe, Fermi and Marshall, and others,[246,247] which expresses the cotangent of the low-energy s-wave scattering phase shift as

$k \cot \delta_{\ell=0} \rightarrow -1/a + r_e k^2/2$ and relates a to the energy of a high-lying bound state when the scattering length is large and positive [see Eq. (3)]

The most systematic separation of short-range and long-range influences on single-channel collisions and spectroscopy in the context of atomic Rydberg states was developed by Seaton,[19] who described his theoretical framework using the term "quantum defect theory" (QDT). An earlier forerunner of these ideas was a derivation by Hartree[89] of the single channel Rydberg formula for any system like an alkali atom with a single electron moving outside of a closed shell ion core. Generalizations exist for effective range theory, e.g., to include energy dependences associated with long range potentials due to polarization[4] or repulsive Coulomb forces relevant in nuclear physics.[90] One key difference between these generalized effective range theories and quantum defect theories is that the former typically concentrate on developing low-energy expansions of phase shift behavior, whereas quantum defect theory normally relies on exact solutions as a function of energy and has a wider range of applicability as a result. Quantum defect theory was also generalized to treat attractive or repulsive charge–dipole potentials $\pm 1/r^2$ in Refs. 12, 13 and polarization potentials that are attractive[10,14,18] or repulsive.[91]

Typically quantum defect theories have a wider energy range of validity than the simpler effective range theories, because the latter rely on Taylor expansions of the long-range field parameters in powers of the threshold energy whereas the former utilize exact (often analytically known) properties. Subsequent generalizations to treat cold collisions began in the 1980s in the context of chemical physics.[15,92] Later when ultracold collisions became crucial to understand for the burgeoning field of dilute quantum gases, theories aiming more concretely at describing two-body interactions with long range van der Waals tails were developed in earnest,[6,8,9,11,16,93] and they all demonstrated great efficiency in characterizing the energy dependence of single-channel and multichannel solutions to the coupled equations whose solutions govern the properties of Fano-Feshbach resonances. One of the most recent generalizations and improvements to the theory by Ruzic _et al._[17] has been applied to the ^{6}Li-^{133}Cs collision system[94] where multichannel QDT (MQDT) predictions are compared with other treatments such as the full close-coupling solutions as well as the simpler and more approximate "asymptotic bound state model" (ABM).[95]

The key idea of all these theories is that beyond some range $r > r_0$, it is usually an excellent approximation to treat the motion of separating particles as though they are moving in a simple, one-dimensional long-range potential $V_{lr}(r)$. In a typical system involving cold collisions of ground state atoms, when particles separate beyond a distance of around $r_0 \approx 30a.u.$, the potential energy can be accurately approximated by a van der Waals potential $-C_6/r^6$. The solutions to the corresponding radial Schrödinger equation have been worked out in fully analytical form by Gao,[8] but it is often simpler to rely on numerical QDT solutions to the radial differential equation. Two linearly independent solutions in this theory are denoted (f^{sr}, g^{sr}) where the superscript "sr" is meant to indicate that these are characterized by "short range" boundary conditions that guarantee these solutions are smooth and analytic functions of energy at small distances $r \leq r_0$.

In the case of long range attractive single-power-law potentials $-C_N/r^N$ with $N \geq 3$, all linearly-independent solutions at small distance are oscillatory all the way down to $r \to 0$ even for nonzero centrifugal potentials $\ell(\ell+1)/2\mu_2 r^2$ with $\ell > 0$. Two energy-analytic solutions $\hat{f}, \hat{g}$ in the notation of Ref. 17 can be identified in such cases by giving them equal amplitude near the origin and choosing a 90 degree phase difference. For instance, in the WKB approximation which is often adequate to specify boundary conditions in cold collision problems, but not essential, these analytic solutions have the following behavior in channel i at small r:

$$\hat{f}_i(r) = \frac{1}{\sqrt{k_i(r)}} \sin\left(\int_{r_x}^{r} k_i(r')dr' + \phi_i\right) \quad \text{at } r = r_x, \tag{8a}$$

$$\hat{g}_i(r) = -\frac{1}{\sqrt{k_i(r)}} \cos\left(\int_{r_x}^{r} k_i(r')dr' + \phi_i\right) \quad \text{at } r = r_x, \tag{8b}$$

where the point r_x is some fixed small distance, and the phase ϕ_i is in general independent of energy and r and is chosen according to some convenient criterion. A choice particularly advantageous for ϕ_i at $\ell_i > 0$ is the one suggested by Ruzic et al.[17] which picks the unique phase such that at zero channel energy, the exact zero-energy independent solution $\hat{g}_i(r)$ approaches $r^{-\ell_i}$ at $r \to \infty$ at channel energy $\epsilon_i = 0$. Because the potential energy is deep in the region where the boundary conditions are chosen, far deeper than the energy range of interest in cold or ultracold collisions for practically any diatomic system, $\hat{f}$, $\hat{g}$ will be extremely smooth functions of energy over a broad range from well above a dissociation threshold to well below. The usual set of real regular f and irregular g functions from scattering theory that oscillate 90 degrees out of phase at infinity with equal amplitudes are more convenient to use for defining physical reaction matrices. For any long range field one must find the energy- and ℓ-dependent constant coefficients A, $\mathcal{G}$ relating f, g to $\hat{f}$, $\hat{g}$, a relationship written as:

$$f_i(r) = A_i^{1/2} \hat{f}_i(r), \tag{9a}$$
$$g_i(r) = A_i^{-1/2} [\mathcal{G}_i \hat{f}_i(r) + \hat{g}_i(r)]. \tag{9b}$$

The physical meaning of these parameters A, $\mathcal{G}$ has been expounded in various references, although the notations and conventions are not always uniform among the various QDT references. For instance, the long range QDT parameter $A^{1/2}$ can be viewed as containing the energy-rescaling factor needed such that the solution f has the usual energy-normalized amplitude at $r \to \infty$. As such, it contains the threshold law physics, i.e., the Wigner threshold laws and their generalizations appropriate to different long range potentials. The parameter A in this notation connects with the parameter $C(E)^{-1}$ in the notation of Ref. 15, where also the long range parameter $\mathcal{G}$ is written as $\tan \lambda(E)$. The significance of the $C(E)$ and $\tan \lambda(E)$ functions for cold atomic collisions is discussed by Julienne and Mies[93] as well as Julienne and Gao[96] along with illustrative figures of their behavior. The parameter $\lambda(E)$ is interpreted qualitatively as reflecting the fact that two linearly independent solutions that oscillate 90 degrees out of phase at small-r generally cannot retain this phase difference all the way to ∞. The asymptotic solutions of scattering theory (reaction matrices, etc.) require solutions 90 degrees out of phase at infinity, and relevant phase correction parameter $\lambda(E)$ or equivalently $\mathcal{G}(\mathcal{E})$ must be determined for each long range field of interest.

A major difference between MQDT and ordinary scattering theory is that in multichannel quantum defect theory, one initially postpones enforcement of large-r boundary conditions in the closed channels, and works with an enlarged channel space that includes closed (Q) as well as open (P) channels. The linearly-independent short-range (sr) reaction matrix solutions in this enlarged channel space have the following form outside the radius r_0 beyond which the potential assumes its purely long-range form:

$$\Psi_{i'} = \sum_i \Phi_i [\hat{f}_i(r)\delta_{ii'} - \hat{g}_i(r) K_{ii'}^{\text{sr}}]. \tag{10}$$

These solutions are sometimes called "unphysical" because they diverge exponentially in the closed (Q) channels at any given energy, and to obtain the physical wavefunctions one must find superpositions of these linearly independent solutions that eliminate the exponentially growing terms. This is straightforward once one writes the coefficient of the exponentially growing terms, namely

$$[\hat{f}_i(r)\delta_{ii'} - \hat{g}_i(r) K_{ii'}^{\text{sr}}] \rightarrow (\cos \gamma_i \delta_{ii'} + \sin \gamma_i K_{ii'}^{\text{sr}}) \exp(\kappa_i r). \tag{11}$$

And the linear algebra that produces the physical reaction is simple, as has been derived in many other references, giving

$$K = K_{oo}^{\text{sr}} - K_{oc}^{\text{sr}} (K_{cc}^{\text{sr}} + \cot \gamma)^{-1} K_{co}^{\text{sr}}. \tag{12}$$

In this form, the appearance of Fano-Feshbach resonances is manifestly clear through the matrix inversion that becomes singular near closed-channel resonances. The smoothness of K^{sr} implies that it is often energy-independent and even field-independent to an excellent approximation over a broad range, i.e., over 0.1-1 K in energy[a] and over hundreds of gauss. Moreover, the eigenvectors are expected on physical grounds to be approximately given by the recoupling transformation coefficients connecting short-range channels $|(i_1 i_2) I (s_1 s_2) S F M_F\rangle$ to the different coupling scheme appropriate at long range. Specifically, in this short range coupling scheme, the nuclear spins i_1, i_2 and electronic spins s_1, s_2 of the individual atoms are first coupled to total nuclear and electronic spins I and S in the presence of strong electronic interaction, then couple to the total hyperfine angular momentum F and its projection m_F for the much weaker hyperfine interactions. At zero B-field, the asymptotic coupling that is relevant has good atomic hyperfine quantum numbers $|(i_1 s_1) f_1 m_1 (i_2 s_2) f_2 m_2\rangle$. But when $B \neq 0$ one must

---

[a]Here the unit of energy (E) is converted to Kelvin by E/k_B, where k_B is the Boltzmann constant.

diagonalize the Breit-Rabi Hamiltonian[97,98]:

$$\hat{h}_j = \zeta \hat{i}_j \cdot \hat{s}_j + g_e \mu_B \mathbf{B} \cdot s_j - g_n \mu_N \mathbf{B} \cdot i_j \tag{13}$$

to find the atomic dissociation thresholds and eigenvectors $\langle (i_1 s_1) f_1 | \lambda_1 \rangle^{m_1}$ and $\langle (i_2 s_2) f | \lambda_2 \rangle^{m_2}$ as well, in order to construct a first order approximation to the asymptotically correct channel states. In Eq. (13), g_e, g_n are electron and nuclear g factors, μ_B is the Bohr magneton, and μ_N is the nuclear magneton. The projection of these onto the short-range eigenchannels gives a unitary matrix $\mathbf{X}$ that plays the role of a frame transformation which can give an effective approximation scheme that does not require solution of any coupled equations. Example applications of this multichannel quantum defect theory with frame transformation (MQDT-FT) are discussed by Burke et al.,[6] Gao et al.,[11] and Pires et al..[94] Note that similar content is expressed in the "three-parameter" description of two-body Fano-Feshbach resonances as developed by Hanna et al.[99]

2.3. Numerical predictions of Feshbach resonances

Although the MQDT has proved to be a very powerful tool to study low-energy scattering problems and to make quantitative predictions in many systems, in heavier atomic systems such as Cs, the importance of dipolar interaction,[100,101] second order spin-orbit coupling,[102–105] and higher order dispersive potential $(-1/r^8, -1/r^{10}, ...)$[106] make the Feshbach spectrum more complicated and difficult to be treated by the MQDT via frame transformation. It is therefore essential to numerically integrate the Schrödinger equation to study the Feshbach physics in such systems.

Numerical studies of atomic collisions relevant to ultracold experiments were firstly done for hydrogen atoms.[101,107] Later studies then focused on alkali atoms[63,100,102,108–112] with their cooling down to quantum degeneracy, and more recently on ultracold lanthanizes — Er and Dy.[113–115] When including the spin degrees of freedom, ultracold atomic collisions can be solved as a standard coupled channel problem with typically a few to tens of channels. In the cases where the magnetic dipole–dipole interaction and/or the anisotropy of electronic interactions are not negligible, orbital angular momentum ceases to be a good quantum number and therefore needs to be coupled. This can lead to thousands of coupled channels in case of highly magnetic atoms like Dy or Er[113–116] and therefore makes scattering studies quite numerically demanding, if scattering over a wide range of magnetic field is to be investigated.

For collisions between alkali atoms that have been widely studied, only the lowest two Born-Oppenheimer potentials $^1\Sigma_g^+$ and $^3\Sigma_u^+$ are involved. Although these potentials are labeled by the total electronic spin S, it is more convenient to expand the scattering wave function in individual atom's electronic and nuclear spin basis where the Hamiltonian matrix can be efficiently evaluated (see, for example, Refs. 60, 87, 117, 118 for implementations). In the asymptotic region where interactions between atoms vanish, the scattering solution is projected to the eigenbasis of individual atoms — the atomic hyperfine states (see Sec. 2.2) — where the scattering matrix is extracted.

The main challenge in the numerical study of ultracold collisions is the large integration range in the atomic separation, since the range should be at least a few times the de Broglie wavelength. This typically means a range over 10^4 Bohr and could be even larger when close to a Feshbach resonance. At small inter-atomic distance the Born-Oppenheimer potentials are generally deep enough for the scattering wave function to have hundreds of nodes, which require a good number of radial points or radial elements to represent. Although it is not a real issue for modern computers to solve such a problem in the simplest scenario, the calculations start becoming cumbersome when different partial waves are coupled by, for instance, the magnetic dipole–dipole interaction.

To improve numerical efficiency, propagation methods are widely used in such scattering studies and also used in finding weakly-bound state energies of Feshbach molecules. For scattering calculations, a practical choice is the eigen-channel R-matrix propagation method,[119] which is numerically stable and has the flexibility of being adaptive to different radial representations.

3. van der Waals Physics for Three Atoms

3.1. *Universal three-body parameter*

The three-body parameter is central in studies of low-energy three-body physics. The origin of the parameter can be traced back to the early discovery of a peculiar quantum behavior for three particles — the Thomas collapse.[120] In 1935, Thomas discovered that the ground state energy of three identical bosons diverges to $-\infty$ as the range of pairwise interactions r_0 shrinks to zero — even if one keeps the two-body binding energy E_{2b} a constant in this limiting process.

The three-body parameter is essentially introduced to "regularize" the unphysical divergence of the three-body spectrum. In the effective field theory (EFT)[71] or in the momentum-space Faddeev equation,[121] the Thomas collapse is manifested as a logarithmic divergence of the three-body spectrum when the inter-particle momentum is taken to the "ultraviolet" limit.[32,122,123] This divergence can be formally avoided by adding a three-body force that effectively introduces a cutoff in the inter-particle momentum, which assumes one form of the three-body parameter.

The three-body parameter can be more intuitively understood in the hyperspherical coordinates.[124] In fact, this was the way used by Efimov to discover the Efimov effect.[37] The hyperspherical coordinates are defined for many particles in analogy to the spherical coordinates for one particle, where only one of the coordinates — the hyperradius

$$
R = \sqrt{\frac{m_1 m_2 r_{12}^2 + m_2 m_3 r_{23}^2 + m_3 m_1 r_{31}^2}{\mu_3 (m_1 + m_2 + m_3)}} \tag{14}
$$

— represents distance, whereas all other coordinates are defined as "angles". The three-body mass μ_3 can be defined by single-particle masses as

$$
\mu_3 = \sqrt{\frac{m_1 m_2 m_3}{m_1 + m_2 + m_3}}. \tag{15}
$$

The advantage of the hyperspherical coordinates is primarily from R being a universal breakup coordinate for all defragmentation processes. The hyperangles, represented by a collective of coordinates Ω, can have different definitions according to the choice of orthogonal coordinates.[124]

Fixing E_{2b} in the limit $r_0 \to 0$ is essentially equivalent to keeping the two-body scattering length a unchanged. To study the behavior of three particles in the universal limit where $|a| \gg r_0$, it is easier to consider the special case with $|a| \to \infty$ and $r_0 \to 0$. In this case it is easy to show[37] that the hyperadial motion separates from the hyperangular motions. The hyperangular motions can be solved analytically and lead to the following hyperadial equation:

$$\frac{\hbar^2}{2\mu_3}\left(-\frac{d^2}{dR^2} + \frac{s_\nu^2 - 1/4}{R^2}\right)F_\nu(R) = E_{3b}F_\nu(R), \qquad (16)$$

where s_ν are universal constants.[37] For identical bosons and some combinations of mass-imbalanced three-body systems, the lowest s_ν^2 is negative — we denote it as $-s_0^2$ with $s_0 \approx 1.00624$. Equation (16) then leads to the well-known "fall-to-the-center" problem[125] — the three-body energy E_{3b} has no lower bound. The problem comes, obviously, from the effective hyperadial potential

$$U_\nu(R) = -\frac{s_0^2 + 1/4}{2\mu_3 R^2}\hbar^2 \qquad (17)$$

being too singular near the origin. To avoid this three-body collapse, some regularization is needed such that the $-1/R^2$ potential does not extend all the way to $R = 0$. A simple cure is to give a cutoff to the $-1/R^2$ potential, which is equivalent to giving an earlier mentioned momentum-space cutoff in the EFT. In either case, a cutoff leads to a finite three-body ground state energy E_0, or a binding wavenumber $\kappa^* = \sqrt{2\mu_3 E_0}/\hbar$ which is another form of the three-body parameter.

Although the $r_0 \to 0$ situation is not realized in nature, the unitarity condition where $|a| \to \infty$ is now routinely realized in ultracold atoms by using Feshbach resonances.[60] In this case, the $-1/R^2$ long-range potential leads to an infinite number of three-body bound states — Efimov states — with energies following a geometric scaling $E_{n+1}/E_n = e^{-2\pi/s_0}$, which can happen even when none of the two-body subsystems is really bound ($|a| \to \infty$ implies $E_{2b} \to 0$). This is known as the Efimov effect[37,52] in quantum physics.

In case of finite $|a|$, the $-1/R^2$ potential still exists but only extends to a distance $R \sim |a|$, so there will be only a finite number of Efimov states. As demonstrated in Fig. 2, however, when $|a|$ increases more Efimov states become bound either below the three-body breakup threshold or the atom-dimer threshold, for $a < 0$ or $a > 0$, respectively. The values of a where the first Efimov state becomes bound, a_-^* ($a < 0$) or a_+^* ($a > 0$), as well as the position of the first three-body recombination minimum a_0^{*}[71,74] at $a > 0$,

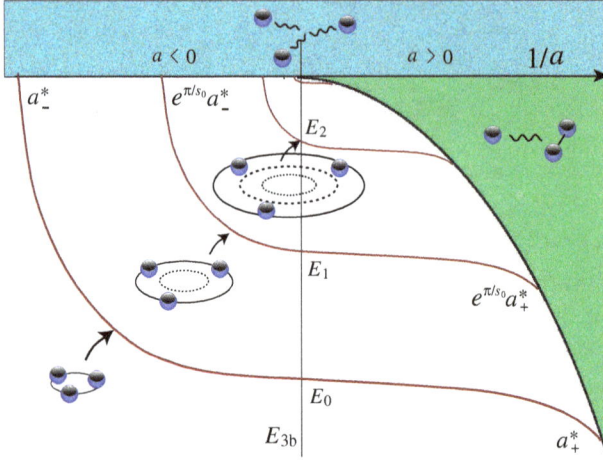

Fig. 2.   The Efimov spectrum in a three-body system. On the negative side of a, the Efimov trimer states are born from the three-body continuum (top blue region) as $|a|$ increases. The first state is born at $a = a_-^*$ and consecutive ones are born at $a = e^{n\pi/s_0}a_-^*$. On the positive side of a where a weakly-bound two-body state exists, the Efimov states merge into the atom-dimer continuum each time when a changes by a factor of e^{π/s_0}. The lowest merging position is denoted as a_+^*.

can also represent the position of the ground Efimov state and are therefore often referred to as the three-body parameter as well.

Generally speaking, the above-mentioned representations of the three-body parameter are related by the universal relations. For identical bosons these are known as[71,126]

$$a_-^* \approx -1.501/\kappa^*, \ a_+^* \approx 0.0708/\kappa^*, \ \text{and} \ a_0^* = 0.316/\kappa^*. \quad (18)$$

In realistic systems such as ultracold atoms, the ground Efimov state is often "contaminated" by the details of short-range interactions that deviate from the $-1/R^2$ effective potential, so large deviations from the universal relations are often reported in the experimental observations at low scattering lengths.[54,127–130]

With the concept that the $-1/R^2$ Efimov potential extends to small R and keeps attractive even when it is modified by short-range interactions, it is natural to expect that the ground Efimov state energy depends strongly on how this potential behaves from $R = 0$ all the way up to $R \gtrsim r_0$, or in the very least, on the semiclassical WKB phase Φ in this range of the potential. It can be shown[124] that when Φ changes by $\pm\pi$ the whole Efimov spectrum is overall shifted by $n \to n \mp 1$, whereas the low-energy three-body observables are kept unchanged. In atomic systems as short-range interactions are strong enough to hold hundreds to thousands of ro-vibrational states already in the two-body level, a less than 1 percent change in the interactions would be sufficient to change the short-range phase by many multiples of π, which makes the prediction of the three-body parameter practically almost impossible. In ultracold experiments, when the magnetic field is scanned through different Feshbach resonances, the number of two-body bound states changes so that the WKB phase in the two-body interactions (at zero energy) changes through multiple of π's. An obvious guess on the change of three-body phase is in a range greater than the change in the two-body phase. Similarly, if experiments are done in different atomic hyperfine states it had been expected that the change in the interactions is also significant enough to give completely different three-body phases.

Another layer of complication is that in realistic three-body systems there are usually many atom–dimer breakup thresholds. In the hyperspherical representation they are manifested by atom–dimer channels, whose potential energies go asymptotically to the dimer energies. These potentials should in principle couple to the Efimov potential when $R \lesssim r_0$, which seems to make the prediction of a three-body parameter even less likely.

Finally as shown in Fig. 3, the non-additive three-body forces for alkali atoms, which differ drastically for different atomic species or different atomic spin states, are often stronger than the sum of two-body interactions near the distance of chemical bond ($\lesssim 30$ Bohr). If three-body forces do contribute to the three-body short-range phase, there would be even less hope for the prediction of a three-body parameter due to the finite precision at which the potential surfaces can be calculated.

Based on the above considerations, it has been generally believed[71,72] that the three-body parameter should not carry any universal properties. However, the experiments with ^{7}Li atoms[131,132] showed the first contradiction to this expectation: the three-body parameter was measured to be independent of the atoms' hyperfine state. More surprisingly, the Innsbruck group later measured a_-^* for Cs near Feshbach resonances with quite different characters but found a persistent value of[75]

$$a_-^* \approx -10 r_{\mathrm{vdW}}. \tag{19}$$

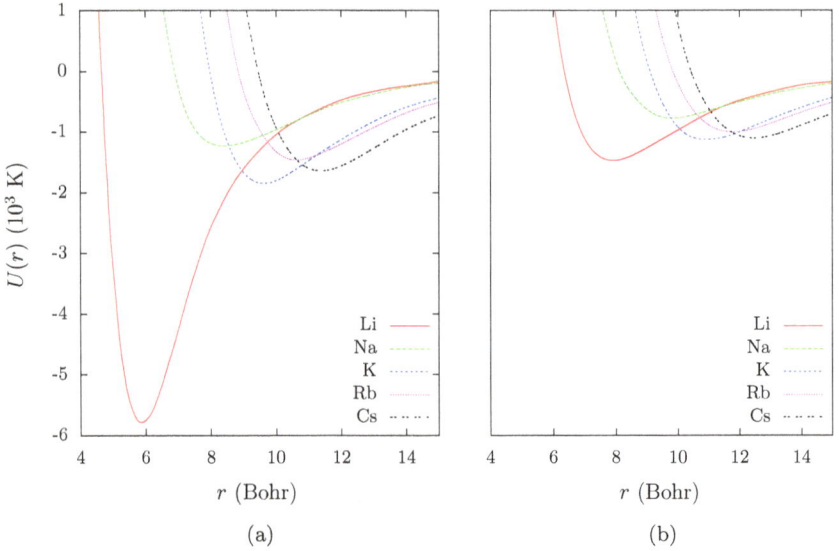

Fig. 3. (a) Three-body full potential surfaces for alkali atoms that include three-body non-additive potentials.[133] (b) Three-body potentials with only the pair-wise sum of two-body potentials. The units of potentials are converted into Kelvin by U/k_B, where k_B is the Boltzmann constant. In both cases the atoms are spin-polarized (quartet state) and are in the the D_{3h} (equilateral triangle) geometry.

Moreover, the magnitude of the three-body recombination induced loss rate L_3 has been observed to be similar near the Efimov resonances in different regions of magnetic fields.[128] After the Innsbruck experiment, more experimental results including the earlier ones (shown in Table 1), are found to be consistent with this "universal" value of a_-^* when it is cast in the unit of r_{vdW}.

The clear disagreement between the experiments and the theoretical expectation triggered a good deal of interest in understanding the physics behind this universality. A key question that needs to be addressed here is the insensitivity of the three-body parameter to the complexity of the three-body dynamics — vibrational and rotational excitations — at small distances, which occurs in the case of alkali atoms where the interactions are strong enough to have hundreds or thousands of ro-vibrational states already on the two-body level.

This rather mysterious universality was soon understood by an effective three-body barrier in the hyperspherical coordinates whose repulsive wall

Table 1. The positions of the first Efimov resonances observed in experiments. The positions of the Feshbach resonances (B_0) are quoted from the corresponding experiments. The values of s_{res} are also quoted from the corresponding experiments unless otherwise stated.

	r_{vdW} (Bohr)	Hyperfine state	B_0 (G)	s_{res}	a_-^*/r_{vdW}
^{7}Li	32.5	$f = 1, m_f = 0$	894.63(24)	0.493[b]	$-8.12(34)$[c]
		$f = 1, m_f = 1$	738.3(3)	0.54[d]	-9[e]
			737.69(12)		$-7.75(31)$[d]
^{39}K	64.5	$f = 1, m_f = 0$	58.92(3)	0.11	$-14.7(23)$[f]
			65.67(5)	0.11	$-14.7(23)$
			471.0(4)	2.8	$-9.92(155)$
		$f = 1, m_f = -1$	33.64(15)	2.6	$-12.9(22)$
			162.35(18)	1.1	$-11.3(19)$
			560.72(20)	2.5	$-9.92(140)$
		$f = 1, m_f = 1$	402.6(2)	2.8	$-10.7(6)$
^{85}Rb	82.1	$f = 2, m_f = -2$	155.04	28[g]	$-9.24(7)$[h]
Cs	101	$f = 3, m_f = 3$	7.56(17)	560	$-8.63(22)$[i]
			553.30(4)	1.6[b]	$-10.19(57)$
			554.71(6)	160[b]	$-9.48(79)$
			818.89(7)	16[b]	$-13.86(149)$
			853.07(56)	1480[b]	$-9.46(28)$

is located near $R = 2r_{vdW}$. This was demonstrated by Wang *et al.*[48] where all the two-body ro-vibrational states from rather deep van der Waals potentials are included in their calculations. Naidon, *et al.*[45] also show such a "universal" barrier by including only the vibrational states. Intuitively, this barrier gives a short-range cutoff to the Efimov potential and "protects" the Efimov states from the influence of three-body forces at small distances. This barrier leads to the suppression of the three-body amplitude inside the

---

[b][134]
[c][131]
[d][127, 130]
[e][132]
[f][135]
[g][136]
[h][75, 128]

barrier, which is in fact not a necessary condition for a universal three-body parameter — as we will discuss below for the heteronuclear systems with extreme mass ratios. The crux for the universality lies in the universal position of the hyperradial node or quasi-node (a node-like structure where the amplitude is small but finite) that can be either understood by the rise of the effective barrier or by the sharp increase in the interaction strength.[48, 137] This nodal structure has also been interpreted as a consequence of universal three-body correlation for van der Waals interactions, and can be generalized to other types of short-range interactions except for square well potentials.[45]

Similar to the homonuclear three-body systems, a heteronuclear three-body system can also have the Efimov effect, particularly when two of the particles are heavy (identical) and the scattering length between heavy (H) and light (L) particles a_{HL} is large.[37] In such a system the Efimov scaling factor $e^{2\pi/s_0}$ is small,[37, 138] excited Efimov states are therefore easy to form and this is often called an "Efimov-favored" scenario.

Since the scaling behavior of the aboved-mentioned hyperradial potentials depends on the mass ratio,[37, 138] the universality in the three-body parameter found in homonuclear systems cannot be readily carried over to heteronuclear systems. Nevertheless, Wang, et al. have shown[77] that the three-body parameter in a heteronuclear atomic system is also universally determined by r_{vdW} without being affected by the details of short-range interactions. In the "Efimov-favored" systems, the universality of the three-body parameter is particularly understood in a simple picture when the motion of the light atom is treated by the Born-Oppenheimer approximation.[77]

In this case the Efimov physics is dictated by the Born-Oppenheimer potential $U^{\text{BO}}(r)$, where r is the distance between the H atoms. Here the origin of the universal three-body parameter is a combination of the universal property of the Efimov behavior in $U^{\text{BO}}(r)$ at large r and that of van der Waals behavior at small r. This can be easily seen by writing $U^{\text{BO}}(r)$ as

$$U_\nu^{\text{BO}}(r) = V_{HH}(r) + V_\nu^{\text{BO}}(r), \tag{20}$$

where $V^{HH}(r)$ is the "bare" $H-H$ interaction and $V_\nu^{\text{BO}}(r)$ is the interaction induced by L. The index ν labels the quantum states of L when H atoms are fixed in space. The channel ν relevant to the Efimov effect is the highest state with σ_g symmetry (zero angular momentum projection along the $H-H$ axis and the L wave function is symmetric upon inversion of its coordinates). It is well-known that in this channel, $V_\nu^{BO}(r)$ has the behavior

$$V_\nu^{BO}(r) \approx -\frac{\chi_0^2}{2m_L r^2}\hbar^2 \tag{21}$$

in the range $r_{\text{vdW},HL} \ll r \ll |a_{HL}|$, where m_L is the mass of L, $r_{\text{vdW},HL}$ is the van der Waals length between H and L, $\chi_0 \approx 0.567143$ is a universal constant. At $r < r_{\text{vdW},HL}$, $V_\nu^{BO}(r)$ behaves in a complicated way due to avoided crossings with other channels, but has a magnitude on the order of the van der Waals energy $E_{\text{vdW},HL} = \hbar^2/(2\mu_{HL} r_{\text{vdW},HL}^2)$, where μ_{HL} is the reduced mass of H and L.

On the other hand, $V_{HH}(r) \approx -C_{6,HH}/r^6$ in the range $r \gg r_{0,HH}$, where $r_{0,HH}$ is the distance where the electronic exchange interaction starts to become significant, with $r_{0,HH} < r_{\text{vdW},HL} < r_{\text{vdW},HH}$ often satisfied in atomic systems. In such a scenario, the potential $U_\nu^{BO}(r)$ is dominated by the $-1/r^2$ behavior when $r \gg r_{\text{vdW},HH}$ but is otherwise dominated by the van der Waals interaction $-C_{6,HH}/r^6$ when $r_{0,HH} \ll r \ll r_{\text{vdW},HH}$. Importantly, thanks to the universal properties of van der Waals interaction (see Sec. 2) the nodal structure of the Born-Oppenheimer radial wave function $F_\nu(r)$ in the van der Waals dominant region is completely determined by $C_{6,HH}$, or $r_{\text{vdW},HH}$, without refering to the interactions at distances near or smaller than $r_{0,HH}$. Moreover, this nodal structure "anchors" the position of the next node in the Efimov region, which determines E_0, or the three-body parameter. Such "universal" nodal determination is demonstrated in Fig. 4 for a three-atom system with extreme mass ratio — YbYbLi. In Fig. 4, the formation of a regular nodal structure in the region $r \lesssim r_{\text{vdW},HH}$ is clearly shown when the van der Waals interaction $-C_{6,HH}/r^6$ extends to smaller distances but with a_{HL} fixed. In fact, this nodal structure closely follows that in the zero-energy wave function in a pure $-C_{6,HH}/r^6$ potential with the same a_{HL}.

Briefly speaking, the universality of the three-body parameter in an "Efimov-favored" system can be understood as the following: all the non-universal ingredients in $V_\nu^{BO}(r)$ near $r \lesssim r_{\text{vdW},HL}$ are "dissolved" by the strong van der Waals interaction in $V_{HH}(r)$, whereas the non-universal ingredients in $V_{HH}(r)$ are completely "absorbed" in a_{HH}. Using the above picture, a universal three-body parameter can be obtained analytically, which depends only on a_{HH}, $r_{\text{vdW},HH}$, and the mass ratio.[77]

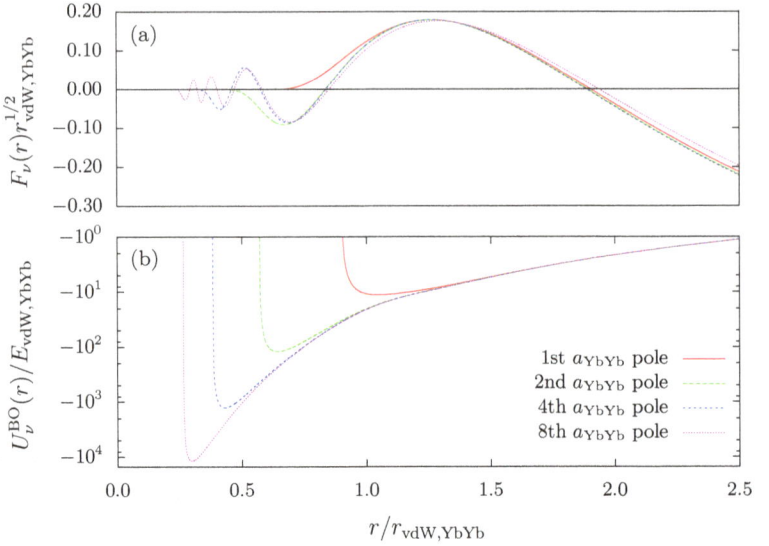

Fig. 4. (a) The Born-Oppenheimer radial wave function $F_\nu(r)$ of the first excited Efimov for YbYbLi. The shot-range cutoff of the Yb-Yb Lennard-Jones potential[139] is tuned at the first a few scattering length poles. (b) The Born-Oppenheimer potential $U_\nu^{BO}(r)$ in the Efimov channel. The short-range parameter corresponds to those used in (a).

In the "Efimov-unfavored" heteronuclear systems where $m_H/m_L \lesssim 1$, Wang *et al.*[77] show that the three-body parameter is still universal, but with a mechanism similar to the identical boson case, i.e., with the existence of a universal repulsive barrier.

Finally, it should be noted that the above theoretical analyses are based on single-channel atomic interactions and are therefore not expected to be applicable to narrow Feshbach resonances. Experimental observations on identical bosons, however, have shown consistency with the universal three-body parameter for relatively narrow ones in ^{7}Li and ^{39}K (see Table 1). These interesting observations are yet to be understood by future studies. In addition, in recent ultracold experiments on the CsCs^{6}Li system[140, 141] where the Cs-Li Feshbach resonance is relatively narrow, the observed three-body parameter is also consistent with the theory for broad Feshbach resonance — a prediction based on Ref. 77. The physics behind these unexpected agreements is still under investigation, and may be understood by the three-body spinor models introduced in Sec. 3.2, where the multichannel character of the Feshbach resonances is properly represented.

3.2. *Three-body physics with spinors — multichannel models*

Although the situation of large scattering length is almost always produced by multichannel phenomena in nature, the majority of studies on Efimov physics so far are performed by single-channel interactions. Such treatment is adequate near broad, or open-channel dominant Feshbach resonances where studies[117] have shown that the two-body physics in those cases can be well represented by models with single-channel interactions.

Generally, an *isolated* Feshbach resonance can be characterized by two parameters: the background scattering length a_{bg} and the s_{res} parameter. A broad or narrow Feshbach resonance can be distinguished by $s_{res} \gg 1$ or $s_{res} \ll 1$. Today the question on Efimov physics and three-body scaling properties is more focused on the cases of non-broad Feshbach resonances where $s_{res} \gg 1$ is not satisfied so that a single-channel treatment is expected to fail. As such Feshbach resonances are commonly accessed in ultracold experiments, it is particularly desirable to understand the three-body physics in such cases.

The success of the zero-range interaction model in predicting the Efimov effect[52] makes it attractive to extend the model to a multichannel version. Efimov made an early effort in this direction to understand how the Efimov effect can manifest in the presence of the nuclear spin.[142] Some other recent efforts focused more on reproducing the energy dependence of two-body scattering near narrow Feshbach resonances by generalizing the zero-range boundary conditions or by introducing a two-body "molecular state" into the three-body Hamilton.[126,143–147] These theories all introduce a parametrization by the effective range r_e or equivalently the R^* parameter.[126,146] For atoms with van der Waals interactions, R^* is defined as $R^* = \bar{a}/s_{res} \approx 0.955978 r_{vdW}/s_{res}$; near narrow Feshbach resonances there is further relation $r_e \approx -2R^*$. Such parametrization, though simple and elegant and is expected to work in the extreme limit $|a| \gg r_0$ and $s_{res} \ll 1$, is not sufficiently capable of representing van der Waals physics for three atoms.

More sophisticated multichannel zero-range models treat the closed-channel Hilbert space in equal footing with the open-channel space, rather than simply a bound "molecular state". A relatively straightforward way to do this is to write the three-body wave function Ψ as a superposition of open- and closed-channel components ψ_{open} and ψ_{close}, and match Ψ separately in the open- and closed-channel zero-range boundary conditions.[148–150] A more rigorous way is to build the internal degrees of freedom — such as spins — directly to individual atoms, as the multichannel zero-range model by Mehta et al.[151]

Thanks to their simplicity, the multichannel zero-range models have recently been applied to fit three-body recombination losses in ultracold experiments[149] and to study spinor condensates.[152] It should be noted, however, that the zero-range boundary condition for the closed-channel wave function implies a weakly-bound two-body Feshbach state in the closed channel, which is rarely the case for realistic systems. In addition, the multichannel zero-range models still have the Thomas collapse,[149] which also limits their predictive power for ultracold experiments.

To better understand the role of van der Waals physics in ultracold collisions of three atoms, it is desirable to have a multichannel model that has finite-range van der Waals interactions built in. Moreover, the connection found by Wang et al.[153] between a d-wave two-body van der Waals state and a three-body state suggests that a correct characterization of the high-lying two-body van der Waals spectrum is also important for predicting ultracold three-body physics in atoms.

Recently, significant progress has been made by Wang and Julienne et al.[51,154] down this road, where the authors have developed a three-body multichannel model with van der Waals interactions. By allowing atoms in their model to carry multiple spin states and interact via spin-coupled van der Waals potentials, the authors have successfully reproduced/predicted many experimentally observed three-body loss features without fitting parameters, particularly in the case of Cs where three-body physics has been extensively studied in Innsbruck experiments.[128] After the groundbreaking discovery of a universal three-body parameter, this theoretical work shows that without addressing the unknown short-range details, ultracold three-body physics in atoms can be quantitatively predicted. This pushes the front of few-body research to a much further position.

In Wang, et al.'s work, three atoms are represented by the Hamiltonian[51]

$$H = T_3 - (\mu_{s_i} + \mu_{s_j} + \mu_{s_k})B + u_{s_i} + u_{s_j} + u_{s_k}$$
$$+ \sum_{s_i,s_j,s_k} |s_k\rangle\langle s_k| \otimes |s_i s_j\rangle \mathcal{S}^\dagger v_{s_i s_j, s_i' s_j'} \mathcal{S}\langle s_i' s_j'|. \qquad (22)$$

Here T_3 is the kinetic energy for the relative motion of three atoms, $|s_{i,j,k}\rangle$ is the spin state of atom i, j, or k, which can take $|a\rangle$, $|b\rangle$, $|c\rangle$, ... in a multi-spin model. Also, μ_{s_i} is the magnetic moment for an individual atom, and $u_{s_{i,j,k}}$ is the zero-field, single-atom energies that represent the hyperfine splittings. The pairwise two-body interactions $v_{s_i s_j, s'_i s'_j}$ are in a symmetrized two-body spin basis $\mathcal{S}|s_i s_j\rangle$ ($\mathcal{S}$ is the symmetrization operator), with 6–12 Lennard-Jones potentials in the diagonals. Thanks to the universal properties of two-body van der Waals physics,[155] the near-threshold two-body spectrum can be reasonably reproduced by the Lennard-Jones potentials that are deep enough for a few bound states. By including more spin states this Hamiltonian is capable of representing complicated Feshbach physics in cold atoms. For instance, isolated Feshbach resonances can be represented by including two spin states for each atom, whereas overlapping Feshbach resonances can be represented by three spin states.

With this model, Wang, *et al.*[51] have studied an interesting case in Cs where an Efimov doublet was observed near a narrow g-wave Feshbach resonance.[75,128] Their calculations successfully reproduced the positions of the Efimov resonances in the doublet and showed how the magnitude of the loss rates can be predicted correctly when a 2-spin model is replaced by a 3-spin model, where the overlapping character of the g-wave Feshbach resonance is better represented.[51] Moreover, even though the theory agrees with the experiment in that the positions of the Efimov resonances in this case are all near the universal value, Wang, *et al.*'s theory further points out that the universal resonance positions are actually a consequence of a_{bg} being near the universal value ($-10r_{\mathrm{vdW}}$) and should not be considered as general.[154]

The above mentioned multichannel three-body models can also be applied to heteronuclear atomic systems. For instance, to study the CsCsLi Efimov physics in the ultracold Cs-Li admixtures realized in the Chicago group[141] and the Heidelburg group,[140] a 2-spin model can be used to treat the CsLi interactions.[141] With the great variety of heteronuclear combinations that are accessible to future experiments, theoretical studies with multichannel models are just in the beginning. In this context, many questions on the van der Waals universality still remain to be answered.

Finally, it should be mentioned that the magnetic Feshbach resonances are by far not the only place where multichannel physics is important. In the emerging fields of ultracold studies where light is used to manipulate the atomic interactions, such as the optical Feshbach resonance[60] and synthesized spin-orbit interactions,[156] new, interesting few-body phenomena due to multichannel physics has already started being predicted or observed (see Sec. 6).

4. van der Waals Physics in the Chemical Reaction of Molecules

We have been discussing the weakly bound states of three atoms, or the collisions of an atom with a weakly bound dimer. Now we turn to collisions or chemical reactions of ordinary molecules, for example, in their ground vibrational state. Such molecules are much more complex than atoms, having additional degrees of freedom, and their chemical reaction dynamics provides an excellent example of few-body dynamics. While we may expect that molecular dynamics is too complex to show any universality, we will show here that there are some cases, perhaps even widespread, where universality is found, that is, insensitivity of the reaction rate coefficient to the details of short-range chemical interactions.

It is well-known that some chemical reactions are described by a universal classical model when there is a unit probability of a reaction occurring if the two reactant species come in close contact with one another. The collision cross section is then determined by the long range potential: it is only necessary to count those classical trajectories of the colliding species that are captured by the long range potential such that the collision partners spiral in to close contact with each other. The Langevin model[157] is a good example of this for ion-molecule reactions, and the related Gorin model[158,159] describes barrierless reactions between neutral chemical species that collide via a van der Waals potential. Fernández-Ramos, *et al.*[160] describes the Langevin and Gorin models in the context of chemical kinetics.

In the low temperature regime such classical capture models need to be adapted to the quantum mechanical near-threshold dynamics governed by quantum threshold laws and resonant scattering. This can be readily done based on QDT treatments of neutral[161] or ionic reactions.[88] We concentrate on the neutral case here, which has also been reviewed by Quéméner and Julienne.[22] It is important to note that QDT theories are not unique, but can be implemented in different ways, in either their analytic or numerical versions, depending on how the reference solutions to the problem are set up. We concentrate here on the particular QDT implementation of Mies.[15,92,93] Gao[10,162–164] obtains similar results with his implementation of the theory.

As discussed in Sec. 2.2, the essence of QDT for molecular reactions is the separation of the interactions into short range and long range regions, with the former characterized by a few quantum defect parameters that allow for chemical reaction at short range. Generalizing the classical Gorin or Langevin models to the quantum threshold is especially simple in a QDT framework.[165] The approach of the two colliding species is governed by long range quantum dynamics, whereas the short range region is characterized by perfectly absorbing, or "black hole" boundary conditions that allow no backscattering of the incoming partial wave from the short range region. On the other hand, the long range region can quantum reflect the incoming flux in a way strongly dependent on partial wave ℓ, allowing for quantum tunneling to short range in the case of centrifugal barriers when $\ell \geq 1$. Following Gao[163] we will call QDT models with such "black hole" boundary conditions quantum Langevin (QL) models.

In general, the short range region may not exhibit unit reactivity, but may reflect some of the incoming flux back into the long range entrance channel. The interference of incoming and outgoing flux establishes a phase shift of the long range standing wave that is related to the scattering length. In this "grey hole" partially reactive case, two dimensionless quantum defect parameters are needed: one, s, related to the phase shift and one, $0 \leq y \leq 1$, related to the short range reactivity. The probability P of short range reaction is $P = 4y/(1 + y)^2$ in Ref. 88. The rate constants for elastic scattering or for inelastic or reactive scattering loss from the entrance channel are determined from the diagonal S-matrix element $S_\ell(E) = e^{2i\eta_\ell(E)}$, which can be specified for any partial wave and E by the complex phase shift $\eta_\ell(E)$, or alternatively, by a complex energy-dependent scattering length $\tilde{a}_\ell(E)$ defined by Idziaszek and Julienne:[166]

$$\tilde{a}_\ell(E) = -\frac{\tan \eta_\ell(E)}{k} = \frac{1}{ik}\frac{1 - S_\ell(E)}{1 + S_\ell(E)}. \tag{23}$$

The universal van der Waals s- and p-wave $\tilde{a}_\ell(E)$ in QDT form reduces in the $\kappa \to 0$ limit to (in van der Waals $\bar{a}$ units):

(a) $\tilde{a}_{\ell=0} \to s + y\dfrac{1 + (1-s)^2}{i + y(1-s)}$ (b) $\tilde{a}_{\ell=1} = -2a_1\kappa^2\dfrac{y + i(s-1)}{ys + i(s-2)}$,

$$\text{where } a_1 = 1.06428^{166}. \quad (24)$$

In the specific "black hole" case of the QL model, where $P = y = 1$, the complex scattering lengths take on an especially interesting universal threshold form, independent of s and with equal magnitudes for their real and imaginary parts: $\tilde{a}_{\ell=0} = 1 - i$ and $\tilde{a}_{\ell=1} = -a_1\kappa^2(1 + i)$. The corresponding elastic and reactive/inelastic loss rate constants are determined respectively from $|1 - S_\ell(E)|^2$ and $1 - |S_\ell(E)|^2$. The loss rate constants for the lowest contributing partial waves near the $k \to 0$ threshold are

$$\text{(a) } K^{\text{loss}}_{\ell=0} = 4\pi g\frac{\hbar}{\mu}\bar{a} \quad \text{(b) } K^{\text{loss}}_{\ell=1} = 40.122g\frac{\hbar}{\mu}k^2\bar{a}^3, \quad (25)$$

where the symmetry factor $g = 2$ in the case of identical particles in identical internal states and $g = 1$ otherwise; identical bosons only have even ℓ collisions and identical fermions only odd ones, whereas nonidentical particles have both. The numerical factor in the p-wave expression is $\Gamma(\frac{1}{4})^6/(12\pi \Gamma(\frac{3}{4})^2) \approx 40.122$, where all three components of the p wave are summed, assuming them to have identical complex scattering lengths, as would be the case for a rotationless molecule (total angular momentum $J = 0$). Thus, the threshold thermally averaged universal van der Waals rate constants for s-wave collisions is the constant expression in Eq. (25a), and for p wave collisions of identical fermions it is

$$K^{\text{loss}}_{\ell=1} \approx 1512.6\frac{k_B T}{h}\bar{a}^3, \quad (26)$$

where k_B is the Boltzmann constant and T is the temperature. Thus, threshold s-wave collisions for "black hole" collisions are universally specified by the vdW mean scattering length $\bar{a}$ and p-wave ones by the van der Waals volume $\bar{a}^3$.

The predictions of these universal rate constants for the QL model agree well with measurements on threshold chemical reaction rates of ^{40}K^{87}Rb fermions in their $v = 0$, $J = 0$ state. The p-wave expression in Eq. (26) agrees within experimental uncertainty when the fermions are in a single spin state, verifying the linear variation of rate with T. The s-wave expression in Eq. (25a) agrees within a factor of 2 or better when the fermions are in different spin states or when they react with ^{40}K atoms.[166] Ni *et al.*[21,167] developed an approximate threshold model to extend the Gorin model when an electric field is introduced to cause the molecules to have a laboratory-frame dipole moment. Idziaszek *et al.*[165] extended this method using a hybrid numerical quantum defect theory based on a sum of van der Waals and dipolar long range potentials to predict the variation of reaction rates of identical fermions with electric field, in excellent agreement with experiment. These quantum defect treatments can readily be adapted to quasi-1D ("tube" geometry with tight confinement in two directions)[168] or quasi-2D ("pancake" geometry with tight confinement in one direction)[169] collisions of reduced dimensionality D due to quantum confinement by optical lattice structures. The three-dimensional (3D) boundary conditions due to short range interactions is incorporated in the QDT parameters s and y, and the asymptotic boundary conditions can be taken to be appropriate to the situation of reduced dimension due to quantized tight confinement.[170] Micheli *et al.*[171] extend the universal vdW rate constants in Eqs. (25a) and (26) to quasi-1D and quasi-2D geometry, and additionally develop numerical methods for "black hole" collisions in quasi-2D geometry when the molecules have a dipole moment; similar calculations have been given for KRb[172] and for the universal collisions of other polar molecule species containing alkali-metal atoms.[173,174] These calculations explain experiments with ^{40}K^{87}Rb fermions in "pancake" quasi-2D geometry, verifying earlier predictions[175] that aligning the dipoles perpendicular to the plane of the "pancake" suppresses the reaction rate, instead of increasing rapidly with dipole strength as in ordinary three-dimensional collisions.

QDT models can be readily extended to include more partial waves as temperature T increases. This gives universal QL rate constants for both the van der Waals[88,163] and the ion-atom[88,164] cases ($N = 6$ and 4), including the quantum effect of centrifugal barrier tunneling. The universal QL model goes to the classical Gorin or Langevin models when the temperature goes above a characteristic temperature associated with the long range energy scale. The QDT model can also be implemented for the more general "grey hole" case, where both s and y are needed. Jachymski *et al.*[88] apply the general QDT theory to explain the merged beam experiments on the

Penning ionization of $^3S_1$ metastable He atoms with Ar atoms[176] up to collision energies on the order of 10 K. In this case, where both atoms are in S states and there is a single potential, the contributions from all partial waves, including a prominent shape resonance for $\ell = 5$, can be explained by a single s-wave complex scattering length parameterized by $s = 3$, $y = 0.007$. In the more general case of molecules with anisotropic potentials, we can expect to need ℓ-dependent QDT parameters. Gao's discussion of the general case[162,163] also describes the role of resonance states, which allow incoming and outgoing scattering flux to make multiple passes between the inner and outer regions, quite unlike the "black hole" case, where resonances are suppressed by the total absorption at short range.

The review by Quéméner and Julienne[22] examines the applicability of the universal QL model to a variety of molecular collisions governed by the van der Waals potential. They review work on the vibrational quenching collisions of atom A with $AB(v)$ or $A_2(v)$ molecules, where A and B represent alkali-metal atoms and v is an excited vibrational level. Both measured $AB(v)$ quenching in magneto-optical traps around 100 μK (see their Fig. 21) and calculated threshold quenching of $A_2(v)$ (see their Fig. 41) tend to be within around a factor of 2 to 4 from the predictions of universal QL model. This suggests that vibrational quenching in general tends to have a high probability of a short range quenching event, that is, has a relatively large y not far from unity. The question that remains to be answered is why the quenching rate coefficients for different species and v levels fluctuate by several factors from the predictions of the QL model.

Finally, it is important to note that real molecular collisions are likely to involve a very dense set of resonance states associated with the various vibrational and rotational degrees of freedom in the molecule. The Bohn group[177-179] has described how the density of resonances are so high that statistical random matrix theories are needed to describe their effects even at ultralow temperatures. Recent cold atoms experiments with the strongly magnetic dipolar atomic species Er have uncovered a high density of Feshbach resonance states characterized by a Wiger-Dyson distribution of resonance spacings that is a characteristic of random matrix theory;[113] similar experiments are being carried out with Dy atoms.[116] If there are many resonances within the thermal spread $k_B T$, the molecules are predicted to stick together in a long-lived collision complex with a universal collision rate given by the QL model.[178] If non-reactive molecules in such a collision complex should undergo short range loss processes with unit probability and never returned outgoing flux to the entrance channel, then they would undergo loss processes at the same rate as a highly reactive collision. The question of the magnitude and control of collision rates for the quenching of internal vibration, rotation, or spin for ultracold molecules having dense sets of resonance levels, as well as the nature of their three-body collisions, remains an open research area experimentally and theoretically.

5. Dipolar Physics for Two and Three Atoms

As discussed in the previous sections, the recent experimental development in producing ultracold ground-state polar molecules[22] or ultracold highly magnetic atoms[27, 113, 115, 180–191] has stimulated a lot of interest in the study of few-body systems with the long-range dipolar interactions. To the simplest level, dipolar atoms and molecules can both be modeled as point dipoles with a "permanent" dipole moment that can be aligned by external fields. As has been discussed in Sec. 4, however, the elastic properties of such models do not hold for reactive molecules due to the highly inelastic chemical processes that occur when molecules are close together, and the random matrix theory arguments[177–179] predict that even for non-reactive molecules the elasticity is practically non-observable. We therefore restrict the application of our discussion below to atoms only, where the elasticity of the short-range interactions is still preserved at current experimental conditions.

5.1. *Universal dipolar physics for two field-aligned dipoles*

Generally speaking, the dynamics of dipoles include rotation of individual dipoles where the angular momentum of a dipole is exchanged with its environment. In practice, however, it is highly challenging to include this rotational degrees of freedom in studies of few-dipole physics even for point dipoles. Nevertheless, from the point of view of universal physics it is more preferential to "freeze" the rotation of individual dipoles by orienting them with external fields and study the scaling behavior of dipolar systems with the anisotropic, long-range dipolar interactions.

For oriented dipoles, their interaction potential can be simply written as

$$V_{\mathrm{dd}}(\boldsymbol{r}) = \frac{2d_l}{\mu_2} \frac{1 - 3(\hat{e} \cdot \boldsymbol{r})^2}{r^3}, \tag{27}$$

where the dipole length d_l is the characteristic length scale for the dipolar interaction,[23] which connects to the induced dipole moment d_m by

$$d_l = \mu_2 d_m^2 \hbar^2 \tag{28}$$

and is tunable by an external fields.[192] The unit vector $\hat{e}$ is along the direction of the external field. For magnetic dipolar atoms d_l can saturate to as large as a few hundred Bohr, whereas for alkali dipolar molecules the saturation value is between 10^3 Bohr to 10^5 Bohr. Without a short-range cutoff, the dipole potential is too singular at the origin — the system collapses again due to the "fall-to-the-center" problem.[125] In reality the "short-range" interaction always deviates from Eq. (27) and is regular at the origin. A non-universal short-range cutoff funtion $f_c(r)$ needs to be introduced in theoretical studies such that

$$f_c(r)V_d(\mathbf{r}) \to \mathcal{O}(r^0), r \to 0; \quad f_c(r)V_d(\mathbf{r}) \to V_d(\mathbf{r}), r \gg r_0. \tag{29}$$

In spite of the non-universal behavior at small distances, universal scaling properties with the strength of the dipolar interaction d_l are in the center of the study. Reference 29 shows that unlike short-range interactions, the dipole potential leads to a low-energy expansion in the scattering phase shift δ_ℓ, whose real part characterizes the cross-section where ℓ is not changed during a collision and has the form

$$\mathrm{Re}[\delta_\ell(k)] = -a_\ell k - b_\ell k^2 - V_\ell k^3 + O(k^4). \tag{30}$$

for any partial wave ℓ, with the power of wavenumber k increases by 1 for consecutive terms. Moreover, it is found that the expansion coefficients may or may not have short-range dependence according to ℓ.[29] For $\ell = 0$ all the coefficients are short-range dependent. For $\ell = 1$, however, short-range dependence only starts from the k^3 term and beyond. Therefore, for bosons non-universal behavior is generally expected in low-energy scattering, whereas for fermions universal expressions for scattering observeables can be derived.[23,29] The imaginary part of δ_l characterizes ℓ-changing collisions, which has k^2 in the leading order and has the leading coefficient that has been analytically derived.[29]

In analogy to the magnetic Feshbach resonance, formation of dipolar resonances by tuning d_l is also of great interest. It is known that the dipolar couplings between different ℓ lead to both shape and Feshbach resonances[27, 192–196] and the relative positions between these resonances are universal[27, 192, 194] — independent of short-range details, due to the anisotropy of the dipolar interaction. However, the different partial wave characters that dipolar resonances bear make them line up with d_l quite irregularly. Nevertheless, such knowledge is still important for the development of many-body dipolar theory, where the renormalization theory can be readily applied and system independent properties can be derived.[197]

As the connector between microscopic and macroscopic phenomena, few-body studies, in particular two-body studies, play the role of developing simple models that can be used in studies of many-body physics. To this end, zero-range models that can correctly reproduce low-energy scattering properies have been developed in both 2D[25, 198, 199] and 3D.[200–203] In view of stable dipolar gases against inelastic collisions, elastic two-dipole physics in 2D or quasi-2D geometries has been studied in great details in Refs. 25, 28, 204–210.

For studies of Efimov physics, however, zero-range models do not have enough information in the scaling behavior of near-threshold bound states, which is crucial for studies of the scaling laws. Numerical study[29] based on the potential in Eq. (27), on the other hand, indicates that the characteristic size of the near-threshold dipolar states scales like d_l and their energies scales with $-1/d_l^2$.

In studies on the transition from the weak dipole regime ($d_l \lesssim r_0$) to the strong dipole regime ($d_l \gg r_0$), the concept of angular momentum mixture is often used to indicate the level of anisotropy of a dipolar state. Although this characterization is useful in connecting the ordinary molecular physics to the dipolar physics when the dipolar interaction is not very strong (or the dipoles are not strongly aligned), it becomes less useful in the strong dipole regime where lots of angular momenta make contributions and dipolar states take characters of "pendulum" states.[29] Nevertheless, the expectation value of the square of the angular momentum $\langle \hat{L}^2 \rangle$ in this regime follows a universal scaling[29]

$$\langle \hat{L}^2 \rangle \propto \sqrt{\frac{d_l}{r_0}}(2\nu + m_\ell + 1), \tag{31}$$

where ν is the vibrational quantum number of the pendulum states and m_ℓ is the magnetic quantum number.

5.2. Universal three-dipole physics

The universal properties found in two-dipole physics have clearly stimulated interest in the universal physics for three dipoles. To begin with, we would first like to point out the non-trivial aspects in three-dipole physics, particularly those relevant to Efimov physics:

• The prediction of the Efimov effect was based on isotropic interactions where the total orbital angular momentum J is conserved. In the case of dipoles, however, it is not clear how the couplings between partial waves may impact the Efimov effect.

• The Efimov effect also assumes a scattering length *only* in the s-wave collisions, whereas the long-range dipolar interaction brings in equally defined scattering lengths for each partial wave.

With the above question marks, a clear, definitive answer to the three-dipole Efimov effect can only be given by a quantitative three-body theory where dipolar interactions are properly treated. To this end, the hyperspherical method would be a good choice for the study, where the existence of the Efimov effect can be directly read off from the long-range scaling behavior of the hyperradial potentials.[124]

By numerically solving the Schrödinger equation in the hyperspherical coordinates, Wang, *et al.*[78] showed that the Efimov effect does exist for three bosonic dipoles near an s-wave dipolar resonance where the s-wave scattering length a_s goes through a pole. It is also found that the dipolar Efimov states follow the same scaling properties as the non-dipolar states. More interestingly, contrary to a common expectation that the three-body parameter is the most likely non-universal because of the irregular pattern in the two-dipole spectrum (see Sec. 5.1), the ground Efimov state energy is found to be universally determined by d_l without referring to the short-range details. This finding is similar to what has been discussed in Sec. 3.1, but was discovered before the concept of universal three-body parameter being built. In this case, the universal three-body parameter is also manifested by a repulsive barrier in the hyperradial potential, but appears near $R = 0.7d_l$. The dipolar Efimov state is still s-wave dominant, the three dipoles in this state therefore prefer to stay far away so that its size is much bigger than d_l. At the dipolar resonance, the position of the ground Efimov state, given by the real part of its energy, scales like[78]

$$\text{Re}(E_0) \approx 0.03 \frac{\hbar^2}{md_l^2}, \tag{32}$$

where m is the mass of a dipole. Interestingly, applying the universal relation between E_0 and the Efimov resonance position a_-^* leads to

$$a_-^* \approx -9d_l, \tag{33}$$

which has a numerical factor very close to that in the van der Waals case [see Eq. (19)]. In addition to the position of the ground Efimov state, its width also shows an $1/d_l^2$ overall scaling behavior, which leads to a quasi-universal scaling for the decay rate of an Efimov state. This decay is characterized by the η_* parameter[71] which can be expressed as $\eta_* = (s_0/2)[\text{Im}(E_0)/\text{Re}(E_0)]$ at $a = \infty$. Based on the numerical results[78] the dipolar η_* has a value around 0.084. However, the width of the dipolar Efimov state is less universal than its position — with a finite sample of values in $d_l/r_0 \gg 1$ and some variations in the form of short-range cutoff, Wang *et al.*[78] showed that η_* can vary up to 60% depending on which dipole resonance the three-body calculation was carried out for.

The stability of ultracold bosonic dipolar gases strongly depends on the dipolar Efimov physics discussed above. In an early study, Ticknor and Rittenhouse[211] performed scaling analysis on three-dipole loss rates and derived a_s^4 scaling law when $|a_s| \gg d_l$ and d_l^4 scaling law when $|a_s| \ll d_l$. Based on the knowledge in Efimov physics and the scaling behavior of two dipoles discussed in Sec. 5.1, Wang, *et al.*[78] have obtained the same scaling laws. These scaling laws indicate significant losses for a 3D dipolar gas in the strong dipole regime. Nevertheless, a stable dipolar gas can be prepared in optical lattices or reduced-dimension traps.[212–217]

For fermionic dipoles where s-wave scattering is absent, Wang et al.[79] showed that existence of a p-wave scattering length does not modify the long-range scaling behavior. The threshold law for three fermionic dipoles is therefore the same as the non-dipole case.[218] Near a p-wave dipolar resonance, however, different from non-dipole fermions where no three-body state exists, there is one, and just one universal three-dipole state with binding energy E_{3b} that scales with d_l as

$$E_{3b} \approx 160/md_l^2. \tag{34}$$

A three-body resonance is therefore expected between two two-body p-wave resonances. In contrast to the bosonic dipoles, the binding of fermionic dipoles is mainly from the anisotropy of the dipolar interaction and the three-dipole state therefore has a size smaller than d_l. This dipolar state has a preferential spatial distribution as shown in Fig. 5. In the most probable configuration the bond lengths were numerically found to be universally determined by d_l as $0.14d_l$, $0.14d_l$, and $0.26d_l$.[79]

Near a p-wave resonance where the scattering volume V_p goes through a pole, the scaling law of the three-body loss rate has been extracted numerically[79] as $L_3 \propto k^4 V_p^{17/2} d_l^{35/2}$. Even though there is the T^2 suppression at ultracold temperatures, L_3 grows very quickly with V_p

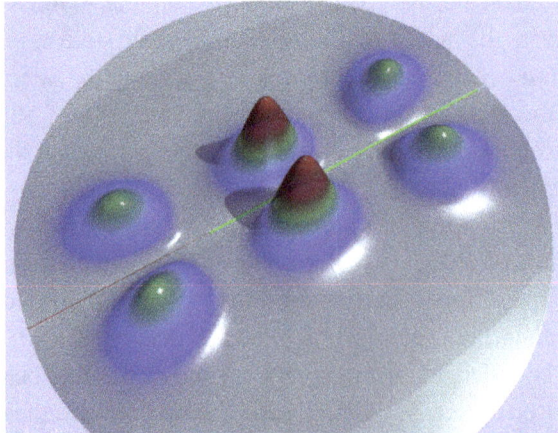

Fig. 5. A cut of the geometrical distribution of the fermionic dipoles in a universal three-body dipolar state along the direction of the external aligning field, indicated by the greern line in the middle of the plane.

and implies short-lived fermionic dipolar gases on resonance. Away from resonance, the loss is expected to have d_l^8 scaling and still implies reduced stability in the strong dipole regime.

6. Fewbody Physics with Other Types of Long-Range Interactions

Long-range interactions bring in long-range correlations, which is the cradle of universal physics. In few-body systems, long-range correlations often lead to binding of exotic quantum states. Such states often have strong impact on the dynamics of atoms and molecules at ultracold temperatures, and provide convenient means in quantum controls in ultracold chemistry. An example is the engineering of molecular states, as demonstrated by Innsbruck group to steer diatomics through high-lying ro-vibrational states using magnetic Feshbach resonances.[219–221] Although the control over triatomics and poly-atomic molecular states is still in the beginning stage, radio-frequency association of Efimov trimers has been experimentally achieved by Lompe *et al.*[222] and Machtey, *et al.*[223] For future experimental development, it is important to understand the properties of exotic few-body states with other types of long-range interactions, as well as possible controls that are realizable. In this section we give a brief review over such systems.

Coulomb interactions. Ultracold few-body systems with Coulomb force is very rich in exotic quantum physics. In a system of identical charges, although interesting phenomena such as ion Coulomb crystal[224] is expected in the many-body level, few-body physics is less intriguing because of the lack of binding. Systems with opposite charges, on the other hand, are much more complicated and have been one of the main subjects in studies of atoms and molecules for many decades. In such systems, the existence of Rydberg series in the two-body sub-systems and the emergence of quantum chaos near a three-body breakup threshold[225] often bring difficulties in the quantum analysis of scaling laws for collisional rates. Such scaling laws are very important for the formation of atoms in cold plasma and has been more widely studied by semi-classical analyses[226–228] and classical Monte-Carlo simulations.[229,230]

Attractive $1/r^2$ potential. Depending on its strength, the attractive $1/r^2$ potential can either be supercritical with infinite number of two-body bound states or subcritical without any of such states. If we write the potential as

$$V(r) = -\frac{\alpha^2 + 1/4}{mr^2}\hbar^2, \tag{35}$$

for two identical bosons of mass m, the super and subcritical behaviors are determined by $\alpha^2 > 0$ and $\alpha^2 < 0$. In the supercritical case, as mentioned earlier the attractive $1/r^2$ potential is too singular at the origin to have a lower bound for the ground state energy — a short-range cutoff is therefore necessary for a system to be physical. In the subcritical case, although the attractive $1/r^2$ singularity does not cause any ill behavior for two bosons, Guevara et al.[231] has numerically showed that the three-boson system suffers a collapse in the ground state energy when α^2 is the range $-0.0072 \lesssim \alpha^2 < 0$. With a two-body short-range cutoff, this collapse is avoided but the three-boson system is no less interesting — it has an infinite number of bound states as a result of the long-range form of a numerically observed, effective hyperradial potential

$$U(R) \approx -\frac{\sqrt{\beta \ln(R/r_0) + \delta}}{2\mu_3 R^2}\hbar^2, \tag{36}$$

with $\beta > 0$ in the range of α^2 where the non-regularized system collapses. The parameter r_0 is the characteristic length scale for the short-range cutoff. Both β and δ depend on α^2, whereas δ depends also on the detail of the short-range cutoff and is therefore non-universal.[231] The collapse of a non-regularized system can be understood by Eq. (36) when the limit $r_0 \to 0$ is taken — $U(R)$ diverges for every R. So unlike the Thomas collapse, this collapse cannot be removed by simply introducing a short-range three-body force.

For fermions, the supercritical behavior occurs when $\alpha^2 > 2$ due to their p-wave barrier. In the subcritical case, it is numerically found[231] that the effective hyperradial potential for three fermions has the asymptotic behavior

$$U(R) \approx -\frac{\alpha_{\text{eff}}^2 + 1/4}{2\mu_3 R^2} + \frac{\gamma}{2\mu_3 R^2 \ln(R/r_0)}, \tag{37}$$

with $\alpha_{\text{eff}}^2 > 0$ when $1.6 \lesssim \alpha^2 < 2$. This again leads to a collapse of the three-body system without proper regularization in $V(r)$. Unlike the case of bosons, though, taking the $r_0 \to 0$ limit here does not lead to a collapse in $U(R)$, therefore a three-body force at small R is sufficient to avoid the collapse of the system. Here the bizarreness is that three fermions have an infinite number of bound states when $1.6 \lesssim \alpha^2 < 1.75$, where the effective two-body interactions (including the p-wave barrier) are all repulsive!

Although the attractive $1/r^2$ interaction doesn't exist between fundamental particles, similar type of interactions can be found between composite particles. One example is the interaction between a charge and dipole, although it is not obvious what changes the anisotropy could make in the three-body physics. Also, in such systems there can be at most two pairs of $1/r^2$ interactions, so that the results discussed above for identical particles do not apply directly. Another possible scenario is a system of three heavy and one light atoms with resonant heavy-light interactions (infinite heavy-light scattering length). In such systems the effective interaction between the heavy ones (induced by the light particle) is also attractive $1/r^2$ [see Eq. (21)]. Here, though, the question is that if the way to get the effective interaction — the Born-Oppenheimer approximation — is good enough to study the new, exotic three-body states in the subcritical regime, where relatively weak effective interactions, or moderate mass ratios, are needed. Another caution is that the effective $1/r^2$ interactions in this system do not exist for all geometries of the heavy particles,[31] where the possible consequences are yet to be investigated.

Ion-atom interactions. The attractive $1/r^4$ interaction between an ion and a neutral atom is of much longer range than the van der Waals interaction between neutral atoms. The direct consequence is that the density of state in the near-threshold two-body spectrum for the ion-atom interaction is much higher than the van der Waals systems, and the characteristic length scale

$$r_4 = (2\mu_2 C_4)^{1/2}/\hbar \tag{38}$$

for the $-C_4/r^4$ interaction is typically on the order of 10^3 Bohr, which is much greater than r_{vdW} in atomic systems. Nevertheless, the near-threshold two-body physics for the attractive $1/r^4$ interaction is essentially the same as that for the van der Waals interaction,[10,14,18,163,232] and so is expected for Efimov physics (assuming attractive $1/r^4$ interactions for all pairs).[45] In realistic systems, however, inelastic collisions due to electron exchange are often non-negligible, which is beyond the physics that single-channel attractive $1/r^4$ interactions describe. Recent experimental studies,[233] though, have suggested that in a system of one Rb^+ ion and two neutral Rb atoms the Rb^+ ion behaves more like a "catalyst" in the process of three-body recombination, and rarely appear in the recombination products. The three-body physics in the ion-neutral hybrid system, including electron exchange and the physics of long-range attractive $1/r^4$ interaction, is therefore an important topic for future studies to understand the dynamics of ion-neutral admixtures.

Spin-orbit interactions. The term "Spin-orbit interaction", or "Spin-orbit coupling", is used both in the context of atomic physics and condensed matter physics. Here we refer to its latter meaning, namely, the interaction between a particle's spin or pseudo-spin s and its *linear* momentum p in the following form:

$$V = (s \cdot a)(p \cdot b), \tag{39}$$

where a and b are some constant vectors that depend on how the spin-orbit interaction is created.[156] In condensed matter systems, the spin-orbit interactions play an important role in many exotic quantum phenomena, such as topological phases and quantum spin Hall effect.[234] Thanks to the recent experimental realization of the spin-orbit interaction, or more generally, the synthetic gauge field in ultracold atomic gases (see a recent review Ref. 235), physics in spin-orbit coupled few-body systems has been a very active research subject in the last few years.

Since particles with the spin-orbit interaction are no longer in their momentum eigenstates even when they are far apart, their low-energy scattering properties, as well as the density of states near the two-body breakup threshold differ dramatically from systems without spin-orbit interactions.[236] New, exotic few-body physics is therefore highly expected in the spin-orbit coupled systems.

Although still in the early stage, understanding in the two-body binding and proper theoretical treatments of two-body scattering have been developed for spin-orbit coupled systems with different identical particle symmetries, spin configurations, and various types of model potentials.[237–242] Such developments have facilitated many experimental observations, for instance, in the anisotropic low-energy scattering properties[243] and in the Feshbach molecules creation controlled by spin-orbit interactions.[244]

Studies on three particles with spin-orbit interactions are just in the beginning. Nevertheless, universal three-body states induced by spin-orbit interactions have been predicted by Shi *et al.*[245] in a mass-imbalanced system with two heavy spinless fermions and one light, spin $1/2$ atom. In this system, extra binding comes from the lift of degeneracy by the spin-orbit interactions in states with the same total orbital angular momentum.

7. Summary

Few-body physics, which connects fundamental quantum physics to macroscopic properties of many-body systems, serves as the foundation of modern quantum physics, quantum chemistry, molecular biology, and many other sciences and technologies. In comparison to larger systems, on the one hand, few-body systems are simple enough so that accurate descriptions and understandings can be obtained; on the other hand they are complex enough for many exotic quantum phenomena to occur and extensive experimental controls to be possible. In our above discussions we have briefly reviewed the universal properties of strongly interacting two- and three-body systems in ultracold atomic gases, where extraordinary level of control and tunability have been achieved over the interaction strength, spatial dimensions, identical particle symmetries, and many other configurations of few-body systems. Here in particular, studies on the van der Waals and dipolar atomic systems have indicated that a full characterization of near-threshold three-body physics is in principle feasible from the relatively simple, yet accurately known two-body properties. Studies on the generalization of these new findings in other ultracold atomic systems, including those with experimentally synthesized interactions, are highly desirable and may lead to another milestone in the quantum few-body research.

Acknowledgments

This work was supported in part by an AFOSR-MURI and by NSF. YW also acknowledges support from Department of Physics, Kansas State University.

References

1. Sadeghpour, H. R., Bohn, J. L., Cavagnero, M. J., Esry, B. D., Fabrikant, I. I., Macek, J. H., and Rau, A. R. P., (2000). Collisions near threshold in atomic and molecular physics, *J. Phys. B*. **33**, 5, R93–R140.

2. Idziaszek, Z. and Karwasz, G. (2006). Applicability of modified effective-range theory to positron-atom and positron-molecule scattering, *Phys. Rev. A.* **73**, 064701. doi: 10.1103/PhysRevA.73.064701.

3. Müller, T.-O., Kaiser, A., and Friedrich, H. (2011). *s*, *Phys. Rev. A.* **84**, 032701. doi: 10.1103/PhysRevA.84.032701.

4. O'Malley, T. F., Spruch, L., and Rosenberg, L. (1961). Modification of effective-range theory in the presence of a long-range (r^{-4}) potential, *J. Math. Phys.* **2**, 4, 491–498. doi: http://dx.doi.org/10.1063/1.1703735.

5. Raab, P. and Friedrich, H. (2009). Quantization function for potentials with $-1/r^4$ tails, *Phys. Rev. A.* **80**, 052705. doi: 10.1103/PhysRevA.80.052705.

6. Burke, J. P., Greene, C. H., and Bohn, J. L. (1998). Multichannel cold collisions: Simple dependences on energy and magnetic field, *Phys. Rev. Lett.* **81**, 3355–3358. doi: 10.1103/PhysRevLett.81.3355.

7. Gao, B. (1998). Quantum-defect theory of atomic collisions and molecular vibration spectra, *Phys. Rev. A.* **58**, 4222–4225. doi: 10.1103/PhysRevA.58.4222.

8. Gao, B. (1998). Solutions of the Schrödinger equation for an attractive $1/r^6$ potential, *Phys. Rev. A.* **58**, 1728–1734. doi: 10.1103/PhysRevA.58.1728.

9. Gao, B. (2001). Angular-momentum-insensitive quantum-defect theory for diatomic systems, *Phys. Rev. A.* **64**, 010701. doi: 10.1103/PhysRevA.64.010701.

10. Gao, B. (2013). Quantum-defect theory for $-1/r^4$-type interactions, *Phys. Rev. A.* **88**, 022701. doi: 10.1103/PhysRevA.88.022701.

11. Gao, B., Tiesinga, E., Williams, C. J., and Julienne, P. S. (2005). Multichannel quantum-defect theory for slow atomic collisions, *Phys. Rev. A.* **72**, 042719. doi: 10.1103/PhysRevA.72.042719.

12. Greene, C., Fano, U., and Strinati, G. (1979). General form of the quantum-defect theory, *Phys. Rev. A.* **19**, 1485–1509. doi: 10.1103/PhysRevA.19.1485.

13. Greene, C. H., Rau, A. R. P., and Fano, U. (1982). General form of the quantum-defect theory. ii, *Phys. Rev. A.* **26**, 2441–2459. doi: 10.1103/PhysRevA.26.2441.

14. Idziaszek, Z., Simoni, A., Calarco, T., and Julienne, P. S. (2011). Multichannel quantum-defect theory for ultracold atom–ion collisions, *New J. Phys.* **13**, 8, 083005.

15. Mies, F. H. (1984). A multichannel quantum defect analysis of diatomic predissociation and inelastic atomic scattering, *J. Chem. Phys.* **80**, 2514–2525.

16. Mies, F. H. and Raoult, M. (2000). Analysis of threshold effects in ultracold atomic collisions, *Phys. Rev. A.* **62**, 012708. doi: 10.1103/PhysRevA.62.012708.

17. Ruzic, B. P., Greene, C. H., and Bohn, J. L. (2013). Quantum defect theory for high-partial-wave cold collisions, *Phys. Rev. A.* **87**, 032706. doi: 10.1103/PhysRevA.87.032706.

18. Watanabe, S. and Greene, C. H. (1980). Atomic polarizability in negative-ion photodetachment, *Phys. Rev. A.* **22**, 158–169. doi: 10.1103/PhysRevA.22.158.

19. Seaton, M. J. (1983). Quantum defect theory, *Rep. Prog. Phys.* **46**, 2, 167–257. ISSN 0034-4885. doi: 10.1088/0034-4885/46/2/002.

20. Blume, D. (2012). Few-body physics with ultracold atomic and molecular systems in traps, *Rep. Prog. Phys.* **75**, 4, 046401. doi: 10.1088/0034-4885/75/4/046401.

21. Ni, K. K., Ospelkaus, S., Wang, D., Quemener, G., Neyenhuis, B., de Miranda, M. H. G., Bohn, J. L., Ye, J., and Jin, D. S. (2010). Dipolar collisions of polar molecules in

the quantum regime, *Nature*. **464**, 7293, 1324–1328. ISSN 0028-0836. doi: 10.1038/nature08953.

22. Quéméner, G. and Julienne, P. S. (2012). Ultracold molecules under control!, *Chemical Reviews*. **112**, 9, 4949–5011.

23. Bohn, J. L., Cavagnero, M., and Ticknor, C. (2009). Quasi-universal dipolar scattering in cold and ultracold gases, *New J. Phys.* **11**, 5, 055039.

24. Cavagnero, M. and Newell, C. (2009). Inelastic semiclassical collisions in cold dipolar gases, *New J. Phys.* **11**, 5, 055040.

25. D'Incao, J. P. and Greene, C. H. (2011). Collisional aspects of bosonic and fermionic dipoles in quasi-two-dimensional confining geometries, *Phys. Rev. A*. **83**, 030702. doi: 10.1103/PhysRevA.83.030702.

26. Giannakeas, P., Melezhik, V. S., and Schmelcher, P. (2013). Dipolar confinement-induced resonances of ultracold gases in waveguides, *Phys. Rev. Lett.* **111**, 183201. doi: 10.1103/PhysRevLett.111.183201.

27. Kanjilal, K. and Blume, D. (2008). Low-energy resonances and bound states of aligned bosonic and fermionic dipoles, *Phys. Rev. A*. **78**, 040703. doi: 10.1103/PhysRevA.78.040703.

28. Ticknor, C. (2011). Two-dimensional dipolar scattering with a tilt, *Phys. Rev. A*. **84**, 032702. doi: 10.1103/PhysRevA.84.032702.

29. Wang, Y. and Greene, C. H. (2012). Universal bound and scattering properties for two dipoles, *Phys. Rev. A*. **85**, 022704. doi: 10.1103/PhysRevA.85.022704.

30. Adhikari, S. K., Delfino, A., Frederico, T., Goldman, I. D. and Tomio, L. (1988). Efimov and Thomas effects and the model dependence of 3-particle observables in 2 and 3 dimensions, *Phys. Rev. A*. **37**, 10, 3666–3673.

31. Amado, R. D. and Greenwood, F. C. (1973). There is no Efimov effect for four or more particles, *Phys. Rev. D*. **7**, 2517–2519. doi: 10.1103/PhysRevD.7.2517.

32. Amado, R. D. and Noble, J. V. (1972). Efimov's effect: A new pathology of three-particle systems. ii, *Phys. Rev. D*. **5**, 1992–2002. doi: 10.1103/PhysRevD.5.1992.

33. Bedaque, P. F., Braaten, E., and Hammer, H.-W. (2000). Three-body recombination in Bose gases with large scattering length, *Phys. Rev. Lett.* **85** 908–911. doi: 10.1103/PhysRevLett.85.908.

34. Deltuva, A. (2010). Efimov physics in bosonic atom-trimer scattering, *Phys. Rev. A*. **82**, 040701. doi: 10.1103/PhysRevA.82.040701.

35. Deltuva, A. and Lazauskas, R. (2010). Breakdown of universality in few-boson systems, *Phys. Rev. A*. **82**, 012705. doi: 10.1103/PhysRevA.82.012705.

36. Efimov, V. (1972). Level spectrum of 3 resonantly interacting particles, *JETP Lett.* **16**, 1, 34.

37. Efimov, V. (1973). Energy levels of three resonantly interacting particles, *Nucl. Phys. A*. **210**, 1, 157–188. ISSN 0375-9474. doi: http://dx.doi.org/10.1016/0375-9474(73)90510-1.

38. Efimov, V. (1979). Low-energy properties of three resonantly interacting particles, *Sov. J. Nuc. Phys.* **29**, 546.

39. Efremov, M. A., Plimak, L., Berg, B., Ivanov, M. Y., and Schleich, W. P. (2009). Efimov states in atom-molecule collisions, *Phys. Rev. A*. **80** 2, 022714. doi: 10.1103/PhysRevA.80.022714.

40. Esry, B. D., Greene, C. H., and Burke, J. P. (1999). Recombination of three atoms in the ultracold limit, *Phys. Rev. Lett.* **83**, 9, 1751–1754. doi: 10.1103/PhysRevLett.83.1751.

41. Esry, B. D., Greene, C. H., Zhou, Y. and Lin, C. D. (1996). Role of the scattering length in three-boson dynamics and Bose-Einstein condensation, *J. Phys. B.* **29**, 2, L51–L57.

42. Esry, B. D., Lin, C. D., and Greene, C. H. (1996). Adiabatic hyperspherical study of the helium trimer, *Phys. Rev. A.* **54**, 1, 394–401.

43. Macek, J. (1986). Loosely bound states of three particles, *Z. Phys. D.* **3**, 1, 31.

44. Macek, J. H., Yu Ovchinnikov, S., and Gasaneo, G. (2006). Exact solution for three particles interacting via zero-range potentials, *Phys. Rev. A.* **73**, 032704. doi: 10.1103/PhysRevA.73.032704.

45. Naidon, P., Endo, S., and Ueda, M. (2014). Microscopic origin and universality classes of the Efimov three-body parameter, *Phys. Rev. Lett.* **112**, 105301. doi: 10.1103/PhysRevLett.112.105301.

46. Nielsen, E., Fedorov, D., Jensen, A. and Garrido, E. (2001). The three-body problem with short-range interactions, *Phys. Rep.* **347**, 5, 373–459. doi: http://dx.doi.org/10.1016/S0370-1573(00)00107-1.

47. Nielsen, E. and Macek, J. H. (1999). Low-energy recombination of identical bosons by three-body collisions, *Phys. Rev. Lett.* **83**, 8, 1566–1569. doi: 10.1103/PhysRevLett.83.1566.

48. Wang, J., D'Incao, J. P., Esry, B. D., and Greene, C. H. (2012). Origin of the three-body parameter universality in Efimov physics, *Phys. Rev. Lett.* **108**, 263001. doi: 10.1103/PhysRevLett.108.263001.

49. Wang, Y., D'Incao, J. P., Nägerl, H.-C., and Esry, B. D. (2010). Colliding Bose-Einstein condensates to observe Efimov physics, *Phys. Rev. Lett.* **104**, 113201. doi: 10.1103/PhysRevLett.104.113201.

50. Wang, Y. and Esry, B. D., (2011). Universal three-body physics at finite energy near Feshbach resonances, *New J. Phys.* **13**, 3, 035025, (2011).

51. Wang, Y. and Julienne, P. S. (2014). Universal van der waals physics for three cold atoms near feshbach resonances, *Nature Phys.* **10**, 10, 768–773. doi: 10.1038/nphys3071.

52. Efimov, V. (1970). Energy levels arising from resonant two-body forces in a three-body system, *Phys. Lett. B.* **33**, 8, 563–564. ISSN 0370-2693. doi: http://dx.doi.org/10.1016/0370-2693(70)90349-7.

53. Efimov, V. (1971). Weakly-bound states of three resonantly-interacting particles, *Sov. J. Nuc. Phys.* **12**, 589–595.

54. Kraemer, T., Mark, M., Waldburger, P., Danzl, J., Chin, C., Engeser, B., Lange, A., Pilch, K., Jaakkola, A., Nägerl, H.-C., et al. (2006). Evidence for Efimov quantum states in an ultracold gas of caesium atoms, *Nature.* **440**, 7082, 315–318.

55. Fano, U. (1935). Absorption spectrum of the noble gases near the limit of the arc spectrum, *Nuovo Cimento.* **12**, 154.

56. Fano, U. (1961). Effects of configuration interaction on intensities and phase shifts, *Phys. Rev.* **124**, 6, 1866–1878.

57. Fano, U., Pupillo, G., Zannoni, A., and Clark, C. W. (2005). On the absorption spectrum of noble gases at the arc spectrum limit, *J. Res. Natl. Inst. Stand. Technol.* **110**, 583–587.

58. Feshbach, H. (1958). Unified theory of nuclear reactions, *Ann. Phys.* **5**, 4, 357. doi: 10.1016/0003-4916(58)90007-1.

59. Feshbach, H. (1962). A unified theory of nuclear reactions. II, *Ann. Phys.* **19**, 2, 287–313. doi: 10.1016/0003-4916(62)90221-X.

60. Chin, C., Grimm, R., Julienne, P., and Tiesinga, E. (2010). Feshbach resonances in ultracold gases, *Rev. Mod. Phys.* **82**, 1225.

61. Inouye, S., Andrews, M. R., Stenger, J., Miesner, H.-J., Stamper-Kurn, D. M., and Ketterle, W. (1998). Observation of Feshbach resonances in a Bose–Einstein condensate, *Nature.* **392**, 6672, 151–154.

62. Stwalley, W. C. (1976). Stability of spin-aligned hydrogen at low temperatures and high magnetic fields: New field-dependent scattering resonances and predissociations, *Phys. Rev. Lett.* **37**, 1628–1631. doi: 10.1103/PhysRevLett.37.1628.

63. Tiesinga, E., Verhaar, B. J., and Stoof, H. T. C. (1993). Threshold and resonance phenomena in ultracold ground-state collisions, *Phys. Rev. A.* **47**, 4114–4122. doi: 10.1103/PhysRevA.47.4114.

64. Clark, C. W. and Siegel, J. (1980). Electron-polar-molecule scattering at intermediate values of J - closed-form treatment, *J. Phys. B.* **13**, 1, L31–L37. doi: 10.1088/0022-3700/13/1/007.

65. Engelking, P. C. and Herrick, D. R. (1984). Effects of rotational doubling on the anomalous photodetachment thresholds resulting from electron-dipole interaction, *Phys. Rev. A.* **29**, 2425–2428. doi: 10.1103/PhysRevA.29.2425.

66. Garrett, W. R. (1970). Critical binding of an electron to a non-stationary electric dipole, *Chem. Phys. Lett.* **5**, 7, 393–397. doi: 10.1016/0009-2614(70)80045-8.

67. Smith, J. R., Kim, J. B., and Lineberger, W. C. (1997). High-resolution threshold photodetachment spectroscopy of OH^-, *Phys. Rev. A.* **55**, 2036–2043. doi: 10.1103/PhysRevA.55.2036.

68. Gailitis, M. and Damburg, R. (1963). The influence of close coupling on the threshold behaviour of cross sections of electron-hydrogen scattering, *Proc. Phys. Soc.* **82**, 2, 192.

69. Greene, C. H. (1980). Interpretation of Feshbach resonances in h- photodetachment, *J. Phys. B.* **13**, 2, L39–L44. doi: 10.1088/0022-3700/13/2/001.

70. Seaton, M. J. (1961). Strong coupling in optically allowed atomic transitions produced by electron impact, *Proc. Phys. Soc.* **77**, 493, 174. doi: 10.1088/0370-1328/77/1/322.

71. Braaten, E. and Hammer, H.-W. (2006). Universality in few-body systems with large scattering length, *Phys. Rep.* **428**, 5–6, 259–390. ISSN 0370-1573. doi: http://dx.doi.org/10.1016/j.physrep.2006.03.001.

72. D'Incao, J. P., Greene, C. H., and Esry, B. D. (2009). The short-range three-body phase and other issues impacting the observation of Efimov physics in ultracold quantum gases, *J. Phys. B.* **42**, 4, 044016.

73. Mehta, N. P., Rittenhouse, S. T., D'Incao, J. P., von Stecher, J. and Greene, C. H. (2009). General theoretical description of N-body recombination, *Phys. Rev. Lett.* **103**, 15, 153201. doi: 10.1103/PhysRevLett.103.153201.

74. Wang, Y., D'Incao, J. P., and Esry, B. D. (2013). Chapter 1 — Ultracold few-body systems. In eds. P. R. B. Ennio Arimondo and C. C. Lin, *Adv. At. Mol. Opt. Phys.*, vol. 62, pp. 1–115. Academic Press. doi: http://dx.doi.org/10.1016/B978-0-12-408090-4.00001-3.

75. Berninger, M., Zenesini, A., Huang, B., Harm, W., Nägerl, H.-C., Ferlaino, F., Grimm, R., Julienne, P. S., and Hutson, J. M. (2011). Universality of the three-body parameter for Efimov states in ultracold cesium, *Phys. Rev. Lett.* **107**, 120401 (Sep, 2011). doi: 10.1103/PhysRevLett.107.120401.

76. Gribakin, G. F. and Flambaum, V. V. (1993). Calculation of the scattering length in atomic collisions using the semiclassical approximation, *Phys. Rev. A.* **48**, 546–553. doi: 10.1103/PhysRevA.48.546.

77. Wang, Y., Wang, J., D'Incao, J. P., and Greene, C. H. (2012). Universal three-body parameter in heteronuclear atomic systems, *Phys. Rev. Lett.* **109**, 243201. doi: 10.1103/PhysRevLett.109.243201.

78. Wang, Y., D'Incao, J. P., and Greene, C. H. (2011). Efimov effect for three interacting bosonic dipoles, *Phys. Rev. Lett.* **106**, 233201. doi: 10.1103/PhysRevLett.106.233201.

79. Wang, Y., D'Incao, J. P., and Greene, C. H. (2011). Universal three-body physics for fermionic dipoles, *Phys. Rev. Lett.* **107**, 233201. doi: 10.1103/PhysRevLett.107.233201.

80. Jones, K. M., Tiesinga, E., Lett, P. D., and Julienne, P. S. (2006). Ultracold photoassociation spectroscopy: Long-range molecules and atomic scattering, *Rev. Mod. Phys.* **78**, 483–535.

81. Gao, B. (2004). Binding energy and scattering length for diatomic systems, *J. Phys. B.* **37**, 4273–4279.

82. Gao, B. (2009). Analytic description of atomic interaction at ultracold temperatures: The case of a single channel, *Phys. Rev. A.* **80**, 012702. doi: 10.1103/PhysRevA.80.012702.

83. Flambaum, V. V., Gribakin, G. F., and Harabati, C. (1999). Analytical calculation of cold-atom scattering, *Phys. Rev. A.* **59**, 1998.

84. Blackley, C. L., Julienne, P. S., and Hutson, J. M. (2014). Effective-range approximations for resonant scattering of cold atoms, *Phys. Rev. A.* **89**, 042701.

85. Julienne, P. S. and Gao, B. (2006).Simple theoretical models for resonant cold atom interactions. In eds. C. Roos, H. Häffner, and R. Blatt, *Atomic Physics 20*, vol. 869, *AIP Conference Proceedings*, pp. 261–268. ISBN 0-7354-0367-8. 20th International Conference on Atomic Physics, Innsbruck, AUSTRIA, JUL 16-21, 2006; arXiv:physics/0609013v1.

86. Julienne, P. S. and Hutson, J. M. (2014). Contrasting the wide Feshbach resonances in ^{6}Li and ^{7}Li, *Phys. Rev. A.* **89**, 052715.

87. Berninger, M., Zenesini, A., Huang, B., Harm, W., Nägerl, H.-C., Ferlaino, F., Grimm, R., Julienne, P. S., and Hutson, J. M. (2013). Feshbach resonances, weakly bound molecular states, and coupled-channel potentials for cesium at high magnetic fields, *Phys. Rev. A.* **87**, 032517. doi: 10.1103/PhysRevA.87.032517.

88. Jachymski, K., Krych, M., Julienne, P. S., and Idziaszek, Z. (2013). Quantum theory of reactive collisions for $1/r^n$ potentials, *Phys. Rev. Lett.* **110**, 213202. doi: 10.1103/PhysRevLett.110.213202.

89. Hartree, D. R. (1928). The wave mechanics of an atom with a non-coulomb central field. part III. term values and intensities in series in optical spectra, *Mathematical Proceedings of the Cambridge Philosophical Society.* **24**, 426–437. ISSN 1469-8064. doi: 10.1017/S0305004100015954.

90. Berger, R. O. and Spruch, L. (1965). Coulombic modified effective-range theory for long-range effective potentials, *Phys. Rev.* **138**, B1106–B1115. doi: 10.1103/PhysRev. 138.B1106.

91. Watanabe, S. (1982). Doubly excited states of the helium negative ion, *Phys. Rev. A.* **25**, 2074–2098. doi: 10.1103/PhysRevA.25.2074.

92. Mies, F. H. and Julienne, P. S. (1984). A multichannel quantum defect analysis of two-state couplings in diatomic molecules, *J. Chem. Phys.* **80**, 2526–2536.

93. Julienne, P. S. and Mies, F. H. (1989). Collisions of ultracold trapped atoms, *J. Opt. Soc. Am. B.* **6**, 2257–2269.

94. Pires, R., Repp, M., Ulmanis, J., Kuhnle, E. D., Weidemüller, M., Tiecke, T. G., Greene, C. H., Ruzic, B. P., Bohn, J. L., and Tiemann, E. (2014). Analyzing Feshbach resonances: A ^{6}Li-^{133}Cs case study, *Phys. Rev. A.* **90**, 012710. doi: 10.1103/PhysRevA. 90.012710.

95. Tiecke, T. G., Goosen, M. R., Walraven, J. T. M., and Kokkelmans, S. J. J. M. F. (2010). Asymptotic-bound-state model for Feshbach resonances, *Phys. Rev. A.* **82**, 042712. doi: 10.1103/PhysRevA.82.042712.

96. Julienne, P. S. and Gao, B. (2006). Simple theoretical models for resonant cold atom interactions, *AIP Conference Proceedings.* **869**, 1, 261–268. doi: http://dx.doi.org/10. 1063/1.2400656.

97. Breit, G. and Rabi, I. I. (1931). Measurement of nuclear spin, *Phys. Rev.* **38**, 2082–2083. doi: 10.1103/PhysRev.38.2082.2.

98. Nafe, J. E. and Nelson, E. B. (1948). The hyperfine structure of hydrogen and deuterium, *Phys. Rev.* **73**, 718–728. doi: 10.1103/PhysRev.73.718.

99. Hanna, T. M., Tiesinga, E., and Julienne, P. S. (2009). Prediction of Feshbach resonances from three input parameters, *Phys. Rev. A.* **79**, 040701. doi: 10.1103/ PhysRevA.79.040701.

100. Moerdijk, A. J., Verhaar, B. J., and Axelsson, A. (1995). Resonances in ultracold collisions of ^{6}Li, ^{7}Li, and ^{23}Na, *Phys. Rev. A.* **51**, 4852–4861. doi: 10.1103/PhysRevA. 51.4852.

101. Stoof, H. T. C., Koelman, J. M. V. A., and Verhaar, B. J. (1988). Spin-exchange and dipole relaxation rates in atomic hydrogen: Rigorous and simplified calculations, *Phys. Rev. B.* **38**, 4688–4697. doi: 10.1103/PhysRevB.38.4688.

102. Chin, C., Vuletić, V., Kerman, A. J., Chu, S., Tiesinga, E., Leo, P. J., and Williams, C. J. (2004). Precision Feshbach spectroscopy of ultracold Cs_2, *Phys. Rev. A.* **70**, 032701. doi: 10.1103/PhysRevA.70.032701.

103. Kotochigova, S., Tiesinga, E., and Julienne, P. S. (2000). Relativistic ab initio treatment of the second-order spin-orbit splitting of the $a^3\Sigma_u^+$ potential of rubidium and cesium dimers, *Phys. Rev. A.* **63**, 012517. doi: 10.1103/PhysRevA.63.012517.

104. Leo, P. J., Tiesinga, E., Julienne, P. S., Walter, D. K., Kadlecek, S., and Walker, T. G. (1998). Elastic and inelastic collisions of cold spin-polarized ^{133}Cs atoms, *Phys. Rev. Lett.* **81**, 1389–1392. doi: 10.1103/PhysRevLett.81.1389.

105. Mies, F. H., Williams, C. J., Julienne, P. S., and Krauss, M. (1996). Estimating bounds on collisional relaxation rates of spin-polarized ^{87}Rb atoms at ultracold temperatures, *Journal of Research-National Institute of Standards and Technology.* **101**, 521–536.

106. Leo, P. J., Williams, C. J., and Julienne, P. S. (2000). Collision properties of ultracold ^{133}Cs atoms, *Phys. Rev. Lett.* **85**, 2721–2724. doi: 10.1103/PhysRevLett.85.2721.

107. Lagendijk, A., Silvera, I. F., and Verhaar, B. J. (1986). Spin exchange and dipolar relaxation rates in atomic hydrogen: Lifetimes in magnetic traps, *Phys. Rev. B.* **33**, 626–628. doi: 10.1103/PhysRevB.33.626.

108. Boesten, H. M. J. M., Vogels, J. M., Tempelaars, J. G. C., and Verhaar, B. J. (1996). Properties of cold collisions of ^{39}K atoms and of ^{41}K atoms in relation to bose-einstein condensation, *Phys. Rev. A.* **54**, R3726–R3729. doi: 10.1103/PhysRevA.54.R3726.

109. Burke, J. P. and Bohn, J. L. (1999). Ultracold scattering properties of the short-lived rb isotopes, *Phys. Rev. A.* **59**, 1303–1308. doi: 10.1103/PhysRevA.59.1303.

110. Houbiers, M., Stoof, H. T. C., McAlexander, W. I., and Hulet, R. G. (1998). Elastic and inelastic collisions of ^{6}Li atoms in magnetic and optical traps, *Phys. Rev. A.* **57**, R1497–R1500. doi: 10.1103/PhysRevA.57.R1497.

111. Moerdijk, A. J., Stwalley, W. C., Hulet, R. G., and Verhaar, B. J. (1994). Negative scattering length of ultracold ^{7}Li gas, *Phys. Rev. Lett.* **72**, 40–43. doi: 10.1103/ PhysRevLett.72.40.

112. Tiesinga, E., Moerdijk, A. J., Verhaar, B. J., and Stoof, H. T. C. (1992). Conditions for bose-einstein condensation in magnetically trapped atomic cesium, *Phys. Rev. A.* **46**, R1167–R1170. doi: 10.1103/PhysRevA.46.R1167.

113. Frisch, A., Mark, M., Aikawa, K., Ferlaino, F., Bohn, J. L., Makrides, C., Petrov, A., and Kotochigova, S. (2014). Quantum chaos in ultracold collisions of gas-phase erbium atoms, *Nature.* **507**, 7493, 475–479. doi: 10.1038/nature13137.

114. Kotochigova, S. and Petrov, A. (2011). Anisotropy in the interaction of ultracold dysprosium, *Phys. Chem. Chem. Phys.* **13**, 19165–19170. doi: 10.1039/C1CP21175G.

115. Petrov, A., Tiesinga, E., and Kotochigova, S. (2012). Anisotropy-induced Feshbach resonances in a quantum dipolar gas of highly magnetic atoms, *Phys. Rev. Lett.* **109**, 103002. doi: 10.1103/PhysRevLett.109.103002.

116. Baumann, K., Burdick, N. Q., Lu, M., and Lev, B. L. (2014). Observation of low-field Fano-Feshbach resonances in ultracold gases of dysprosium, *Phys. Rev. A.* **89**, 020701.

117. Köhler, T., Góral, K., and Julienne, P. S. (2006). Production of cold molecules via magnetically tunable Feshbach resonances, *Rev. Mod. Phys.* **78**, 1311–1361. doi: 10. 1103/RevModPhys.78.1311.

118. Hutson, J. M., Tiesinga, E., and Julienne, P. S. (2008). Avoided crossings between bound states of ultracold cesium dimers, *Phys. Rev. A.* **78**, 052703. doi: 10.1103/ PhysRevA.78.052703.

119. Aymar, M., Greene, C. H., and Luc-Koenig, E. (1996). Multichannel Rydberg spectroscopy of complex atoms, *Rev. Mod. Phys.* **68**, 1015–1123. doi: 10.1103/ RevModPhys.68.1015.

120. Thomas, L. H. (1935). The interaction between a neutron and a proton and the structure of H^3, *Phys. Rev.* **47**, 903–909. doi: 10.1103/PhysRev.47.903.

121. Faddeev, L. D. (1961). Scattering theory for a three-particle system, *Sov. Phys.-JETP.* **12**, 1014–1019.

122. Amado, R. D. and Noble, J. V. (1971). On Efimov's effect: A new pathology of three-particle systems, *Phys. Lett. B.* **35**, 1, 25–27, (1971). ISSN 0370-2693. doi: http: //dx.doi.org/10.1016/0370-2693(71)90429-1.

123. Bedaque, P. F., Hammer, H.-W., and van Kolck, U. (1999). Renormalization of the three-body system with short-range interactions, *Phys. Rev. Lett.* **82**, 463–467. doi: 10.1103/PhysRevLett.82.463.

124. Wang, Y. (2010). *Universal Efimov physics in three- and four-body collisions*. PhD thesis, Kansas State University.

125. Perelomov, A. and Popov, V. (1970). "Fall to the center" in quantum mechanics, *Theoretical and Mathematical Physics.* **4**, 1, 664–677. ISSN 0040-5779. doi: 10.1007/BF01246666.

126. Gogolin, A. O., Mora, C., and Egger, R. (2008). Analytical solution of the bosonic three-body problem, *Phys. Rev. Lett.* **100**, 140404. doi: 10.1103/PhysRevLett.100.140404.

127. Dyke, P., Pollack, S. E., and Hulet, R. G. (2013). Finite-range corrections near a Feshbach resonance and their role in the Efimov effect, *Phys. Rev. A.* **88**, 023625. doi: 10.1103/PhysRevA.88.023625.

128. Ferlaino, F., Zenesini, A., Berninger, M., Huang, B., Nägerl, H.-C., and Grimm, R. (2011). Efimov resonances in ultracold quantum gases, *Few-Body Sys.* **51** (2-4), 113–133. ISSN 0177-7963. doi: 10.1007/s00601-011-0260-7.

129. Knoop, S., Ferlaino, F., Mark, M., Berninger, M., Schöbel, H., Nägerl, H.-C. and Grimm, R. (2009). Observation of an Efimov-like trimer resonance in ultracold atom–dimer scattering, *Nature Phys.* **5**, 3, 227–230.

130. Pollack, S. E., Dries, D. and Hulet, R. G. (2009). Universality in three- and four-body bound states of ultracold atoms, *Science.* **326**, 5960, 1683–1685. doi: 10.1126/science.1182840.

131. Gross, N., Shotan, Z., Kokkelmans, S., and Khaykovich, L. (2009). Observation of universality in ultracold ^{7}Li three-body recombination, *Phys. Rev. Lett.* **103**, 163202. doi: 10.1103/PhysRevLett.103.163202.

132. Gross, N., Shotan, Z., Kokkelmans, S., and Khaykovich, L. (2010). Nuclear-spin-independent short-range three-body physics in ultracold atoms, *Phys. Rev. Lett.* **105**, 103203. doi: 10.1103/PhysRevLett.105.103203.

133. Soldán, P., Cvitaš, M. T., and Hutson, J. M. (2003). Three-body nonadditive forces between spin-polarized alkali-metal atoms, *Phys. Rev. A.* **67**, 054702. doi: 10.1103/PhysRevA.67.054702.

134. Jachymski, K. and Julienne, P. S. (2013). Analytical model of overlapping Feshbach resonances, *Phys. Rev. A.* **88**, 052701. doi: 10.1103/PhysRevA.88.052701.

135. Roy, S., Landini, M., Trenkwalder, A., Semeghini, G., Spagnolli, G., Simoni, A., Fattori, M., Inguscio, M., and Modugno, G. (2013). Test of the universality of the three-body Efimov parameter at narrow Feshbach resonances, *Phys. Rev. Lett.* **111**, 053202. doi: 10.1103/PhysRevLett.111.053202.

136. Wild, R. J., Makotyn, P., Pino, J. M., Cornell, E. A., and Jin, D. S. (2012). Measurements of Tan's contact in an atomic Bose-Einstein condensate, *Phys. Rev. Lett.* **108**, 145305. doi: 10.1103/PhysRevLett.108.145305.

137. Chin, C. (2011). Universal scaling of Efimov resonance positions in cold atom systems, *arXiv preprint arXiv:1111.1484.*

138. D'Incao, J. P. and Esry, B. D. (2006). Enhancing the observability of the Efimov effect in ultracold atomic gas mixtures, *Phys. Rev. A.* **73**, 030703. doi: 10.1103/PhysRevA.73.030703.

139. Jones, J. E. (1924). On the determination of molecular fields. II. from the equation of state of a gas, *Proc. Roy. Soc. London: Ser. A*. **106**, (738), 463–477. doi: 10.1098/rspa. 1924.0082.

140. Pires, R., Ulmanis, J., Häfner, S., Repp, M., Arias, A., Kuhnle, E. D. and Weidemüller, M., (2014). Observation of Efimov resonances in a mixture with extreme mass imbalance, *Phys. Rev. Lett.* **112**, 250404. doi: 10.1103/PhysRevLett.112.250404.

141. Tung, S.-K., Jiménez-García, K., Johansen, J., Parker, C. V., and Chin, C. (2014). Geometric scaling of efimov states in a ^{6}Li-^{133}Cs mixture, *Phys. Rev. Lett.* **113**, 240402. doi: 10.1103/PhysRevLett.113.240402.

142. Bulgac, A. and Efimov, V. (1976). Spin dependence of the level spectrum of three resonantly interacting particles, *Sov. J. Nuc. Phys.* **22**, 2, 153.

143. Jona-Lasinio, M. and Pricoupenko, L. (2010). Three resonant ultracold bosons: Off-resonance effects, *Phys. Rev. Lett.* **104**, 023201. doi: 10.1103/PhysRevLett.104. 023201.

144. Lee, M. D., Köhler, T., and Julienne, P. S. (2007). Excited Thomas-Efimov levels in ultracold gases, *Phys. Rev. A*. **76**, 012720. doi: 10.1103/PhysRevA.76.012720.

145. Massignan, P. and Stoof, H. T. C. (2008). Efimov states near a Feshbach resonance, *Phys. Rev. A*. **78**, 030701. doi: 10.1103/PhysRevA.78.030701.

146. Petrov, D. S. (2004). Three-boson problem near a narrow Feshbach resonance, *Phys. Rev. Lett.* **93**, 143201. doi: 10.1103/PhysRevLett.93.143201.

147. Pricoupenko, L. (2010). Crossover in the Efimov spectrum, *Phys. Rev. A*. **82**, 043633. doi: 10.1103/PhysRevA.82.043633.

148. Sørensen, P. K., Fedorov, D. V., Jensen, A. S., and Zinner, N. T. (2012). Efimov physics and the three-body parameter within a two-channel framework, *Phys. Rev. A*. **86**, 052516. doi: 10.1103/PhysRevA.86.052516.

149. Sørensen, P. K., Fedorov, D. V., and Jensen, A. S. (2013). Three-body recombination rates near a Feshbach resonance within a two-channel contact interaction model, *Few-Body Sys.* **54**, 5–6, 579–590. ISSN 0177-7963. doi: 10.1007/s00601-012-0312-7.

150. Sørensen, P. K., Fedorov, D. V., Jensen, A. S., and Zinner, N. T. (2013). Finite-range effects in energies and recombination rates of three identical bosons, *J. Phys. B*. **46**, 7, 075301.

151. Mehta, N. P., Rittenhouse, S. T., D'Incao, J. P., and Greene, C. H. (2008). Efimov states embedded in the three-body continuum, *Phys. Rev. A*. **78**, 020701. doi: 10.1103/ PhysRevA.78.020701.

152. Colussi, V. E., Greene, C. H., and D'Incao, J. P. (2014). Three-body physics in strongly correlated spinor condensates, *Phys. Rev. Lett.* **113**, 045302. doi: 10.1103/ PhysRevLett.113.045302.

153. Wang, J., D'Incao, J. P., Wang, Y., and Greene, C. H. (2012). Universal three-body recombination via resonant d-wave interactions, *Phys. Rev. A*. **86**, 062511. doi: 10. 1103/PhysRevA.86.062511.

154. Wang, Y., Julienne, P. S., and Greene, C. H. (2014). Unpublished.

155. Gao, B. (2000). Zero-energy bound or quasibound states and their implications for diatomic systems with an asymptotic van der waals interaction, *Phys. Rev. A*. **62**, 050702. doi: 10.1103/PhysRevA.62.050702.

156. Zhai, H. (2012). Spin-orbit coupled quantum gases, *International Journal of Modern Physics B*. **26**, 01, 1230001. doi: 10.1142/S0217979212300010.

157. Langevin, P. (1905). *Ann. Chim. Phys.* **5**, 245.
158. Gorin, E. (1938). *Acta Physicochim. URSS.* **9**, 681.
159. Gorin, E., Kauzmann, W., Walter, J., and Eyring, H. (1939). Reactions involving hydrogen and the hydrocarbons, *J. Chem. Phys.* **7**, 633–645.
160. Fernández-Ramos, A., Miller, J. A., Klippenstein, S. J., and Truhlar, D. G. (2006). Modeling the kinetics of bimolecular reactions, *Chem. Rev.* **106**, 11, 4518–4584. doi: 10.1021/cr050205w.
161. Gao, B. (1996). Theory of slow-atom collisions, *Phys. Rev. A.* **54**, 2022–2039. doi: 10.1103/PhysRevA.54.2022.
162. Gao, B. (2008). General form of the quantum-defect theory for $-1/r^\alpha$ type of potentials with $\alpha > 2$, *Phys. Rev. A.* **78**, 012702.
163. Gao, B. (2010). Universal model for exoergic bimolecular reactions and inelastic processes, *Phys. Rev. Lett.* **105**, 263203.
164. Gao, B. (2010). Universal properties in ultracold ion-atom interactions, *Phys. Rev. Lett.* **104**, 213201.
165. Idziaszek, Z., Quemener, G., Bohn, J. L., and Julienne, P. S. (2010). Simple quantum model of ultracold polar molecule collisions, *Phys. Rev. A.* **82**.
166. Idziaszek, Z. and Julienne, P. S. (2010). Universal rate constants for reactive collisions of ultracold molecules, *Phys. Rev. Lett.* **104**, 11, 113202.
167. Quéméner, G. and Bohn, J. L. (2010). Strong dependence of ultracold chemical rates on electric dipole moments, *Phys. Rev. A.* **81**, 022702.
168. Olshanii, M. (1998). Atomic scattering in the presence of an external confinement and a gas of impenetrable bosons, *Phys. Rev. Lett.* **81**, 938–941.
169. Petrov, D. S. and Shlyapnikov, G. V. (2001). Interatomic collisions in a tightly confined Bose gas, *Phys. Rev. A.* **64**, 012706.
170. Naidon, P., Tiesinga, E., Mitchell, W. F., and Julienne, P. S. (2007). Effective-range description of a Bose gas under strong one- or two-dimensional confinement, *New J. Phys.* **9**, 19.
171. Micheli, A., Idziaszek, Z., Pupillo, G., Baranov, M. A., Zoller, P., and Julienne, P. S. (2010). Universal rates for reactive ultracold polar molecules in reduced dimensions, *Phys. Rev. Lett.* **105**, 073202.
172. Quéméner, G. and Bohn, J. L. (2011). Dynamics of ultracold molecules in confined geometry and electric field, *Phys. Rev. A.* **83**, 012705.
173. Quéméner, G., Bohn, J. L., Petrov, A., and Kotochigova, S. (2011). Universalities in ultracold reactions of alkali-metal polar molecules, *Phys. Rev. A.* **84**, 062703.
174. Julienne, P. S., Hanna, T. M., and Idziaszek, Z. (2011). Universal ultracold collision rates for polar molecules of two alkali-metal atoms, *Phys. Chem. Chem. Phys.* **13**, 19114–19124.
175. Büchler, H. P., Demler, E., Lukin, M., Micheli, A., Prokof'ev, N., Pupillo, G., and Zoller, P. (2007). Strongly correlated 2d quantum phases with cold polar molecules: Controlling the shape of the interaction potential, *Phys. Rev. Lett.* **98**, 060404.
176. Henson, A. B., Gersten, S., Shagam, Y., Narevicius, J., and Narevicius, E. (2012) Observation of resonances in Penning ionization reactions at sub-Kelvin temperatures in merged beams, *Science.* **338**, 6104, 234–238.
177. Croft, J. F. E. and Bohn, J. L. (2014). Long-lived complexes and chaos in ultracold molecular collisions, *Phys. Rev. A.* **89**, 012714. doi: 10.1103/PhysRevA.89.012714.

178. Mayle, M., Quéméner, G., Ruzic, B. P., and Bohn, J. L. (2013). Scattering of ultracold molecules in the highly resonant regime, *Phys. Rev. A.* **87**, 012709. doi: 10.1103/ PhysRevA.87.012709.

179. Mayle, M., Ruzic, B. P., and Bohn, J. L. (2012). Statistical aspects of ultracold resonant scattering, *Phys. Rev. A.* **85**, 062712. doi: 10.1103/PhysRevA.85.062712.

180. Aikawa, K., Frisch, A., Mark, M., Baier, S., Grimm, R., and Ferlaino, F. (2014). Reaching Fermi degeneracy via universal dipolar scattering, *Phys. Rev. Lett.* **112**, 010404. doi: 10.1103/PhysRevLett.112.010404.

181. Aikawa, K., Frisch, A., Mark, M., Baier, S., Rietzler, A., Grimm, R., and Ferlaino, F. (2012). Bose-Einstein condensation of erbium, *Phys. Rev. Lett.* **108**, 210401. doi: 10.1103/PhysRevLett.108.210401.

182. Connolly, C. B., Au, Y. S., Doret, S. C., Ketterle, W., and Doyle, J. M. (2010). Large spin relaxation rates in trapped submerged-shell atoms, *Phys. Rev. A.* **81**, 010702. doi: 10.1103/PhysRevA.81.010702.

183. Griesmaier, A., Werner, J., Hensler, S., Stuhler, J., and Pfau, T. (2005). Bose-Einstein condensation of chromium, *Phys. Rev. Lett.* **94**, 160401. doi: 10.1103/PhysRevLett. 94.160401.

184. Hancox, C. I., Doret, S. C., Hummon, M. T., Luo, L. and Doyle, J. M. (2004). Magnetic trapping of rare-earth atoms at millikelvin temperatures, *Nature.* **431**, 7006, 281–284. doi: 10.1038/nature02938.

185. Koch, T., Lahaye, T., Metz, J., Frhlich, B., Griesmaier, A., and Pfau, T. (2008). Stabilization of a purely dipolar quantum gas against collapse, *Nature Phys.* **4**, 3, 218–222. doi: 10.1038/nphys887.

186. Lahaye, T., Koch, T., Fröhlich, B., Fattori, M., Metz, J., Griesmaier, A., Giovanazzi, S., and Pfau, T. (2007). Strong dipolar effects in a quantum ferrofluid, *Nature.* **448**, 7154, 672–675. doi: 10.1038/nature06036.

187. Lahaye, T., Metz, J., Fröhlich, B., Koch, T., Meister, M., Griesmaier, A., Pfau, T., Saito, H., Kawaguchi, Y., and Ueda, M. (2008). *d*-wave collapse and explosion of a dipolar bose-einstein condensate, *Phys. Rev. Lett.* **101**, 080401. doi: 10.1103/PhysRevLett. 101.080401.

188. Lu, M., Burdick, N. Q., and Lev, B. L. (2012). Quantum degenerate dipolar Fermi gas, *Phys. Rev. Lett.* **108**, 215301. doi: 10.1103/PhysRevLett.108.215301.

189. Lu, M., Burdick, N. Q., Youn, S. H., and Lev, B. L. (2011). Strongly dipolar bose-einstein condensate of dysprosium, *Phys. Rev. Lett.* **107**, 190401. doi: 10.1103/ PhysRevLett.107.190401.

190. Stuhler, J., Griesmaier, A., Koch, T., Fattori, M., Pfau, T., Giovanazzi, S., Pedri, P., and Santos, L. (2005). Observation of dipole-dipole interaction in a degenerate quantum gas, *Phys. Rev. Lett.* **95**, 150406. doi: 10.1103/PhysRevLett.95.150406.

191. Werner, J., Griesmaier, A., Hensler, S., Stuhler, J., Pfau, T., Simoni, A., and Tiesinga, E. (2005). Observation of Feshbach resonances in an ultracold gas of ^{52}Cr, *Phys. Rev. Lett.* **94**, 183201. doi: 10.1103/PhysRevLett.94.183201.

192. Roudnev, V. and Cavagnero, M. (2009). Resonance phenomena in ultracold dipole-dipolescattering, *J. Phys. B.* **42**, 4, 044017.

193. Deb, B. and You, L. (2001). Low-energy atomic collision with dipole interactions, *Phys. Rev. A.* **64**, 022717. doi: 10.1103/PhysRevA.64.022717.

194. Roudnev, V. and Cavagnero, M. (2009). Universal resonant ultracold molecular scattering, *Phys. Rev. A.* **79**, 014701. doi: 10.1103/PhysRevA.79.014701.
195. Ticknor, C. (2007). Energy dependence of scattering ground-state polar molecules, *Phys. Rev. A.* **76**, 052703. doi: 10.1103/PhysRevA.76.052703.
196. Ticknor, C. and Bohn, J. L. (2005). Long-range scattering resonances in strong-field-seeking states of polar molecules, *Phys. Rev. A.* **72**, 032717. doi: 10.1103/PhysRevA. 72.032717.
197. Beane, S. R., Bedaque, P. F., Childress, L., Kryjevski, A., McGuire, J., and van Kolck, U. (2001). Singular potentials and limit cycles, *Phys. Rev. A.* **64**, 042103. doi: 10.1103/PhysRevA.64.042103.
198. Kanjilal, K. and Blume, D. (2006). Coupled-channel pseudopotential description of the Feshbach resonance in two dimensions, *Phys. Rev. A.* **73**, 060701. doi: 10.1103/ PhysRevA.73.060701.
199. Shih S.-M. and Wang, D.-W. (2009). Pseudopotential of an interaction with a power-law decay in two-dimensional systems, *Phys. Rev. A.* **79**, 065603. doi: 10.1103/ PhysRevA.79.065603.
200. Derevianko, A. (2003). Anisotropic pseudopotential for polarized dilute quantum gases, *Phys. Rev. A.* **67**, 033607. doi: 10.1103/PhysRevA.67.033607.
201. Derevianko, A. (2005). Erratum: Anisotropic pseudopotential for polarized dilute quantum gases [Phys. Rev. A 67, 033607 (2003)], *Phys. Rev. A.* **72**, 039901. doi: 10.1103/PhysRevA.72.039901.
202. Yi, S. and You, L. (2000). Trapped atomic condensates with anisotropic interactions, *Phys. Rev. A.* **61**, 041604. doi: 10.1103/PhysRevA.61.041604.
203. Yi, S. and You, L. (2001). Trapped condensates of atoms with dipole interactions, *Phys. Rev. A.* **63**, 053607. doi: 10.1103/PhysRevA.63.053607.
204. Ticknor, C. (2009). Two-dimensional dipolar scattering, *Phys. Rev. A.* **80**, 052702. doi: 10.1103/PhysRevA.80.052702.
205. Ticknor, C. (2010). Quasi-two-dimensional dipolar scattering, *Phys. Rev. A.* **81**, 042708. doi: 10.1103/PhysRevA.81.042708.
206. Armstrong, J. R., Zinner, N. T., Fedorov, D. V., and Jensen, A. S. (2010). Bound states and universality in layers of cold polar molecules, *Europhys. Lett.* **91**, 1, 16001, (2010).
207. Volosniev, A. G., Zinner, N. T., Fedorov, D. V., Jensen, A. S., and Wunsch, B. (2011). Bound dimers in bilayers of cold polar molecules, *J. Phys. B.* **44**, 12, 125301.
208. Ticknor, C. (2012). Finite-temperature analysis of a quasi-two-dimensional dipolar gas, *Phys. Rev. A.* **85**, 033629. doi: 10.1103/PhysRevA.85.033629.
209. Zinner, N. T., Armstrong, J. R., Volosniev, A. G., Fedorov, D. V., and Jensen, A. S. (2012). Dimers, effective interactions, and Pauli blocking effects in a bilayer of cold fermionic polar molecules, *Few-Body Sys.* **53**, 3–4, 369–385. ISSN 0177-7963. doi: 10.1007/s00601-012-0304-7.
210. Koval, E. A., Koval, O. A., and Melezhik, V. S. (2014). Anisotropic quantum scattering in two dimensions, *Phys. Rev. A.* **89**, 052710. doi: 10.1103/PhysRevA.89.052710.
211. Ticknor, C. and Rittenhouse, S. T. (2010). Three body recombination of ultracold dipoles to weakly bound dimers, *Phys. Rev. Lett.* **105**, 013201. doi: 10.1103/ PhysRevLett.105.013201.
212. Chotia, A., Neyenhuis, B., Moses, S. A., Yan, B., Covey, J. P., Foss-Feig, M., Rey, A. M., Jin, D. S., and Ye, J. (2012). Long-lived dipolar molecules and Feshbach molecules

in a 3D optical lattice, *Phys. Rev. Lett.* **108**, 080405. doi: 10.1103/PhysRevLett.108. 080405.

213. Danzl, J. G., Mark, M. J., Haller, E., Gustavsson, M., Hart, R., Aldegunde, J., Hutson, J. M., and Nägerl, H.-C. (2010). An ultracold high-density sample of rovibronic ground-state molecules in an optical lattice, *Nature Phys.* **6**, 4, 265–270. doi: 10.1038/nphys1533.

214. Danzl, J. G., Mark, M. J., Haller, E., Gustavsson, M., Hart, R., Liem, A., Zellmer, H., and Nägerl, H.-C. (2009). Deeply bound ultracold molecules in an optical lattice, *New J. Phys.* **11**, 5, 055036.

215. Hummon, M. T., Yeo, M., Stuhl, B. K., Collopy, A. L., Xia, Y., and Ye, J. (2013). 2D magneto-optical trapping of diatomic molecules, *Phys. Rev. Lett.* **110**, 143001. doi: 10.1103/PhysRevLett.110.143001.

216. Takekoshi, T., Reichsöllner, L., Schindewolf, A., Hutson, J. M., Le Sueur, C. R. Dulieu, O., Ferlaino, F., Grimm, R., and Nägerl, H.-C. (2014). Ultracold dense samples of dipolar RbCs molecules in the rovibrational and hyperfine ground state, *Phys. Rev. Lett.* **113**, 205301. doi: 10.1103/PhysRevLett.113.205301.

217. Yan, B., Moses, S. A., Gadway, B., Covey, J. P., Hazzard, K. R. A., Rey, A. M., Jin, D. S. and Ye, J., (2013). Observation of dipolar spin-exchange interactions with lattice-confined polar molecules, *Nature.* **501**, 7468, 521–525. doi: 10.1038/nature12483.

218. Esry, B. D., Greene, C. H., and Suno, H. (2001). Threshold laws for three-body recombination, *Phys. Rev. A.* **65**, 010705. doi: 10.1103/PhysRevA.65.010705.

219. Lang, F., Straten, P. v. d., Brandstätter, B., Thalhammer, G., Winkler, K., Julienne, P. S., Grimm, R. and Denschlag, J. H. (2008). Cruising through molecular bound-state manifolds with radiofrequency, *Nature Phys.* **4**, 3, 223–226. doi: 10.1038/nphys838.

220. Mark, M., Ferlaino, F., Knoop, S., Danzl, J. G., Kraemer, T., Chin, C., Nägerl, H.-C. and Grimm, R. (2007). Spectroscopy of ultracold trapped cesium Feshbach molecules, *Phys. Rev. A.* **76**, 042514. doi: 10.1103/PhysRevA.76.042514.

221. Zenesini, A., Huang, B., Berninger, M., Nägerl, H.-C., Ferlaino, F., and Grimm, R. (2014). Resonant atom-dimer collisions in cesium: Testing universality at positive scattering lengths, *Phys. Rev. A.* **90**, 022704. doi: 10.1103/PhysRevA.90.022704.

222. Lompe, T., Ottenstein, T. B., Serwane, F., Wenz, A. N., Zürn, G., and Jochim, S. (2010). Radio-frequency association of Efimov trimers, *Science.* **330**, 6006, 940–944. doi: 10.1126/science.1193148.

223. Machtey, O., Shotan, Z., Gross, N., and Khaykovich, L. (2012). Association of Efimov trimers from a three-atom continuum, *Phys. Rev. Lett.* **108**, 210406. doi: 10.1103/PhysRevLett.108.210406.

224. Mavadia, S., Goodwin, J. F., Stutter, G., Bharadia, S., Crick, D. R., Segal, D. M., and Thompson, R. C. (2013). Control of the conformations of ion coulomb crystals in a penning trap, *Nat. Commun.* **4**, 2571. doi: 10.1038/ncomms3571.

225. Xu, J., Le, A.-T., Morishita, T., and Lin, C. D. (2008). Signature of Ericson fluctuations in helium inelastic scattering cross sections near the double ionization threshold, *Phys. Rev. A.* **78**, 012701. doi: 10.1103/PhysRevA.78.012701.

226. Macek, J. H. and Ovchinnikov, S. Y. (1996). Hyperspherical theory of three-particle fragmentation and wannier's threshold law, *Phys. Rev. A.* **54**, 544–560. doi: 10.1103/PhysRevA.54.544.

227. Macek, J. H., Ovchinnikov, S. Y., and Pasovets, S. V. (1995). Hidden crossing theory of threshold ionization of atoms by electron impact, *Phys. Rev. Lett.* **74**, 4631–4634. doi: 10.1103/PhysRevLett.74.4631.

228. Wannier, G. H. (1953). The threshold law for single ionization of atoms or ions by electrons, *Phys. Rev.* **90**, 817–825. doi: 10.1103/PhysRev.90.817.

229. Mansbach, P. and Keck, J. (1969). Monte Carlo trajectory calculations of atomic excitation and ionization by thermal electrons, *Phys. Rev.* **181**, 275–289. doi: 10.1103/PhysRev.181.275.

230. Pohl, T., Vrinceanu, D., and Sadeghpour, H. R. (2008). Rydberg atom formation in ultracold plasmas: Small energy transfer with large consequences, *Phys. Rev. Lett.* **100**, 223201. doi: 10.1103/PhysRevLett.100.223201.

231. Guevara, N. L., Wang, Y. and Esry, B. D. (2012). New class of three-body states, *Phys. Rev. Lett.* **108**, 213202. doi: 10.1103/PhysRevLett.108.213202.

232. Li, M., You, L., and Gao, B. (2014). Multichannel quantum-defect theory for ion-atom interactions, *Phys. Rev. A.* **89**, 052704. doi: 10.1103/PhysRevA.89.052704.

233. Härter, A., Krükow, A., Brunner, A., Schnitzler, W., Schmid, S., and Denschlag, J. H. (2012). Single ion as a three-body reaction center in an ultracold atomic gas, *Phys. Rev. Lett.* **109**, 123201. doi: 10.1103/PhysRevLett.109.123201.

234. Hasan, M. Z. and Kane, C. L. (2010). *Colloquium*: Topological insulators, *Rev. Mod. Phys.* **82**, 3045–3067. doi: 10.1103/RevModPhys.82.3045.

235. Galitski, V. and Spielman, I. B. (2013). Spin-orbit coupling in quantum gases, *Nature.* **494**, 7435, 49–54. doi: 10.1038/nature11841.

236. Vyasanakere, J. P. and Shenoy, V. B. (2011). Bound states of two spin-$\frac{1}{2}$ fermions in a synthetic non-abelian gauge field, *Phys. Rev. B.* **83**, 094515. doi: 10.1103/PhysRevB.83.094515.

237. Cui, X. (2012). Mixed-partial-wave scattering with spin-orbit coupling and validity of pseudopotentials, *Phys. Rev. A.* **85**, 022705. doi: 10.1103/PhysRevA.85.022705.

238. Duan, H., You, L., and Gao, B. (2013). Ultracold collisions in the presence of synthetic spin-orbit coupling, *Phys. Rev. A.* **87**, 052708. doi: 10.1103/PhysRevA.87.052708.

239. Wu, Y. and Yu, Z. (2013). Short-range asymptotic behavior of the wave functions of interacting spin-1/2 fermionic atoms with spin-orbit coupling: A model study, *Phys. Rev. A.* **87**, 032703. doi: 10.1103/PhysRevA.87.032703.

240. Zhang, L., Deng, Y., and Zhang, P. (2013). Scattering and effective interactions of ultracold atoms with spin-orbit coupling, *Phys. Rev. A.* **87**, 053626. doi: 10.1103/PhysRevA.87.053626.

241. Zhang, P., Zhang, L., and Deng, Y. (2012). Modified bethe-peierls boundary condition for ultracold atoms with spin-orbit coupling, *Phys. Rev. A.* **86**, 053608. doi: 10.1103/PhysRevA.86.053608.

242. Zhang, P., Zhang, L., and Zhang, W. (2012). Interatomic collisions in two-dimensional and quasi-two-dimensional confinements with spin-orbit coupling, *Phys. Rev. A.* **86**, 042707. doi: 10.1103/PhysRevA.86.042707.

243. Williams, R. A., LeBlanc, L. J., Jiménez-García, K., Beeler, M. C., Perry, A. R., Phillips, W. D., and Spielman, I. B. (2012). Synthetic partial waves in ultracold atomic collisions, *Science.* **335**, 6066, 314–317. doi: 10.1126/science.1212652.

244. Fu, Z., Huang, L., Meng, Z., Wang, P., Zhang, L., Zhang, S., Zhai, H., Zhang, P., and Zhang, J. (2014). Production of Feshbach molecules induced by spin-orbit coupling in fermi gases, *Nature Phys.* **10**, 2, 110–115. doi: 10.1038/nphys2824.

245. Shi, Z.-Y., Cui, X., and Zhai, H. (2014). Universal trimers induced by spin-orbit coupling in ultracold fermi gases, *Phys. Rev. Lett.* **112**, 013201. doi: 10.1103/PhysRevLett.112.013201.

246. Bethe, H. A. (1949). Theory of the effective range in nuclear scattering, *Phys. Rev.* **76**, 38–50. doi: 10.1103/PhysRev.76.38.

247. Fermi, E. and Marshall, L. (1947). Interference phenomena of slow neutrons, *Phys. Rev.* **71**, 666–677. doi: 10.1103/PhysRev.71.666.

CHAPTER 3

SPIN-ORBIT COUPLING IN OPTICAL LATTICES

Shizhong Zhang*, William S. Cole[†], Arun Paramekanti[‡],
and Nandini Trivedi[§]

*Department of Physics and Center of Theoretical and
Computational Physics, The University of Hong Kong,
Hong Kong, China

[†],[§]Department of Physics, The Ohio State University, Columbus,
Ohio, 43210, USA

[‡]Department of Physics, University of Toronto, Toronto M5S1A7,
and Canadian Institute for Advanced Research, Toronto,
Ontario, M5G 1Z8, Canada

In this review, we discuss the physics of spin-orbit coupled quantum gases
in optical lattices. After reviewing some relevant experimental techniques,
we introduce the basic theoretical model and discuss some of its generic
features. In particular, we concentrate on the interplay between spin-orbit
coupling and strong interactions and show how it leads to various exotic
quantum phases in both the Mott insulating and superfluid regimes. Phase
transitions between the Mott and superfluid states are also discussed.

1. Introduction

Cold atom experiments are performed with charge-neutral atoms.[1,2]
At first sight, this would have precluded the effects of orbital magnetism,
as well as spin-orbit coupling to be studied in these cold atomic gases.
However, in the past few years, by using atom-light coupling (Raman lasers
and shaking optical lattice), it has become possible to simulate these effects

in neutral atomic samples. This provides cold atom experimentalists with the exciting opportunity to produce and investigate several paradigmatic quantum states such as the quantum Hall liquids, topological insulators and superfluids, Dirac and Weyl semimetals, as well as many other exotic phenomena that have recently been predicted for electron systems in external magnetic fields or with strong spin-orbit interaction.[3–5] What is perhaps more interesting is that this capacity would open an entire new vista for the investigation of bosonic topological states that have so far only been subjected to theoretical studies.[6–8] Indeed, the great tunability of cold atom systems provides an avenue to the realization of conceptually important models that may not have a natural correspondence to a condensed matter system,[9,10] as was beautifully demonstrated in the recent implementation of Haldane's honeycomb model of a Chern insulator[11] and measurement of Chern number in a Hofstadter band,[12] in optical lattices, using shaking and Raman technique, respectively.

As a result of this fundamental interest, there has been tremendous effort in realizing synthetic magnetic flux and spin-orbit coupling in neutral atomic gases over the past few years; for recent reviews, see Refs. 13–16. Many schemes have been proposed and several of them have been implemented in experiments. At present, there are two schemes which have accumulated the most substantial experimental success: the Raman scheme[17–21] and shaken optical lattices.[11,22–24] The former has been utilized to produce both synthetic magnetic flux and spin-orbit coupling, while the later has so far only been used to produce synthetic magnetic flux. However, several proposals exist in the literature using shaking optical lattices to produce spin-orbit coupling.[25–28] In this review, we shall concentrate our attention on the experimentally implemented schemes. In Table 1, we give a summary of the experiments conducted so far on spin-orbit coupled quantum gases using Raman scheme.

For a general overview of the subject, we refer readers to Ref. 15. For a more complete introduction to the subject and in particular, on the experimental techniques, see Refs. 13, 27 and 37. Results related to spin-orbit coupled quantum gases in a harmonic trap are reviewed in Ref. 16, which concentrates mostly on the weakly interacting regime. In this review, we focus rather on the interplay between the effects of spin-orbit coupling and strong interactions. In the case of bosons, perhaps the simplest route to this regime is to load bosons in an deep optical lattice with spin-orbit

Table 1. Experiments on spin-orbit coupled quantum gases.

Group	Element	Phenomena	Comments
NIST	^{87}Rb	Structure of BEC;[17] partial wave scattering;[29] spin hall effect;[30] Zitterbewegung[31]	Harmonic trap
USTC	^{87}Rb	Dipole oscillation;[20] finite temperature phase diagram;[32] collective excitations[33]	Harmonic trap
Shanxi	^{40}K	ARPES, Fermi surface transition;[19] spin-orbit coupled molecule[34]	fermion; Harmonic trap
MIT	^{6}Li	Inverse ARPES, Zeeman Lattice[18]	Fermion
Purdue	^{87}Rb	Landau-Zener transitions[35]	Harmonic trap
WSU	^{87}Rb	Dynamical instability of spin-orbit BEC,[21] collective excitations[36]	Moving optical lattice

coupling generated either by the Raman scheme or by shaking, as we shall review below in Sec. 2. In Sec. 3 we then move on to review the basic theoretical models that describe spin-orbit coupled bosons in the optical lattice and discuss the resulting band structure in Sec. 3.1. In Sec. 3.2 we discuss some general themes that emerge from the interplay between spin-orbit coupling and strong interactions, specifically identifying a few of the more novel aspects. Finally, in Sec. 4, we conclude our review and offer some perspectives on the subject.

2. Experimental Realizations

So far, several ways of generating spin-orbit coupling in optical lattices have been proposed and, in some cases, implemented. Building on the Raman scheme which realizes the effect of spin-orbit coupling in the uniform system, an extra pair of lattice beams can be added. This has been achieved in a recent experiment[21] with a moving optical lattice. Other schemes for generating spin-orbit coupling include Raman-assisted tunneling[38] and shaking the optical lattice.[25–28] A particularly interesting idea is the so-called "Zeeman" lattice, in which the spin-orbit coupling and optical lattice are generated at the same time by a combination of radio-frequency beams and Raman beams.[18,39] We shall discuss each of these techniques in turn.

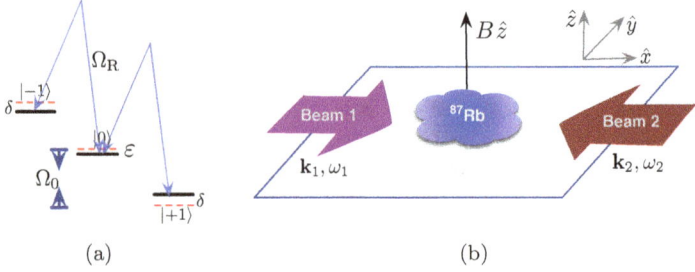

(a)   (b)

Fig. 1. Schematic setup of the NIST experiment. (a) The level diagram of ^{87}Rb in its ground state $F = 1$ manifold. The linear Zeeman splitting is given by $\hbar\Omega_0$. $\delta = \omega - \Omega_0$ is the detuning of the beams from the Raman resonance. Ω_R is the two-photon Rabi frequency. ε-term is the quadratic Zeeman effect which shifts the $F_z = 0$ state downwards by amount $\hbar\varepsilon$. (b) Two counter-propagating laser beams impinge on a cloud of ^{87}Rb atoms along the $\pm\hat{x}$-axis. An external magnetic field is applied along the $\hat{z}$-direction. In the resulting adiabatic states, the atoms behave as charged particles in an external gauge field. Figure adapted from Ref. 42.

2.1. *Raman scheme with an optical lattice*

As a first step, we outline the Raman scheme for generating spin-orbit coupling in a uniform system.[40,41] As shown in Fig. 1, a pair of Raman beams with frequencies $\omega_{1,2}$, wave vectors $\mathbf{k}_{1,2}$ and polarization $\hat{\lambda}_{1,2}$ are applied to the atomic ^{87}Rb vapor. An external magnetic field $\mathbf{B}$ is applied along the $\hat{z}$-axis and sets the quantization axis of the hyperfine spin $\mathbf{F}$. The Raman lasers transfer momentum $2\mathbf{q} \equiv \mathbf{k}_1 - \mathbf{k}_2$, which we take to be along the $\hat{x}$-direction, to the atom and at the same time flip its spin, depending on the polarizations of the two laser beams.

The single particle Hamiltonian has the form,

$$H(t) = \frac{\mathbf{p}^2}{2m} - \hbar\Omega_0 F_z + \hbar\varepsilon F_z^2 - \frac{\hbar\Omega_R}{2}[e^{i(2qx-\omega t)}(F_x + iF_y) + \text{H.c.}], \quad (1)$$

where Ω_0 is the Larmor frequency associated with the uniform external magnetic field along $\hat{z}$ direction and m is the mass of the atom under consideration. The ε-term arises from the quadratic Zeeman effect which shift the $F_z = 0$ state downwards by amount $\hbar\varepsilon$ (we have neglect a constant term $-\hbar\varepsilon$ in the Hamiltonian). The Rabi frequency Ω_R describes the coupling between different spin states due to the laser fields and is referred to as the two-photon Rabi frequency. The momentum and energy transfer between the atom and laser field are given by $2\mathbf{q} = \mathbf{k}_1 - \mathbf{k}_2 \equiv 2q_x\hat{x}$

(say along $\hat{x}$-direction) and $\omega = \omega_1 - \omega_2$. The explicit time-dependence of $H(t)$ can be eliminated by performing a unitary transformation $U(t) = \exp(i\omega t F_z)$, then $\tilde{H} = U^\dagger H(t) U$ is time-independent

$$\tilde{H} = \frac{\mathbf{p}^2}{2m} + \exp(-i2qx F_z)\left[-\hbar(\Omega_0 - \omega)F_z + \hbar\varepsilon F_z^2 - \hbar\Omega_R F_x\right]$$
$$\exp(i2qx F_z) \quad (2)$$

This transformation describes the spins spiraling around the $\hat{z}$-axis with a period π/q. It is then possible to eliminate the spatial dependence $\exp(-i2qx F_z)$ by a similar unitary transformation on the Hamiltonian $\tilde{H}$ and one then ends up with a spin-orbit coupled Hamiltonian of the form[17,42]

$$H_{\text{so}} = \frac{(\mathbf{p} + q F_z \hat{x})^2}{2m} - \hbar(\Omega_0 - \omega)F_z + \hbar\varepsilon F_z^2 - \hbar\Omega_R F_x. \quad (3)$$

In the ground state manifold of ^{87}Rb atoms with $F = 1$, depending on the choice of various parameters, this apparently simple Hamiltonian contains both the abelian synthetic gauge field and spin-orbit coupling as limiting cases.

(1) When $\Omega_R \gg \varepsilon, q^2/2m$ and $\omega \approx \Omega_0$, the single lowest spin state is given by $F_x = -1$. In this case, one can project the Hamiltonian to this single state, and with a magnetic field gradient, one realizes the traditional $U(1)$ abelian gauge field. The formation of superfluid vortices has indeed been observed in experiment[43] in this regime.

(2) When $\varepsilon \gg \Omega_R, q^2/2m$ and $\omega \approx \Omega_0 - \varepsilon$ the two states with $F_z = 0$ and $F_z = 1$ are nearly degenerate and upon projecting the Hamiltonian to the space spanned by these two spin states, which we shall denote as (pseudo-spin) σ, a spin-orbit coupling is realized, with the Hamiltonian taking on the following form

$$H_{\text{so}} = \frac{(p_x + q\sigma_z)^2}{2m} + \frac{\delta}{2}\sigma_z + \frac{\Omega_R}{2}\sigma_x, \quad (4)$$

where $\delta = \omega - \Omega_0$ is the detuning from Raman resonance.

Now let us introduce an one-dimensional optical lattice with optical potential given by $V(x) = s E_R \sin^2(Kx)$, where K is the wave vector of the optical lattices and $E_R \equiv \hbar^2 K^2/2m$ is the recoil energy and s characterizes the depth of the potential. The single particle Hamiltonian becomes

$$H_0 = \frac{(p_x + q\sigma_z)^2}{2m} + \frac{\delta}{2}\sigma_z + \frac{\Omega_R}{2}\sigma_x + sE_R \sin^2(Kx). \tag{5}$$

While the discussion below will be for one-dimensional case, it is straightforward to generalize it to higher dimensions.

It is helpful to construct an appropriate tight-binding (TB) model to describe this system in the limit of deep optical lattices $s \gg 1$. To do this, in principle one should solve for the band spectrum of the Hamiltonian in Eq. (5) and then construct the appropriate Wannier states. When the spin-orbit coupling is weak, there will be two nearly degenerate bands (corresponding to, roughly, the two spin components) with a large band gap $\Delta \propto \sqrt{s}$ to all other higher bands. As a result, we can concentrate on the lowest two nearly degenerate bands. Thus, we expect to find a TB model with two spin-resolved orbitals associated with every lattice site $W_i^I(\mathbf{r})$ and $W_i^{II}(\mathbf{r})$, each in general a superposition of spin-up and spin-down states. The appropriate hopping constant can be calculated as

$$T_{ij}^{\tau\tau'} = \langle W_i^\tau(\mathbf{r})|H_0|W_j^{\tau'}(\mathbf{r})\rangle; \quad \tau, \tau' = I, II, \tag{6}$$

where τ and τ' label the two Wannier states. Writing this in terms of the original spin degree of freedom, one obtains the appropriate hopping Hamiltonian. While this work may be necessary for detailed quantitative comparisons between theory and experiments, we can reason from very general considerations to determine the general structure of the hopping model. In particular, by interpreting the spin-orbit coupling itself as a gauge field which enters the Hamiltonian in the minimal coupling form, we can invoke the Peierls' substitution to provide the correct lattice model. Thus, we can write the hopping matrix along $\hat{x}$-direction as[44–46]

$$T_{\hat{x}} = t_{\hat{x}} \exp(-i\alpha A_{\hat{x}}) = t_{\hat{x}} \exp\left[i\frac{\pi q_x}{K}\sigma_z\right], \tag{7}$$

where we have used the fact that lattice constant $a = \pi/K$ and the gauge field along $\hat{x}$-direction is given by $A_{\hat{x}} = -q_x\sigma_z$. $t_{\hat{x}}$ is the hopping parameter in the absence of spin-orbit coupling. Since one can arrange the direction of momentum transfer $2\mathbf{q}$ to be different from the lattice direction, a similar term can be generated along $\hat{y}$-direction. Note that the coupling term would be $p_y\sigma_z$ instead of the isotropic Rashba form $p_x\sigma_y - p_y\sigma_x$. The strength of the spin-orbit coupling can be tuned by changing the value of $2\mathbf{q}$. In general, microscopic calculations with the correct Wannier states find that spin-orbit coupling modifies *both* $t_{\hat{x}}$ and the phases factors, but the general structure (σ_z dependences) will be left un-modified as long as the two-band approximation remains appropriate. A careful comparison of the Peierls' substitution with a numerical calculation of the Wannier functions and hopping matrix elements was carried out in Ref. 46. We note in passing that in the experiments conducted so far, the momentum transfer is typically comparable to the lattice wavevector K and are in the regime where Peierls' substitution begins to become quantitatively inaccurate.

2.2. *Laser assisted tunneling*

The physics of laser-assisted hopping can be illustrated most easily with a double-well potential.[47,48] Consider an atom with two internal states (i.e., spin-$1/2$) in a spin-independent double-well potential. For simplicity, let us further assume that around each minimum of the double-well, the oscillator frequencies are identical and are given by ω_0. The associated localized wave functions are given by $\varphi_{\mathbf{R}_i}(\mathbf{r})$ and $\varphi_{\mathbf{R}_j}(\mathbf{r})$, where $\mathbf{R}_i$ and $\mathbf{R}_j$ label the two wells along $\hat{x}$-direction with $a \equiv |\mathbf{R}_i - \mathbf{R}_j|$. Now, in the absence of a tilting potential, $\Delta = 0$, the normal tunneling between these two sites is diagonal in spin space and is given by J_0; its value depends on the overlap of the two Wannier wave functions $\varphi_{\mathbf{R}_i}(\mathbf{r})$ and $\varphi_{\mathbf{R}_j}(\mathbf{r})$. When $\Delta \neq 0$, normal tunneling between the two sites $\mathbf{R}_i$ and $\mathbf{R}_j$ is suppressed due to the energy mismatch. To restore hopping, a pair of far-off resonant laser beams are applied with wave vectors $\mathbf{k}_1$ and $\mathbf{k}_2$, frequencies ω_1 and ω_2. This induces a coupling term of the following form,

$$V_{\text{laser}} = \hbar\Omega_R \left[\exp(i2\mathbf{q} \cdot \mathbf{r} - i\omega t)\hat{S} + \exp(-i2\mathbf{q} \cdot \mathbf{r} + i\omega t)\hat{S}^\dagger \right], \quad (8)$$

where Ω_R is the two-photon Rabi frequency and recall that $2\mathbf{q} = \mathbf{k}_1 - \mathbf{k}_2 \equiv 2q_x\hat{x}$ and $\omega = \omega_1 - \omega_2$ are the momentum and energy transfer between the atom and the laser field. $\hat{S}$ describes the action of the two laser beams on the spin states of the atom and depends on the polarization of the two laser beams. If $\hat{S}$ is diagonal in spin space, the result is two decoupled optical lattices for each spin component. Each copy can feature a complex hopping amplitude, which in general can give rise to abelian gauge fields. On the

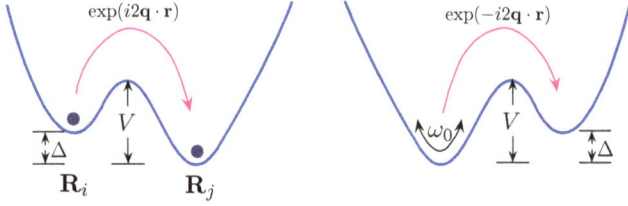

Fig. 2. Laser assisted hopping between two wells with energy offset Δ and a potential barrier V. The harmonic frequency at the bottom of the well is ω_0. When an atom from the higher energy well at $\mathbf{R}_i$ tunnels to the lower energy well at $\mathbf{R}_j$, an energy of Δ must be absorbed by the radiation field with a spatially-varying phase factor $\exp(i2\mathbf{q}\cdot\mathbf{r})$. The hopping from $\mathbf{R}_j$ to $\mathbf{R}_i$, on the other hand, will be associated with a phase factor $\exp(-i2\mathbf{q}\cdot\mathbf{r})$.

other hand, if $\hat{S}$ is non-diagonal in spin space, this may be used to realize a non-abelian gauge field in an optical lattice, of which spin-orbit coupling is a special case. It is worthwhile to point out that a non-diagonal $\hat{S}$ relies on the internal atomic spin-orbit coupling, and this can lead to siginificant spontaneous emission in alkali atoms.[38]

Resonant hopping can be restored when $\hbar\omega = \Delta$. In the dressed atom picture, this means that an atom initially residing at the higher potential $\mathbf{R}_i$ with n_1-photon in mode $(\omega_1, \mathbf{k}_1)$ and n_2-photon in mode $(\omega_2, \mathbf{k}_2)$, is resonant with a state in which the atom is at the lower potential $\mathbf{R}_j$ with $(n_1 + 1)$-photon in mode $(\omega_1, \mathbf{k}_1)$ and $(n_2 - 1)$-photon in mode $(\omega_2, \mathbf{k}_2)$. Crucially, the spatial phase associated with hopping from $\mathbf{R}_i$ to $\mathbf{R}_j$ is given by $\exp(i2\mathbf{q}\cdot\mathbf{r})$, while that from $\mathbf{R}_j$ to $\mathbf{R}_i$ is given by $\exp(-i2\mathbf{q}\cdot\mathbf{r})$. Thus the laser beams imprint a complex Peierls' phase during the hopping process. It can be shown that the effective hopping amplitude as modified by the laser beams is given by

$$J_{\text{eff}} \exp(-i\mathbf{q} \cdot (\mathbf{R}_i + \mathbf{R}_j)), \tag{9}$$

where $J_{\text{eff}} = J_0 \mathcal{J}_1(\kappa)$, where $\kappa = 4\Omega \sin(q_x a)/\Delta$, depending on the details of the laser arrangements and the distance between the neighboring sites. $\mathcal{J}_1(\kappa)$ is the first order Bessel function.

The idea of Raman assisted tunnelling in optical lattices is proposed in Ref. 49 and later extended in Ref. 50. Within the Raman scheme, two types of magnetic flux patterns have been realized to date. In the first experiment from Munich, a staggered magnetic flux along one-direction was realized with a superlattice potential.[51,52] The magnetic flux per plaquette can be tuned easily by changing the angle between the two laser beams. In later experiments from both Munich[53] and MIT,[54] uniform flux is realized with a linear potential generated by the magnetic field gradient. Extension of the Raman scheme to create spin-orbit coupling is discussed in Ref. 38.

2.3. *Periodically driven lattice*

Another way of generating artificial magnetic fields and spin-orbit coupling is to use periodically driven systems.[27,55] In this scheme, a time-periodic Hamiltonian is considered $\hat{H}(t) = \hat{H}(t + T)$. The time evolution of the system is described by the evolution operator $\hat{U}(t) = \mathcal{T} \exp[-i \int_0^t \hat{\mathcal{H}}(t')dt']$. Because of the periodicity of the problem, one looks for the evolution operator over a period $\hat{U}(T)$ and defines an effective Hamiltonian

$$\hat{U}(T) = \mathcal{T} \exp[-i \int_0^T \hat{\mathcal{H}}(t')dt'] \equiv \exp[-iT\hat{H}_{\text{eff}}]. \tag{10}$$

The form of $\hat{H}_{\text{eff}}$ can be very complicated and no closed form exists in general. However, provided that the modulation frequency $\omega = 2\pi/T$ is large compare with typical energy scales in the problem and the modulation amplitude is small, it is possible to develop a formal expansion in $1/\omega$. Let us write $\hat{H}_{\text{eff}}$ as a Fourier series

$$\hat{H}_{\text{eff}} = \sum_{n=-\infty}^{\infty} \hat{H}_n \exp[in\omega t], \tag{11}$$

then the effective Hamiltonian up to first order in $1/\omega$ is given by[11,27]

$$\hat{H}_{\text{eff}} = \hat{H}_0 + \frac{1}{\omega} \sum_{n=1}^{\infty} \frac{1}{n} [\hat{H}_n, \hat{H}_{-n}]. \tag{12}$$

We note that the zeroth order term is just the time average of the Hamiltonian over a period T. Typically, this modulation is applied in combination with an optical lattice ("shaking lattice") and the time dependence enters into the Hamiltonian by coupling to a term of the form

$$\sum_{\sigma\sigma'\mathbf{r}} \hat{H}_{\text{mod}}(\mathbf{r}, t) a_{\sigma\mathbf{r}}^\dagger a_{\sigma'\mathbf{r}}, \tag{13}$$

where the modulation coupling $\hat{H}_{mod}(\mathbf{r}, t)$ can be spatially varying and, furthermore, can be a matrix in spin space. A few examples that have been realized in recent experiments are given in Table 2. Extensions of the shaking scheme to generate spin-orbit coupling are discussed in Refs. 25–28.

2.4. Zeeman lattice

The concept of a "Zeeman lattice" was introduced in Ref. 39, where a combination of Raman and radio-frequency laser beams produce an effective magnetic field that varies periodically both in its *magnitude* and *direction*. This is related to the more general idea of optical flux lattices, introduced in Ref. 57. As before, consider ^{87}Rb atoms with the $F = 1$ ground state split due to a Zeeman field, as shown in Fig. 1. In addition to the Raman beams, one adds an extra radio-frequency (RF) beam which drive direct transitions between hyperfine-Zeeman levels. The coupling strength and frequency of the RF field is given by Ω_{rf} and ω_{rf}, respectively. One also defines the detuning as $\delta = \omega_{rf} - \Omega_0$. It is necessary that the energy transfer from the Raman beams be the same as the RF beam, in order that one can go to a common rotating frame. As a result, the effective Hamiltonian is given by

$$\hat{H}_{rf+Raman} = \frac{p^2}{2m}\hat{I} + \mathbf{\Omega}(x) \cdot \hat{\mathbf{F}} + \hat{H}_Q, \qquad (14)$$

where $\hat{I}$ is a 3×3 identity matrix. $\hat{H}_Q = -\varepsilon(\hat{I} - F_z^2)$ is the quadratic Zeeman shift. $\hat{\mathbf{F}} = (F_x, F_y, F_z)$ and the effective magnetic field is given by[39]

$$\mathbf{\Omega}(x) = \frac{1}{\sqrt{2}}(\Omega_{RF} + \Omega_R \cos(2qx), -\Omega_R \sin(2qx), \sqrt{2}\delta), \qquad (15)$$

where $2q$ is the momentum transfer from the Raman beams. Without the RF-field, one finds a Zeeman field whose direction is rotating in the x-y plane, but the amplitude stays the same. This realizes the standard spin-orbit coupling in the uniform system. With additional rf-field, the magnitude of the Zeeman coupling, $|\Omega(x)|$, is changing periodically and provides a one-dimensional "Zeeman lattice". When an atom hops from one minimum of the lattice to its nearest neighbors, the effective magnetic field winds in the Bloch sphere and generate a geometric Berry phase.[39] An experiment using ^{87}Rb has measured the Peierls' phase generated and also the effective mass close to the band minimum.[39] The spin resolved band structure in a

Table 2. Different lattices and physical models realized so far with the periodic driven lattice technique. In the table, ω is the driving frequency and T is the period of the driving. $\hat{e}_{1,2}$ are two orthonormal vectors in the xy-plane. $\delta\mathbf{k}$ is the momentum transfer from the Raman beams. Ω_R is the Rabi frequency. F and $F_{1,2}$ are the amplitude of the driving.

Group	Underlying lattice	Driven term ($\hat{H}_{\mathrm{mod}}(\mathbf{r}, t) = \mathbf{F}(t) \cdot \mathbf{r}$)	Physical models
Hamburg	1D lattice	$\mathbf{F}(t) = F \sin(\omega t), t < T_1; \mathbf{F}(t) = 0, T_1 < t < T$	Peierls phase[23]
Hamburg	triangular lattice	$\mathbf{F}(t) = F_1 \cos(\omega t)\hat{e}_1 + F_2 \sin(\omega t)\hat{e}_2$	frustrated spin model[22]
Hamburg	triangular lattice	$\mathbf{F}(t) = F_1 \cos(\omega t)\hat{e}_1 + F_2[\sin(\omega t) + \delta\sin(2\omega t)]\hat{e}_2$	Ising-XY spin model[24]
Munich	superlattice	$\Omega_R \sin(\delta\mathbf{k} \cdot \mathbf{r} - \omega t)$	staggered flux[51]
Munich MIT	optical lattice+ linear potential	$\Omega_R \sin(\delta\mathbf{k} \cdot \mathbf{r} - \omega t)$	uniform flux[53,54]
Chicago	1D lattice	$U_0 \sin^2(k(x - x_0(t)))$	ferromagnetic domain[56]
ETH	honeycomb lattice	$\mathbf{F}(t) = F[\cos(\omega t)\hat{e}_1 + \cos(\omega t - \varphi)\hat{e}_2]$	Haldane model[11]
Munich	square lattice	$V(t) = \kappa \sum_{mn,i}|f_i(m)\cos(\omega t + g_i(n))|$	Hofstadter model[12]

"Zeeman lattice" has been mapped out using fermionic ^{6}Li with a novel spin injection spectroscopy technique in Ref. 18.

3. Basic Theoretical Model

Having now discussed several experimentally viable routes to implementing spin-orbit coupling in an optical lattice, we next turn our attention to new many-body physics which results from the interplay of spin-orbit coupling, lattice environment, and interactions. We do not tie ourselves to any specific experimental realization, assuming that specific model Hamiltonians we consider can be realized using schemes discussed above or suitable variants. The appropriate degrees of freedom will be boson or fermion operators associated with Wannier states localized near the lattice sites, which also carry a (typically two-component) pseudospin degree of freedom. For a sufficiently deep lattice potential that the occupation of excited bands can be neglected, the resulting system is well-described by a tight-binding Hamiltonian

$$H_0 = -t \sum_{\langle ij \rangle} a_{i\sigma}^\dagger \mathcal{R}_{ij}^{\sigma\sigma'} a_{j\sigma'} + \text{h.c.}, \tag{16}$$

where t sets an overall hopping amplitude (and natural energy scale), while $\mathcal{R}_{ij}^{\sigma\sigma'}$ characterizes the hopping of an atom with spin σ' from site j to the spin σ state on a neighboring site i. Here we have dropped the Zeeman terms associated with Rabi frequency and detuning in order to define a minimal model to investigate the interplay between spin-orbit coupling and strong interactions. Because of the lattice-translation invariance of this Hamiltonian, it can be Fourier transformed into $\mathbf{k}$-space, and the Hamiltonian can be generically written as

$$H_0 = \sum_{\mathbf{k}} \psi_{\mathbf{k}}^\dagger \mathcal{H}(\mathbf{k}) \psi_{\mathbf{k}}, \tag{17}$$

where $\mathcal{H}(\mathbf{k}) = d_0(\mathbf{k}) + \mathbf{d}(\mathbf{k}) \cdot \sigma$, is written in terms of its expansion in the Pauli matrices $\sigma = (\sigma_x, \sigma_y, \sigma_z)$. The topological properties of any Hamiltonian expressed in this way can be obtained from a straightforward computation of the Berry curvature (for two spatial dimensions)

$$F_{ij} = \frac{1}{2} \epsilon_{abc} \hat{d}_a \partial_i \hat{d}_b \partial_j \hat{d}_c, \tag{18}$$

where $\hat{d}_a = d_a/|\mathbf{d}|$. The Chern number can then be computed by integrating the Berry curvature over the occupied states.[58]

In the following, we shall explicitly work through two important examples: (1) Anisotropic spin-orbit coupling (e.g., $p_x\sigma_y$) in a 1D optical lattice and (2) isotropic Rashba spin-orbit coupling in 2D optical lattices. The first case has been realized in an experiment which explores the stability of the spin-orbit coupled condensate in a moving 1D optical lattice.[21] The second case has not yet been realized, but is the subject of a substantial experimental effort.

3.1. *Band structure*

(1) *One-dimensional spin-orbit coupling in a one dimensional lattice.* This is the case with the present implementation of spin-orbit coupling, where only one component of the momentum (say k_x) is coupled nontrivially to the spin. The hopping matrix is then given by

$$\mathcal{R}_{ij}^{\hat{x}} = \cos\alpha \pm i\sin\alpha\,\sigma_y, \tag{19}$$

where $\pm$ refers to hopping along the $+\hat{x}$ or $-\hat{x}$ directions, and reveals the explicitly broken spatial inversion symmetry. The single-particle spectrum is given by two bands with $\epsilon_\pm(k) = -2t\cos(k \pm a)$. There are two degenerate minima at $k = \pm a$ and the associated spinor wave functions are $\chi_{\pm,a}(x) = \frac{1}{\sqrt{2}}\exp(\pm iax)(1, \pm i)^T$. These two states form a pair, $\chi_{+,a}(x) = -i\sigma_x\chi_{-,a}(-x)$, related to each other by inversion followed by a spin rotation by π about the σ_x axis. At zero momentum ($k = 0$) the two bands have a level crossing, leading to a doublet protected by time-reversal symmetry. This Kramers degeneracy is broken in the presence of a symmetry-breaking Zeeman term, as appears in present experiments with Raman-induced spin-orbit coupling. Near this avoided crossing, the energy spectrum is described by a one-dimensional massive Dirac equation.

(2) *Rashba spin-orbit coupling in a two-dimensional square lattice.* In this case, in addition to the hopping along the $\hat{x}$-direction, Eq. (19), there appears an additional contribution from hopping along the $\hat{y}$ direction, given by the matrix

$$\mathcal{R}_{ij}^{\hat{y}} = \cos\beta \pm i\sin\beta\,\sigma_x, \tag{20}$$

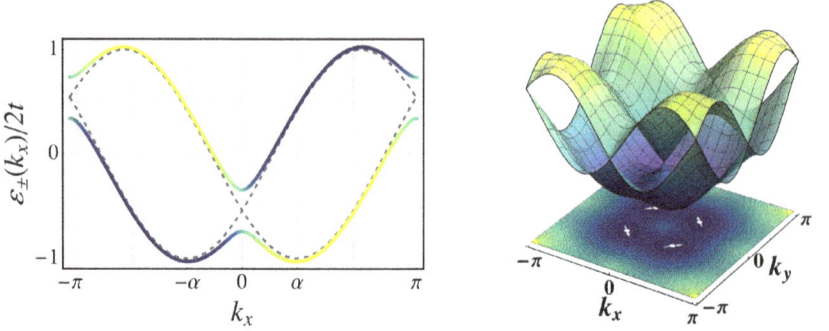

Fig. 3. (Left) Energy eigenvalues $\epsilon_{\pm}(k_x) = -2t\cos(k_x \pm \alpha)$ resulting from the one-dimensional spin-orbit coupling Eq. (19). Additionally introducing a small Zeeman coupling splits the two bands, generating avoided crossings at $k_x = 0, \pi$. The color corresponds to the y-component of spin $\langle\sigma_y\rangle$ in that state. Dark blue corresponds to states where the spin is locked to the $-y$ direction, yellow to $+y$. (Right) Energy bands arising from Rashba spin-orbit coupling in a square lattice. Here, the color simply tracks the energy. The momenta corresponding to the four lowest-energy states are marked with white dots and arrows which represent the spin wavefunction associated with those states. The right-side figure is adapted from Ref. 45.

which links motion along the $\hat{y}$ direction to the spin projection along $\hat{x}$. In general, α and β can be different, resulting in an arbitrary linear combination of the linear Rashba and linear Dresselhaus spin-orbit couplings.

Using the expansion in Pauli matrices, the tight-binding Hamiltonian can be characterized by the d-vector

$$d_0(\mathbf{k}) = -2t\left(\cos\alpha\cos k_x + \cos\beta\cos k_y\right), \qquad (21)$$

$$d_x(\mathbf{k}) = -2t\sin\beta\sin k_y, \qquad (22)$$

$$d_y(\mathbf{k}) = -2t\sin\alpha\sin k_x, \qquad (23)$$

$$d_z(\mathbf{k}) = 0. \qquad (24)$$

The energy spectrum is easily evaluated as $\epsilon_{\pm}(k) = d_0(k) \pm |d(k)|$, while the spin eigenstates with $d_z = 0$ are

$$\chi_{\pm}(k) = \begin{pmatrix} 1 \\ \mp i e^{i\varphi_k} \end{pmatrix}, \quad \varphi_k = \arctan\left(d_x/d_y\right), \qquad (25)$$

which has singularities whenever $d_x = d_y = 0$. In the first Brillouin zone, this occurs at the four time-reversal invariant momenta $(k_x, k_y) \in \{(0, 0), (0, \pi), (\pi, 0), (\pi, \pi)\}$, and signals the locations of Dirac points. The circulation of the spin wavefunction is counter-clockwise for contours that enclose $k = (0, 0)$ or (π, π) (in the usual counter-clockwise sense), and clockwise for those that surround $k = (0, \pi)$ or $(\pi, 0)$. These two species of Dirac points are therefore topologically distinct from one another, with winding numbers ± 1.

3.2. Strong interaction physics

One of the primary reasons for interest in optical lattices is that they provide a route to tunable strong interactions between particles. Working in the regime where the interaction scale $U \ll \Delta$, with Δ the gap to the lowest excited band, it is possible to write the interaction contribution to the Hamiltonian as

$$H_{\text{int}} = \sum_i \left(\frac{U}{2} \sum_\sigma [n_{i\sigma}(n_{i\sigma} - 1)] + U' n_{i\uparrow} n_{i\downarrow} \right) + \cdots, \quad (26)$$

where ... accounts for (typically negligible) further neighbors interactions. U and $U' \equiv \lambda U$ describe the intra- and inter-species interactions. Considering only the on-site interaction, the full model Hamiltonian has the form

$$H = -t \sum_{\langle ij \rangle} a_{i\sigma}^\dagger \mathcal{R}_{ij}^{\sigma\sigma'} a_{j\sigma'} + \sum_i \left(\frac{U}{2} \sum_\sigma [n_{i\sigma}(n_{i\sigma} - 1)] + U' n_{i\uparrow} n_{i\downarrow} \right).$$
$$(27)$$

We shall refer to this as spin-orbit coupled Bose-Hubbard model (SOBHM). In the following, we discuss a few special features that occur in the combined presence of spin-orbit coupling and strong inter-particle interaction.

3.3. General Discussions

There are already several new physical effects associated with spin-orbit coupling that have been demonstrated in recent experiments. In the case of bosons, for example, depending on the parameters, the single particle ground state can be degenerate and the Bose condensate can feature novel density and spin density patterns. A further consequence of spin-orbit coupling is the lack of Galilean invariance, as demonstrated in the recent moving optical lattice experiment. Furthermore, in a harmonic trap, spin-momentum locking provides a way to couple the dipole oscillation to magnetic oscillations. In this review, we focus on interesting new features that are brought about by spin-orbit coupling in connection with strong interaction effects.

Exotic magnetic structures in Mott insulators. A natural question regarding SOBHM is the magnetic phases deep in the Mott insulating regime and this has been addressed in several recent works.[45,46,59,60] Similar problems in spin-orbit coupled fermions has been discussed in, for example, Refs. 61 and 62. For the standard Bose-Hubbard model with spinless bosons, the Mott insulating state is a featureless Mott insulator with a charge gap and zero compressibility. However, introducing a spin degree of freedom as well as the spin-orbit coupling present in H_0 allows for the realization of a rich class of magnetically ordered Mott insulators, similar to the generic form introduced by Moriya[63] for electronic Mott insulators with spin-orbit coupling. To $\mathcal{O}(t^2/U)$, this hamiltonian is

$$H_{\text{mag}} = \sum_{i,\mu} \left[J S_i \cdot S_{i+\mu} + D_\mu \cdot \left(S_i \times S_{i+\mu} \right) + \sum_{a,b} S_i^a \Gamma_\mu^{ab} S_{i+\mu}^b \right], \quad (28)$$

where μ represents the spatial direction of a bond of the lattice. For concreteness, in later sections we will take a 1D chain along $\hat{x}$ and a 2D square lattice with lattice vectors $\hat{x}$ and $\hat{y}$. A detailed derivation of the above Hamiltonian and the appropriate coefficients for these cases is provided in Appendix.

The natural energy scale here is given by $\mathcal{J} = \frac{4t^2}{\lambda U}$. In terms of this scale, we can write the exchange constant $J = -\mathcal{J} \cos 2\alpha$ which accompanies the spin-isotropic Heisenberg interaction; this is clearly ferromagnetic in the limit of vanishing spin-orbit coupling, as it must be for bosons. The remaining terms account for anisotropies in spin space that are generated either by the explicitly spin-anisotropic interactions of the bosons ($U' \neq U$) or by the explicit coupling of spin to orbital motion ($\alpha \neq 0$). The vectors $D_x = -(\mathcal{J}\lambda \sin 2\alpha)\hat{y}$ and $D_y = -(\mathcal{J}\lambda \sin 2\alpha)\hat{x}$ arise purely from SOC, and characterize the antisymmetric Dzyaloshinsky-Moriya interaction[63,64] which generically leads to long-wavelength magnetic spirals in solid-state materials. The SOC also generates symmetric anisotropic interactions of a "compass model" type, $\Gamma_y^{xx} = \Gamma_x^{yy} = -\mathcal{J}(1 - \cos 2\alpha)$, while an out-of-plane anisotropy $\Gamma_x^{zz} = \Gamma_y^{zz} = -2\mathcal{J}(\lambda - 1)$ arises from the original spin-anisotropy in the interactions (all other components of the tensor Γ^{ab}_{μ} are zero). Compass model interactions have become a major research topic of late for their role in Kitaev's exactly-solvable "honeycomb model" of a spin liquid,[65] which itself might describe the magnetism of certain transition metal oxides with strong spin-orbit coupling.[66]

In the combined presence of these terms, the magnetic Hamiltonian is generically frustrated and can support a wide variety of complex magnetic structures, including spiral and skyrmion states. This is discussed further in the following sections.

The Mott transition from a non-uniform superfluid state. In the standard BHM with only on-site interactions, both the superfluid state and the Mott state have uniform density, and the Mott-superfluid transition is accompanied by the breaking of $U(1)$ symmetry. On the other hand, when one considers the SOBHM, the superfluid state can exhibit spin density wave, while the spin density of the magnetically ordered Mott states is also generally inhomogeneous. Thus, apart from the usual broken $U(1)$ symmetry of the Mott-superfluid transition, there are order parameters associated with broken lattice translation symmetry in the Mott and superfluid states. As we shall discuss later, the magnetic structure can also persist across the Mott-superfluid transition.

These considerations imply that some generalizations to the standard treatment of BHM need to be made when considering the quantum phase transitions in a SOBHM. At the most naïve level, since the superfluid state is no longer uniform, it is not possible to use the uniform Gutzwiller approximation, as is often done to describe mean-field properties of the BHM. It is at least necessarily to perform the mean-field analysis in a finite cluster. This has been explored in some detail by Refs. 45, 67 and 68. Beyond this level of mean-field theory, recently a bosonic variant of the *dynamical* mean-field theory (BDMFT) has been applied.[69] Other, more exact, numerical methods are available in one dimension. The presence of an extra spin-density wave order parameter implies that the effective field theory of the Mott-superfluid transition could be quite different from the standard one and further research in this direction is worthwhile.

Topological states in the SOBHM. In the simplest case of a square lattice with Rashba spin-orbit coupling, the resulting single-particle band structure is topologically trivial, in the sense that the bands have zero Chern number. However even (or especially) in the absence of a nonzero single-particle Chern number, it is very interesting to ask if interaction effects can lead to nontrivial topological properties. The most natural place to look for such non-trivial topological properties is the Mott insulating state, where the single particle excitation spectrum is gapped.

Having pointed out several new features that are likely to be encountered with spin-orbit coupling in the presence of Hubbard-type interactions, we proceed in the following sections with a few illustrative examples.

3.3.1. One-dimensional lattice with spin-orbit coupling

Let us consider first the case of a one-dimensional SOBHM, for which more exact treatment using density matrix renormalization group is possible. For simplicity, we shall neglect altogether the Zeeman terms and concentrate on the interplay between spin-orbit coupling and interactions. The Hamiltonian is given by Eq. (27) with the hopping matrix $\mathcal{R}$ given by Eq. (19). Using intra species interaction U to set the energy scale, we have $\alpha, t/U$ and $\lambda \equiv U'/U$ as three independent dimensionless parameters.

In the strong coupling limit $t/U \ll 1$ with unit filling, the system enters Mott state with one boson per site and one obtains the effective magnetic Hamiltonian by the standard perturbation theory. In order to put the magnetic Hamiltonian in the standard form, we rotate the spin around $\hat{x}$-axis by $\pi/2$, such that the Dzyaloshinsky-Moriya vector is along $\hat{z}$-axis,

$$H_{\mathrm{mag}} = -\frac{4t^2}{U} \sum_{\langle ij \rangle} \left[\frac{\cos(2\alpha)}{\lambda} s_i^x s_j^x + \frac{\cos(2\alpha)}{\lambda}(2\lambda - 1)s_i^y s_j^y + \frac{1}{\lambda}s_i^z s_j^z \right.$$
$$\left. + \sin(2\alpha)(s_i^x s_j^y - s_i^y s_j^x) \right]. \tag{29}$$

We note the following features: (1) The overall exchange energy scale is given by t^2/U as usual, but the sign can be tuned by changing α and can be both ferromagnetic or antiferromagnetic; (2) There appears the Dzyaloshinsky-Moriya term, in addition to the standard Heisenberg coupling, with Dzyaloshinsky-Moriya vector given by $\mathbf{D} = \sin(2\alpha)\hat{z}$. H_{mag} cannot be solved exactly for general α and λ, but in various limits, it can be reduced to known models or exactly solvable.[70–74] The full phase diagram of H_{mag} is given in Fig.4. In the following, we consider a few special cases of H_{mag}.

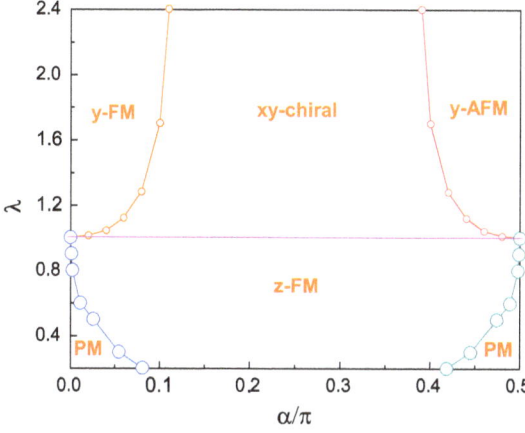

Fig. 4. Phase diagram of the effective spin model H_{mag} in the λ-α plane. In addition to the weak coupling magnetic phases, $\hat{z}$-FM and xy-chiral, one finds three additional magnetic phases: $\hat{y}$-FM, $\hat{y}$-AFM and PM states. Figure adapted from Ref. 72.

(1) For $SU(2)$ invariant interaction, i.e., $\lambda = 1$, H_{mag} reduces to

$$H_{mag} = -\cos(2\alpha)\frac{4t^2}{U} \sum_{\langle ij \rangle} \left[s_i^x s_j^x + s_i^y s_j^y + \frac{1}{\cos(2\alpha)} s_i^z s_j^z \right.$$

$$\left. + \tan(2\alpha)(s_i^x s_j^y - s_i^y s_j^x) \right]. \tag{30}$$

This Hamiltonian can be transformed to an isotropic Heisenberg model if we make the following transformation.[75] At each site, the spin is rotated around $\hat{z}$-axis by an angle θ_i, $\tilde{s}_i^+ \equiv \exp(-i\theta_i s_z) s_i^+ \exp(i\theta_i s_z) = \exp(-i\theta_i) s_i^+$, where $s_i^+ = s_x + i s_y$ is the spin raising operator, while $\tilde{s}_i^z = s_i^z$. Choosing $\theta_{i+1} - \theta_i = -2\alpha$, the Hamiltonian Eq. (29) reduces to an isotropic ferromagnetic Heisenberg model in terms of the $\tilde{s}_i$ spins for any α. That is, $H_{mag} = -\frac{4t^2}{U} \sum_{ij} [\tilde{s}_i^x \tilde{s}_j^x + \tilde{s}_i^y \tilde{s}_j^y + \tilde{s}_i^z \tilde{s}_j^z]$. The exact ground state is a ferromagnet and the elementary excitations are spin waves with quadratic dispersion. In terms of the original spin s, this corresponds to an exact spiral ground state with wave vector 2α along the chain.

(2) When $\alpha = \frac{\pi}{4}$, $H_{mag} = -\frac{4t^2}{U} \sum_{ij} [\frac{1}{\lambda} s_i^z s_j^z + (s_i^x s_j^y - s_i^y s_j^x)]$. This is a one-dimensional Ising model with DM interactions and has been studied in the literature.[76] It has two phases: for $\lambda > 1$, the DM term dominates and the system is in a chiral phase in which the spin spirals around the $\hat{z}$-axis along the chain. We refer to this as the chiral xy-magnet since the interactions will kill long range spin order but preserve the chirality. For $\lambda < 1$, the

ferromagnetic term dominates and the system is in a ferromagnetic state, pointing along the $\hat{z}$ direction. The ferromagnet has a twofold ground state degeneracy.

(3) A particularly interesting limit corresponds to taking $\lambda \to \infty$. In this case, the spin model is given by

$$H_{\text{mag}} = -\frac{4t^2}{U} \sum_{\langle ij \rangle} \left[2\cos(2\alpha)s_i^y s_j^y + \sin(2\alpha)(s_i^x s_j^y - s_i^y s_j^x) \right]. \quad (31)$$

This can be solved by the standard Jordan-Wigner transformation. The final result is a Bogoliubov-de Genne type of Hamiltonian

$$H_{\text{fermion}} = \sum_{k>0} [c_k^\dagger, c_{-k}] \begin{bmatrix} \epsilon(k) & \Delta(k) \\ \Delta^*(k) & -\epsilon(-k) \end{bmatrix} \begin{bmatrix} c_k \\ c_{-k}^\dagger \end{bmatrix}, \quad (32)$$

where $\Delta(k) = i\cos(2\alpha)\sin k$ and $\epsilon(k) = -\cos(k-2\alpha)$. In terms of Nambu spinor $\Psi_k^\dagger \equiv [c_k^\dagger, c_{-k}]$, $H_{\text{fermion}} = \sum_{k>0} \Psi_k^\dagger \hat{\mathcal{H}}(k)\Psi_k$, with $\hat{\mathcal{H}}(k) = d_0(k)\hat{I} + \sum_{i=x,y,z} d_i(k)\hat{\sigma}_i$, where $\hat{I}$ is the 2×2 identity matrix and $\hat{\sigma}_{x,y,z}$ are the Pauli matrices. $d_0(k) = -\sin 2\alpha \sin k$, $d_x(k) = 0$, $d_y(k) = -\cos 2\alpha \sin k$ and $d_z(k) = -\cos 2\alpha \cos k$. The spectrum of fermion modes is given by $E_\pm(k) = -\sin(2\alpha)\sin k \pm |\cos(2\alpha)|$. The critical values for α where the spectrum $E_\pm(k)$ becomes gapless are given by $\alpha = \frac{1}{8}\pi, \frac{3}{8}\pi$. For $\alpha < \frac{1}{8}\pi$, the system is a $\hat{y}$-ferromagnet, while for $\alpha > \frac{3}{8}\pi$, it is a $\hat{y}$-anti-ferromagnet; Both phases are gapped. In the intermediate region, $\frac{1}{8}\pi < \alpha < \frac{3}{8}\pi$, it is in the xy-chiral phase with gapless excitations.

What is interesting is that the Hamiltonian Eq. (32) describes p-wave pairing in one dimension, analogous to the Kitaev chain model.[77] The Hamiltonian obeys the following symmetry: $\hat{\mathcal{H}}(k) = -\sigma_x \hat{\mathcal{H}}(-k)^* \sigma_x$ and belongs to the "D" symmetry class, characterised by a $\mathbb{Z}_2$ invariant.[78] In the special case when $\alpha = 0, \frac{\pi}{2}$, Eq. (32) reduces to the standard Kitaev model. We note that the magnetic transitions described by JW fermions in the limit $\lambda \to \infty$ occur also for finite values of $\lambda > 1$, as shown in Fig. 5. It is thus tempting to conclude that the phase boundaries between the xy-chiral and $\hat{y}$-ferromagnetic or $\hat{y}$-antiferromagnetic, to be described by the same topological transitions. Further investigations are necessary in this direction.

Now, let us turn to the question of Mott-superfluid transition and in particular, how the magnetic phases obtained above evolve into the superfluid phase. For more detailed discussion, see Refs. 70–72. It is

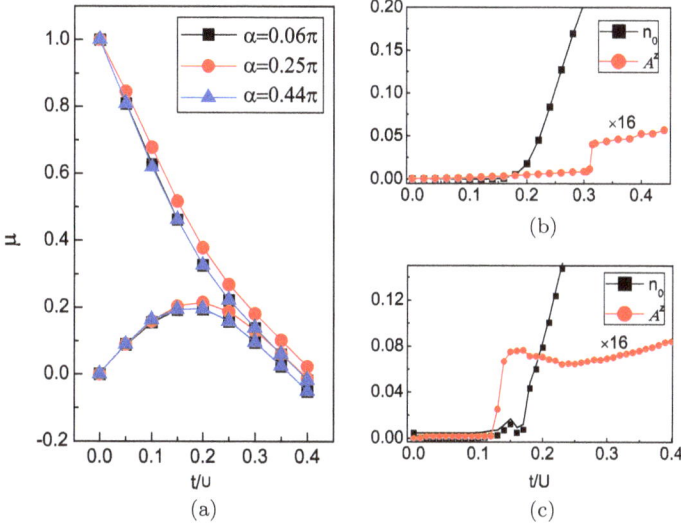

Fig. 5. (a) Phase diagram of the one-dimensional SOBHM in the μ-t/U plane. Note that for different values of α, corresponding to various strength of spin-orbit coupling, the superfluid-Mott boundaries are only slightly modified. (b) and (c) show the magnetic transition from $\hat{y}$-FM to xy-chiral phase as one increase the hopping amplitude t/U. For (b) $\alpha = 0.08\pi$ and $\lambda = 1.5$ and for (c) $\alpha = 0.07\pi$ and $\lambda = 1.2$. n_0 can be regarded as the superfluid order parameter and describes the superfluid-Mott transition, while A^z is the chiral order parameters. Note that the sequence of transition depends on the values of α and λ. Figure adapted from Ref. 72

instructive to look first at the weak coupling limit when $U \rightarrow 0$. In this case, the single particle spectrum has two degenerate ground states at $k = \pm \alpha$ with the corresponding wave function given by $\Psi_{\pm}(k) = \exp(\pm iax)(1, \pm i)$. The superfluid order parameter is a superposition of the two states $\Psi_{\pm}(x)$ for $\lambda > 1$, which leads to an order parameter of the form $(\langle a_{x\uparrow} \rangle, \langle a_{x\downarrow} \rangle) = (\cos \alpha x, -\sin \alpha x)$. This corresponds to spin spiraling around the $\hat{y}$-axis with wave vector 2α. This is the xy-chiral phase found in the magnetic Hamiltonian, after rotating the spin around $\hat{x}$ by $\pi/2$. When $\lambda < 1$, the system breaks the $\mathbb{Z}_2$ symmetry and chooses one of the $\Psi_{\pm}$ as its order parameter. The superfluid state is a ferromagnetic state along $\hat{y}$-direction, which, after rotating around $\hat{x}$-axis by $\pi/2$, is consistent with what is found in the strong coupling limit.

However, strong interaction leads to more magnetic phases as is evident in the Mott regime, where additional paramagnetic, $\hat{y}$-ferromagnetic and $\hat{y}$-anti-ferromagnetic are found. The interesting question is whether these new magnetic phases, not found in the weak coupling limit, arise concomitant with emergence of Mott insulating phases, or they develop either before or after Mott-superfluid transition. To investigate this question, we first establish that the superfluid-Mott transition is only slightly modified by the presence of spin-orbit coupling. As an example, we calculate the μ-t/U phase diagram for three values of spin-orbit coupling $\alpha = 0.06\pi, 0.25\pi, 0.44\pi$ by identifying values of μ and t/U where single particle $(E_+ \equiv E(N + 1) - E(N))$ or hole excitation $(E_+ \equiv E(N - 1) - E(N))$ energies approach zero. As can be seen from Fig. 5(a), the phase boundary is only slightly modified.

On the other hand, the magnetic phases depend crucially on the value of spin-orbit coupling. In Fig. 5(b) and (c), we show how magnetic phases in the strong coupling limit evolve into the superfluid phases, for two sets of parameters $(\alpha = 0.08\pi, \lambda = 1.5)$ and $(\alpha = 0.07\pi, \lambda = 1.2)$. We calculate the one-body density matrix $\langle a_{i\alpha}^\dagger a_{j\sigma'} \rangle$ and extract its maximal eigenvalues n_0 whose eigenfunction decays algebraically. n_0 is non-zero only in the superfluid state. We also define the chiral correlation function $\mathcal{A}^\gamma (i, j) \equiv \langle A_i^\gamma A_j^\gamma \rangle$, where $\gamma = x, y, z$. In the Mott regime, $A_i^\gamma = \varepsilon^{\gamma \mu \nu}(s_i^\mu s_{i+1}^\nu - s_i^\nu s_{i+1}^\mu)$, describing the chirality of the spins in the ground state, while in the superfluid state, we replace $s_i^\gamma = \frac{1}{2}a_{i\alpha}^\dagger \sigma_{\alpha\beta}^\gamma a_{i\beta}$, with underlying boson operator. In the chiral state, one expects that the asymptotic value $\mathcal{A}^\gamma \equiv \lim_{|i-j|\to\infty} \mathcal{A}^\gamma (i, j)$ to remain finite. As can be seen from Figs. 5(b) and (c), depending on the values of (α, λ), the magnetic transition can occur either before or after the superfluid transition.

3.3.2. Two-dimensional lattice with Rashba spin-orbit coupling

Let us now turn to the two dimensional case with Rashba SOC. The Hamiltonian is given by Eq. (27) with the hopping matrices given in Eq. (19) and Eq. (20). As before, we have $\alpha, t/U$ and $\lambda \equiv U'/U$ as three independent dimensionless parameters.

It was noted previously that the single-particle spectrum has four degenerate lowest energy states. As a result, any state where N bosons are distributed among these minima is a valid ground state in the absence of interactions. It is expected that when weak interactions are taken into account, a unique ground state will be selected. To explore this, we first assume that the bosons condense into one single-particle state with the generic wavefunction $\varphi(r) = \sum_{m=1}^{4} c_{\mathbf{k}_m} e^{i\mathbf{k}_m \cdot r} \chi_{\mathbf{k}_m}$, where the c_m are a set of normalized complex variational parameters and $\chi_{\mathbf{k}m}$ are the associated spin wave functions at the four minima. The optimal set of c_m minimizes the interaction energy $E_{\mathrm{int}}[\{c_m\}] \equiv \langle \Phi | H_{\mathrm{int}} | \Phi \rangle$, and fully characterizes the properties of the condensate. It is convenient to rewrite H_{int} in terms of the operators that diagonalize H_0, and then the interaction energy only receives contributions from terms where all 4 operators correspond to the 4 minima. This process yields an expression

$$E_{\mathrm{int}} \propto \sum_{\alpha\beta} \sum_{kpq}{}' U_{\alpha\beta}(c_{p+q-k}\chi_{p+q-k,\alpha}^{-})^*(c_k\chi_{k,\beta}^{-})^*(c_p\chi_{p,\beta}^{-})(c_q\chi_{q,\alpha}^{-}). \quad (33)$$

The primed sum indicating that we only consider terms where all momentum indices correspond to energy minima.

Minimizing this quantity, we find — similar to studies in the absence of an optical lattice[79–82] — that either a single minimum is occupied (leading to a "plane wave" condensate) or two opposite momenta are equally occupied (leading to a uniform density but spin-polarization-striped condensate). Which state is chosen depends on the deviation from a totally spin-isotropic interaction $U = U'$, with the striped condensate being energetically favorable when $U' > U$.

The conceptual explanation of this result is rather straightforward, but it is useful to first write down the wavefunctions which are macroscopically occupied. We have, for the plane wave phase,

$$\Psi_{\mathrm{PW}}(r) = \frac{1}{\sqrt{2}} \exp(ik_1 \cdot r) \begin{pmatrix} 1 \\ e^{i\pi/4} \end{pmatrix}. \quad (34)$$

For the stripe phase,

$$\Psi_{\mathrm{stripe}}(r) = \frac{1}{2} \left[\Psi_{\mathrm{PW},k_1}(r) + \Psi_{\mathrm{PW},k_3}(r) \right] = \begin{pmatrix} \cos(k_1 \cdot r) \\ e^{i3\pi/4} \sin(k_1 \cdot r) \end{pmatrix}. \quad (35)$$

We may also consider, although it fails to appear as a ground state in this weak-coupling approach, a Skyrmion state where an equal-weight combination of all four minima is occupied,

$$\Psi_{\text{Skyrmion}}(r) = \frac{1}{2\sqrt{2}} \sum_{m=1}^{4} \exp{(ik_m \cdot r)} \begin{pmatrix} 1 \\ e^{i(2m-1)\pi/4} \end{pmatrix}. \tag{36}$$

The stripe and plane wave solutions are the only two states that can be constructed in this way which have a spatially uniform number density, which is favored by the spin-isotropic part of the interaction. The stripe phase additionally has a sort of "phase separation" into regions where the two spin densities minimize their spatial overlap. This is favored when the interspecies interaction U' is dominant. The Skyrmion state describes a *local* minimum in energy, but is never a global minimum. It does not have a uniform number density, and is less efficient than the stripe phase at minimizing the spatial overlap of the two spin components. It is interesting to note, however, that other authors have observed that density-modulated condensates, including quasicrystals, can be stabilized by *long-ranged dipolar* interactions, even in the weak-coupling limit.[83,84]

    In the weakly interacting limit, the interactions are responsible for supporting a unique ground state and the structure of the ground state emerges from interference between the spinor wavefunctions describing the single-particle minima. In the opposite limit where $U, U' \gg t$, we begin, however, with the single-site spectrum of H_{int}, as the band structure and low-energy states of H_0 are less relevant. In the following, we restrict our attention to unit filling. Then, reintroducing H_0 to second-order in

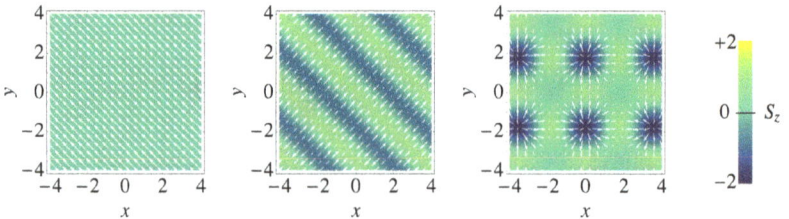

Fig. 6. Spin densities of the various ordered condensates: (left) single plane wave, (middle) stripe, (right) Skyrmion. The z projection of the spin density is indicated by color, while the x and y projections are indicated by the white arrows. The plane wave and stripe solutions have uniform total number density, while the Skyrmion has a density wave, with peaks in the dark blue regions and vanishing density in the interstitial regions where all spin components are zero.

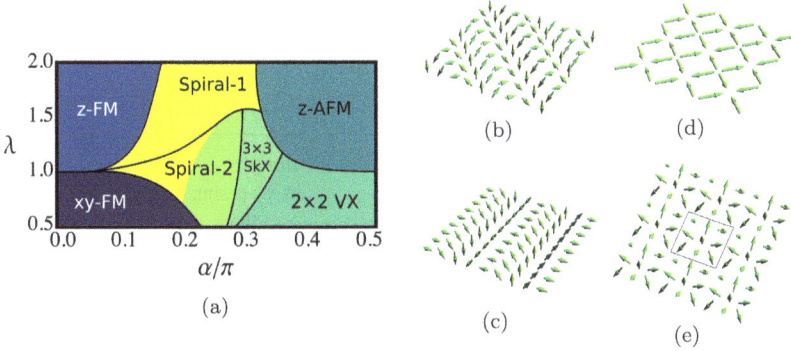

Fig. 7. (a) Magnetic phase diagram in the deep Mott limit determined by Monte Carlo annealing on a 36 × 36 site square lattice. Although the spiral states are generally incommensurate, the shaded green area in the Spiral-2 region corresponds to a likely commensurate state. In this region the spiral unit cell contains 4 sites, with the spin winding by $\pi/2$ along each bond in the spiral direction. This pattern maximizes the cross-product of neighboring spins, and is therefore quite favorable in the region $\theta \sim \pi/4$, where the isotropic ferromagnetic interaction vanishes. On the right side, we show several classical spin configurations. (b) Spiral-1 state; coplanar spin-orientation rotating along (11). (c) Spiral-2 state; coplanar spin-orientation rotating along (10). (d) Vortex crystal; 2 × 2 unit cell with $\pi/2$ rotation along each bond. (e) Skyrmion crystal; the 3 × 3 unit cell is highlighted with a gray box. The central spin points in the positive z direction, while the remaining spins tumble outward toward $-z$. Figure adapted from Ref. 45

perturbation theory yields the model given by Eq. (28) and the paragraph that follows it.

These exchange interactions are frustrated even on the square lattice, and finding ground states is quite challenging. To gain some insight into the possible states supported by such a model, we revert to classical Monte Carlo simulations, in which we treat the spins S as classical variables. The resulting phase diagram is shown in the $\lambda \equiv U'/U$ and α plane in Fig. 7, together with a few selected spin configurations. We characterize the different phases through the magnetic structure factor $S_q = |\sum_x S_x e^{iq \cdot r_x}|$. The peaks in this structure factor tell us about the magnetic ordering vectors (as shown in the inset of Fig. 7). A summary of various magnetic phases is given in Table 3, and we now proceed to point out some interesting features of the various phases.

(I) The existence of the two ferromagnetic phases which occupy the small α region can be understood as follows. In the limit $\alpha \to 0$, the magnetic

Table 3. Summary of the classical spin states supported by the effective Hamiltonian.

Phase	Location of peaks in $S_q = \left\| \sum_x S_x e^{iq \cdot r_x} \right\|$	Spin orientation
zFM	$(0, 0)$	along z
xyFM	$(0, 0)$	in xy plane, at angle $(2n + 1)\pi/4$ to the x axis
zAFM	(π, π)	along z
Spiral-1	$(q, \pm q)$	in z-q plane
Spiral-2	$(q, 0)$ or $(0, q)$	in z-q plane
Vortex Crystal (VX)	$(\pi, 0)$ and $(0, \pi)$	in xy plane, spin components: $S^x = (-1)^{r_x}/\sqrt{2}$, $S^y = (-1)^{r_y}/\sqrt{2}$
Skyrmion Crystal (SkX)	$(2\pi/3, 0)$ and $(0, 2\pi/3)$	non-coplanar

Hamiltonian Eq.(28) reduces to the standard Heisenberg XXZ model, with the only anisotropy in the exchange coming from $\lambda \neq 1$. For $\lambda > 1$, the $\hat{z}$-component of the exchange interaction is larger than the in-plane component, and one then expects ordering along $\hat{z}$. When $\lambda < 1$, we have the opposite case. For small nonzero α, these phases survive but with additional Ising anisotropies that pin the direction of the xy-ferromagnetism.

(II) The existence of the two magnetic phases near $\alpha = \pi/2$ can likewise be understood in the limiting case. The D vectors again vanish and the λ dependence that leads to z-axis or xy-plane anisotropy is identical to the previous case. Now, though, the sign has switched so that the z-component of the exchange is antiferromagnetic. For the vortex crystal (VX) phase, the exchange along x and y directions are of different sign and additionally the exchange in spin space is of different sign for the x and y components. Ordering is therefore frustrated. To understand the classical phase that emerges here, a variational solution is useful. We propose a state $S_i^x = (-1)^{x_i} \sin \varphi$, $S_i^y = (-1)^{y_i} \cos \varphi$ with a uniform φ. Now, plugging this state into the full Hamiltonian, we find that the energy is independent of φ; that is, the VX phase has a $U(1)$ degeneracy. For illustrative purposes, we settle on the choice $\varphi = \pi/4$, as this state emerges in our Monte Carlo annealing. We conjecture that this is because the degeneracy is broken slightly above $T = 0$ by thermal and quantum fluctuations. Finally, because the state is coplanar, it gains no energy from the DM term. As α is reduced, the DM term grows and the coplanar state becomes unstable giving way to the non-coplanar Skyrmion crystal (SkX).

(III) For $\lambda > 1$ and intermediate values of the spin-orbit coupling, we recover a magnetic phase reminiscent of the "stripe" condensate described previously. Here we have an incommensurate spin spiral along the (11) or $(1\bar{1})$ direction of the lattice. This sort of spiral state results quite generically from the combination of ferromagnetic exchange with any nonzero DM interaction. At weak coupling, the pitch of the stripe condensate was determined solely by the location of the energy minima k_m, while at strong coupling, it is determined by the ratio of the DM interaction to the spin-isotropic interaction. Thus, even though the magnetic structure is similar, the underlying mechanism — interference in the condensate and superexchange in the insulator — is quite different. For $\lambda < 1$, coplanar spiral order is also found, but the spiral vector is along the (10) or (01) direction. This kind of order does the most to compromise between the DM term (tumbling the spins in one direction so that the cross product between spins in that direction does not vanish) while also satisfying the large compass interaction (by aligning spins along the other direction).

(IV) In a small parameter regime, we find that the energy is minimized by a superposition of stripes in the (10) and (01) directions. This superposition leads to a magnetic texture that again is reminiscent of the Skyrmion condensate described above. In this case it has a unit cell of 3×3 lattice sites. The central spin in the unit cell points in either the positive or negative z direction, while the remaining spins tumble outward toward the opposite z direction. For $\lambda < 1$, the extra planar anisotropy prevents these off-center spins from having a significant z-component, however. As the only non-coplanar arrangement of spins, this state also carries a non-zero spin chirality $\sum_i S_i \cdot \left(S_{i+\hat{x}} \times S_{i+\hat{y}} \right)$. In this lattice discretization of the spin chirality, there is no need for the result to be quantized, however it is useful to note that this definition is inspired by a continuum formulation, wherein this spin chirality *is* topologically quantized, and simply counts the number of Skyrmions present in the texture.

In between the two limits so far described, a transition must occur from a Mott insulating phase supporting various magnetic structures to a superfluid phase. Unlike the one-dimensional case, where exact numerical methods

(DMRG, for example) can be applied, here we must resort to mean field theory. In the presence of spin-orbit coupling, the order-parameter is multicomponent and may vary from site to site to incorporate inhomogeneous spin-density and phase structure. This requires us to extend the standard homogeneous Gutzwiller mean field theory for the Bose-Hubbard model to consider a more general Gutzwiller ansatz:

$$|\Psi\rangle = \prod_i \left(f_{i,0} + f_{i,1,1} b_{i\uparrow}^\dagger + f_{i,1,-1} b_{i\downarrow}^\dagger + \right.$$

$$\left. + f_{i,2,2} b_{i\uparrow}^\dagger b_{i\uparrow}^\dagger + f_{i,2,0} b_{i\downarrow}^\dagger b_{i\uparrow}^\dagger + f_{i,2,-2} b_{i\downarrow}^\dagger b_{i\downarrow}^\dagger \ldots \right) |0\rangle, \quad (37)$$

where the coefficients $\{f_{i,2S,2m_S}\}$ can be determined by diagonalization in the local Hilbert space of each lattice site, where each site is coupled to its neighbors through the whole set of $\varphi_{i\sigma} \equiv \langle b_{i\sigma} \rangle$, and the set of coefficients are sought which satisfy the self-consistency condition at each site. By starting with different initial states we can then search for global energy minima in the space of such self-consistent solutions. From Eq. 37, it is clear that any number of terms can be added (the local Hilbert space is infinite), but to study the $n = 1$ Mott insulator to superfluid transition, the six terms written are typically sufficient since higher occupancies are strongly suppressed by the Hubbard repulsion near the Mott transition. Finally, due to the inhomogeneity expected in the solution, calculations must be carried out on a finite cluster of linear dimension L, checking for the stability of the ground state as L is varied.

Similar questions about the Mott-superfluid transition in two dimensions arise as those from the one-dimensional case. (1) Does the superexchange-induced magnetic order in the insulator persist across the Mott transition? (2) If so, how does this impact the nature of the transition, compared to the traditional Bose-Hubbard model? (3) Just above the Mott transition, can there exist spin-ordered superfluid phases which have no weak-coupling analog?

Some representative results of the Gutzwiller approach are shown in Fig. 8, where we plot the Mott lobes for filling $n = 1$ and $\alpha = 0, \pi/4, \pi/2$. In general, increasing α frustrates the hopping so that a larger bare t is required to support a superfluid phase. This result was first obtained, absent the possibility of a spatially varying order parameter, by Graß, et al. using

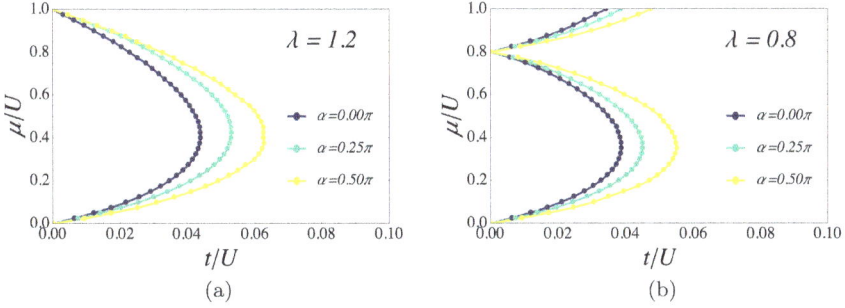

Fig. 8. Phase diagrams of the spin-orbit coupled Bose-Hubbard model in μ/U vs. t/U plane, showing Mott lobes and superfluid states. (a) phase diagram with $\lambda = 1.2$ and $\alpha = (0, 0.25, 0.5)\pi$ and (b) $\lambda = 0.8$ and $\alpha = (0, 0.25, 0.5)\pi$. The width of the $n = 1$ lobe is given by λU and the critical value $(t/U)_c$ increases with λ.

a series expansion approach.[85] This series expansion method has also been used to give some indication of the excitations of the model,[67] as well as the very intriguing proposal by Wong and Duine[86] of the possibility of quasiparticle excitation bands that carry nontrivial Chern number, which will be discussed shortly.

In addition to merely locating the phase transition, the Gutzwiller approach also allows for a spatially varying order-parameter, and it is interesting to investigate the strong-coupling superfluid states by looking at the spatial dependence of the solutions. One way to characterize these states is through local spin-densities and bond currents

$$\mathbf{m}_i = \left\langle b_{i\mu}^\dagger \sigma_{\mu\mu'} b_{i\mu'} \right\rangle, \tag{38}$$

$$J_{ij}^{\mu\nu} = -it(\mathcal{R}_{ij}^{\mu\nu} \langle b_{i\mu}^\dagger b_{j\nu}\rangle - c.c.). \tag{39}$$

The latter quantity describes the current flow from site j to i, with $\mu\nu$ indicating that it is a tensor in the spin space. In Fig. 9, we show the $\hat{z}$-component of the magnetization $\mathbf{m}_{iz}$ and the number current along the bond $\tilde{J}_{ij} = \sum_\mu J_{ij}^{\mu\mu}$ for a variety of mean-field states. All of these states have uniform number density, but exhibit different magnetic order. For example, with $\lambda = 1.5$ and $\alpha = \pi/2$, the superfluid state exhibits zAF magnetic order, consistent with the magnetic phase in the Mott insulating regime, while at $\lambda = 0.5$ the magnetization is in the plane and adopts the VX structure. In addition, however, plaquette currents

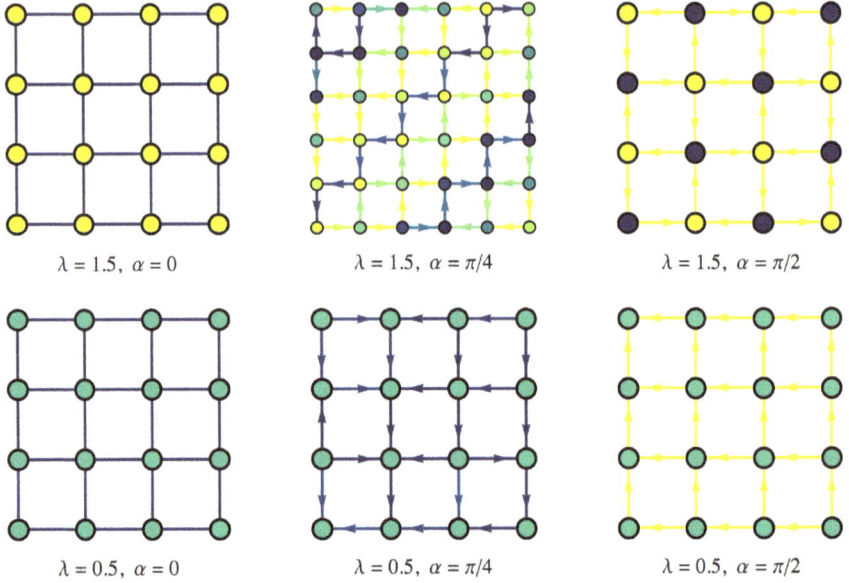

Fig. 9. Magnetic structure and currents in the strong coupled superfluid state close to the Mott-superfluid phase boundary. Blue (yellow) dots denote spin up (down), while green ones indicate that the magnetization is ordered in the xy-plane. Number currents are plotted as arrows on the bonds, with the brightness of the color indicating the magnitude of the current.

are generated in the superfluid. We note that such current patterns can be observed in experiments using quantum quenches.[87] Similar results to those outlined here have also been obtained by other authors at this level of approximation.[68] Additionally, recent calculations using a much more sophisticated bosonic DMFT approach have led to similar conclusions.[69]

Insights into the nature of the bond currents in the ground state may be obtained using a slave boson approach, which has been formulated for spinor bosons in the context of the SOBHM.[45] For magnetically ordered superfluids, this approach is particularly simple to understand, and it amounts to freezing the spinor part of the boson wavefunction while allowing for superfluidity and currents to be determined by the charge sector of the Hamiltonian. Schematically, we can set $b_{i\mu}^\dagger = a_i^\dagger z_{i\mu}$ where the spinor wavefunction $z_{i\mu}$ is chosen to correspond to the spin structure of the ground state, and the a-boson simply carries a 'charge' quantum number. Explicitly, denoting angles (θ_i, ϕ_i) to refer to the local spin direction, we arrive at

$$z_{i\uparrow} = \cos(\theta_i/2)e^{-i\phi_i/2} \tag{40}$$

$$z_{i\downarrow} = \sin(\theta_i/2)e^{+i\phi_i/2}, \tag{41}$$

so that the effective Hamiltonian for the a-bosons takes the form,

$$H_a = -t\sum_{i\delta}(a_i^\dagger a_{i+\delta}[z_{i\alpha}^* R_{i,i+\delta}^{\alpha\beta} z_{i+\delta,\beta}] + \text{h.c.}) + \frac{U}{2}\sum_i a_i^\dagger a_i^\dagger a_i a_i$$

$$+ (\lambda - 1)U \sum_i |z_{i\uparrow}|^2 |z_{i\downarrow}|^2 a_i^\dagger a_i^\dagger a_i a_i. \tag{42}$$

It then becomes clear that $[z_{i\alpha}^* R_{i,i+\delta}^{\alpha\beta} z_{i+\delta,\beta}] \sim e^{iA_{i,i+\delta}}$ leads to an effective $U(1)$ gauge field (a 'synthetic magnetic field') for the charge bosons, and different magnetic orders imprint different background gauge fields, which allows us to understand the novel bond current patterns found in the mean field theory. For instance, an Ising antiferromagnetic order, observed at $\alpha = \pi/2$ and $\lambda > 1$, imprints a π-flux through each plaquette for the a-bosons, leading to a checkerboard pattern of current order, spontaneously breaking the translationally symmetry of the lattice as shown in Fig. 9. Furthermore, the Hamiltonian in Eq. (42) also allows us to describe the superfluid to Mott transition of bosons with a given magnetic order, as varying U can lead to Mott localization of the a-bosons which describes charge localization in the SOBHM.

Finally, let us discuss the possible topological phases in the spin-orbit coupled Bose-Hubbard models, as proposed by Wong and Duine.[86] Unlike the Fermi system, where the topology of the band structure can be readily explored with free fermions, free bosons, even in a nontrivial band, will automatically condense into the lowest single particle state. Free bosons are thus insensitive to the topology of the band structure. In addition, for weak interactions, the Bose-condensed system exhibits gapless bulk phonon excitations, in contrast to the bulk-gapped topological insulators. As a result, a natural place to look for the possible emergence of topological properties for SOBHM is in the Mott insulating regime, where the single particle excitations are gapped due to strong interactions.

For simplicity, let us consider a ferromagnetic Mott insulating state, as done in Ref. 86. This simplifies the discussion considerably, as the Mott insulating state respects the lattice translational invariance. As a result, we can write the inverse of the single particle Green function in the Mott regime as

$$-\hat{G}^{-1}(\omega, \mathbf{k}) = \tilde{d}_0(\omega, \mathbf{k}) + \tilde{d}_i(\omega, \mathbf{k}) \cdot \sigma. \tag{43}$$

The quasi-particle (hole) excitations are determined by $\det[\hat{G}^{-1}(\omega, \mathbf{k})] = 0$ and we shall denote the excitation energy as $\omega_0(\mathbf{k})$. The quasi-particle can then be regarded as moving in an effective magnetic field with an effective $\mathbf{d}$-vector given by

$$\tilde{d}_i(\omega_0(\mathbf{k}), \mathbf{k}). \tag{44}$$

In the ground state, all the quasi-hole excitations are occupied and the topological character of the Mott insulating state can be determined by integrating the Berry curvature of the quasi-hole excitations over the Brillouin Zone. Within the random phase approximation for the SOBHM, the quasi-particle $\tilde{\mathbf{d}}$ assumes a particularly simple form

$$\tilde{d}_x(\omega_0(\mathbf{k}), \mathbf{k}) = d_x(\mathbf{k}),$$

$$\tilde{d}_y(\omega_0(\mathbf{k}), \mathbf{k}) = d_y(\mathbf{k}),$$

$$\tilde{d}_z(\omega_0(\mathbf{k}), \mathbf{k}) = d_z(\mathbf{k}) + \frac{1}{2}(g_{\downarrow\downarrow}^{-1}(\omega_0(\mathbf{k})) - g_{\uparrow\uparrow}^{-1}(\omega_0(\mathbf{k}))),$$

where $g_{\sigma\sigma}(\omega)$ is the onsite Green function. Thus interactions enter only through the modification of the $\hat{z}$-component of the effective magnetic field d. In the cases investigated in Ref. 86, such a modification can lead to an integer Hall conductivity even though the underlying free-particle band structure is trivial. In the language of the slave-boson picture discussed above, this means the spontaneous ordering of spins in the Mott insulator leads to time-reversal breaking, and an extra added charge boson (particle or hole in this case), which senses local orbital magnetic fields on plaquettes, can develop a gapped, topologically nontrivial, band structures with nonzero Chern numbers.

4. Future prospects

In this article, we first reviewed several experimental schemes currently employed to study the spin-orbit coupling in cold atoms in the continuum as well as in optical lattices. In the later case, we concentrate on the interplay between the strong onsite interaction and the spin-orbit coupling and point out several novel phenomena associated with them. Theoretically, a few outstanding questions remain to be understood in SOBHM.

- What is the nature of the superfluid to Mott insulator phase transitions in the presence of spin-orbit coupling? How does the density or spin modulation in the superfluid state modify the critical properties of the transition?
- It is necessary to better characterize the superfluid state by, for example, calculating and measuring the superfluid and spin superfluid density. The same problem remains to be done in the case of spin-orbit coupled quantum gases in the absence of an optical lattice.
- Investigate the possible band topology and edge states in the Mott insulating regime where interesting magnetic phases (SkX and VX) are present.

So far, only spin-orbit coupling along one direction is realized in actual experiments and there exists proposals to engineer Rashba spin-orbit coupling using extension of Raman scheme[88–90] and pulsed inhomogeneous magnetic fields using atomic chip.[91] The realisation of Rashba spin-orbit coupling would enable the study of the remarkable spin-textured Mott insulators and superfluids as unveiled theoretically. Theoretical work using a finite temperature generalization of the Gutzwiller approach also suggests the appearance of an unusual normal stripe fluid of lattice bosons with spin orbit coupling.[92]

Currently, the major obstacle with the Raman scheme is heating due to spontaneous emission and this concern seems to be less severe in the case of shaking lattice. The problem of heating could be mitigated by using atoms with a long-lived electronic excited state such as Yb[93–96] and Sr,[97] or Lanthanide atoms like Dy[98] and Er.[99] These atoms offer, in addition to the possible spin-orbit coupling induced by Raman lasers, a larger manifold of spin states which could open the gateway towards new exotic quantum states in cold atoms.[100] On the other hand, with the recently realized Haldane model using shaking lattice,[11] an obvious next step would be to investigate the interaction effects and search for fractional Chern insulators.[101,102]

Finally, there may be interesting directions to explore by putting spinor atoms in close proximity to the surfaces of cryogenic materials.[103] Letting the atoms interact with surfaces of complex oxides which can support novel magnetic textures may lead to novel gauge field configurations.

Acknowledgement

We would like to thank Xu Zhihao and Subroto Mukerjee for discussions. S.Z. is supported by a startup grant from the University of Hong Kong and a grant from the Research Grants Council of the Hong Kong Special Administrative Region, China (Grant No. HKUST3/CRF/13G, and RGC/HKU-709313P). WSC acknowledges NSF grant DMR139461 and NT was supported under ARO Grant No. W911NF-13-1-0018 with funds from the DARPA OLE program. AP acknowledges support from NSERC of Canada, and thanks the Aspen Center for Physics (Grant No. NSF PHY-1066293) for hospitality during completion of this manuscript.

Appendix. Effective Magnetic Hamiltonian Derived from Two-Site Perturbation theory

Taking the limit $U, U' \gg t$ for the Hamiltonian Eq. (27), particle number fluctuations are effectively blocked, leaving only *virtual* hopping processes to reduce the ground-state degeneracy. An effective magnetic Hamiltonian governing these residual spin fluctuations can be derived for the Mott insulator by ordinary second-order perturbation theory. The starting point is to consider a restriction to two lattice sites, and to decompose the Hamiltonian as $H = H_0 + H_1$ where H_0 includes only the onsite (i.e., the interaction) terms and

$$H_1 = \sum_{\alpha\beta} b_{1\alpha}^{\dagger} h_{\alpha\beta} b_{2\beta} + \text{h.c.} \qquad (A.1)$$

On these two sites, the set of lowest energy eigenstates of H_0 is spanned by

$$\mathcal{G} = \{|\uparrow_1, \uparrow_2\rangle, |\uparrow_1, \downarrow_2\rangle, |\downarrow_1, \uparrow_2\rangle, |\downarrow_1, \downarrow_2\rangle\}, \qquad (A.2)$$

which all have eigenvalue -2μ. The perturbative shifts to the eigenvalues and eigenstates can be combined into an *effective* magnetic Hamiltonian in this basis, whose matrix elements between states $|s_1\rangle, |s_2\rangle \in \mathcal{G}$ are

$$(H_{\text{mag}})_{s_1 s_2} = -\sum_{\gamma} \frac{\langle s_1| H_1 |\gamma\rangle \langle \gamma| H_1 |s_2\rangle}{E_\gamma - \frac{1}{2}\left(E_{s_1} + E_{s_2}\right)}, \qquad (A.3)$$

where $|\gamma\rangle$ are the six available excited states which can be obtained by acting on the four ground states with H_1. The algebra that goes into constructing this spin model is tedious, but is included here for completeness. To start, we can calculate

$$H_1 |\mu_1, \nu_2\rangle = \sum_{\alpha\beta} \left(b_{1\alpha}^\dagger h_{\alpha\beta} b_{2\beta} + b_{2\beta}^\dagger h_{\alpha\beta}^* b_{1\alpha} \right) b_{1\mu}^\dagger b_{2\nu}^\dagger |0\rangle \tag{A.4}$$

$$= \sum_{\alpha\beta} \left(b_{1\alpha}^\dagger h_{\alpha\beta} b_{1\mu}^\dagger \delta_{\beta,\nu} + b_{2\beta}^\dagger h_{\alpha\beta}^* b_{2\nu}^\dagger \delta_{\alpha,\mu} \right) |0\rangle \tag{A.5}$$

$$= \sum_{\alpha} h_{\alpha\nu} b_{1\alpha}^\dagger b_{1\mu}^\dagger |0\rangle + \sum_{\beta} h_{\mu\beta}^* b_{2\beta}^\dagger b_{2\nu}^\dagger |0\rangle . \tag{A.6}$$

While substituting in the four ground-state spin orientations yields

$$H_1 |\uparrow_1, \uparrow_2\rangle = \sqrt{2} h_{\uparrow\uparrow} |(\uparrow\uparrow)_1, 0_2\rangle + h_{\downarrow\uparrow} |(\uparrow\downarrow)_1, 0_2\rangle$$
$$+ \sqrt{2} h_{\uparrow\uparrow}^* |0_1, (\uparrow\uparrow)_2\rangle + h_{\uparrow\downarrow}^* |0_1, (\uparrow\downarrow)_2\rangle , \tag{A.7}$$

$$H_1 |\uparrow_1, \downarrow_2\rangle = \sqrt{2} h_{\uparrow\downarrow} |(\uparrow\uparrow)_1, 0_2\rangle + h_{\downarrow\downarrow} |(\uparrow\downarrow)_1, 0_2\rangle$$
$$+ \sqrt{2} h_{\uparrow\downarrow}^* |0_1, (\downarrow\downarrow)_2\rangle + h_{\uparrow\uparrow}^* |0_1, (\uparrow\downarrow)_2\rangle , \tag{A.8}$$

$$H_1 |\downarrow_1, \uparrow_2\rangle = \sqrt{2} h_{\downarrow\uparrow} |(\downarrow\downarrow)_1, 0_2\rangle + h_{\uparrow\uparrow} |(\uparrow\downarrow)_1, 0_2\rangle$$
$$+ \sqrt{2} h_{\downarrow\uparrow}^* |0_1, (\uparrow\uparrow)_2\rangle + h_{\downarrow\downarrow}^* |0_1, (\uparrow\downarrow)_2\rangle , \tag{A.9}$$

$$H_1 |\downarrow_1, \downarrow_2\rangle = \sqrt{2} h_{\downarrow\downarrow} |(\downarrow\downarrow)_1, 0_2\rangle + h_{\uparrow\downarrow} |(\uparrow\downarrow)_1, 0_2\rangle$$
$$+ \sqrt{2} h_{\downarrow\downarrow}^* |0_1, (\downarrow\downarrow)_2\rangle + h_{\downarrow\uparrow}^* |0_1, (\uparrow\downarrow)_2\rangle . \tag{A.10}$$

The factors of $\sqrt{2}$ arise from normalization, e.g., $|(\uparrow\uparrow)_1, 0_2\rangle = \frac{1}{\sqrt{2}} b_{1\uparrow}^\dagger b_{1\uparrow}^\dagger |0\rangle$. At this point it is convenient to rewrite this as a table of the matrix elements which go into Eq. (A.3), provided in Table A.1.

At this point we may insert the matrix elements in Table A.1 into the expression Eq. (A.3). The result is a 4×4 matrix of couplings which is not particularly illuminating. However we will need these matrix elements for an alternative representation with a more physical character, which we now construct.

Each term in the effective Hamiltonian can be replaced by a boson operator expression, since H_{mag} can be expanded in the set of projection

Table A.1. Virtual state matrix elements which enter into Eq. (A.3) for calculating the low-energy effective spin Hamiltonian deep in the Mott insulating limit. The rightmost column gives the energy gap to the corresponding virtual excitation.

$\langle \gamma\,|\,H_1\,|s\rangle$	$|{\uparrow}_1,{\uparrow}_2\rangle$	$|{\uparrow}_1,{\downarrow}_2\rangle$	$|{\downarrow}_1,{\uparrow}_2\rangle$	$|{\downarrow}_1,{\downarrow}_2\rangle$	$E_\gamma - \frac{1}{2}\left(E_{s_1}+E_{s_2}\right)$
$\langle(\uparrow\uparrow)_1,0_2|$	$\sqrt{2}h_{\uparrow\uparrow}$	$\sqrt{2}h_{\uparrow\downarrow}$	0	0	$U_{\uparrow\uparrow}$
$\langle(\uparrow\downarrow)_1,0_2|$	$h_{\downarrow\uparrow}$	$h_{\downarrow\downarrow}$	$h_{\uparrow\uparrow}$	$h_{\uparrow\downarrow}$	$U_{\uparrow\downarrow}$
$\langle(\downarrow\downarrow)_1,0_2|$	0	0	$\sqrt{2}h_{\downarrow\uparrow}$	$\sqrt{2}h_{\downarrow\downarrow}$	$U_{\downarrow\downarrow}$
$\langle0_1,(\uparrow\uparrow)_2|$	$\sqrt{2}h^*_{\uparrow\uparrow}$	0	$\sqrt{2}h^*_{\downarrow\uparrow}$	0	$U_{\uparrow\uparrow}$
$\langle0_1,(\uparrow\downarrow)_2|$	$h^*_{\uparrow\downarrow}$	$h^*_{\uparrow\uparrow}$	$h^*_{\downarrow\downarrow}$	$h^*_{\downarrow\uparrow}$	$U_{\uparrow\downarrow}$
$\langle0_1,(\downarrow\downarrow)_2|$	0	$\sqrt{2}h^*_{\uparrow\downarrow}$	0	$\sqrt{2}h^*_{\downarrow\downarrow}$	$U_{\downarrow\downarrow}$

operators into $\mathcal{G}$

$$\left|\sigma_1\sigma_2'\right\rangle\left\langle\tau_1\tau_2'\right| = b^\dagger_{1\sigma}b^\dagger_{2\sigma'}b_{2\tau'}b_{1\tau}, \tag{A.11}$$

which can then be expressed in local spin operators through the transformations

$$b^\dagger_{i\uparrow}b_{i\uparrow} = \frac{1}{2} + S^z_i, \quad b^\dagger_{i\downarrow}b_{i\downarrow} = \frac{1}{2} - S^z_i, \quad b^\dagger_{i\uparrow}b_{i\downarrow} = S^+_i, \quad b^\dagger_{i\downarrow}b_{i\uparrow} = S^-_i. \tag{A.12}$$

In the remainder of this Appendix, we carry out the expansion of H_{mag} and rearrangement of terms that give the more familiar magnetic model presented in the main text. First the expansion, which is simply writing the sum $H_{\text{mag}} = \sum_{\alpha\beta}(H_{\text{mag}})_{\alpha\beta}\,|\alpha\rangle\,\langle\beta|$ out explicitly. We simplify the notation by taking $V_{\alpha\beta} \equiv \left(H_{\text{mag}}\right)_{\alpha\beta}$

$$\begin{aligned}
H_{\text{mag}} = {}& V_{11}b^\dagger_{1\uparrow}b_{1\uparrow}b^\dagger_{2\uparrow}b_{2\uparrow} + V_{22}b^\dagger_{1\uparrow}b_{1\uparrow}b^\dagger_{2\downarrow}b_{2\downarrow} \\
& + V_{33}b^\dagger_{1\downarrow}b_{1\downarrow}b^\dagger_{2\uparrow}b_{2\uparrow} + V_{44}b^\dagger_{1\downarrow}b_{1\downarrow}b^\dagger_{2\downarrow}b_{2\downarrow} \\
& + \Big(V_{12}b^\dagger_{1\uparrow}b_{1\uparrow}b^\dagger_{2\uparrow}b_{2\downarrow} + V_{13}b^\dagger_{1\uparrow}b_{1\downarrow}b^\dagger_{2\uparrow}b_{2\uparrow} \\
& \quad + V_{14}b^\dagger_{1\uparrow}b_{1\downarrow}b^\dagger_{2\uparrow}b_{2\downarrow} + V_{23}b^\dagger_{1\uparrow}b_{1\downarrow}b^\dagger_{2\downarrow}b_{2\uparrow} \\
& \quad + V_{24}b^\dagger_{1\uparrow}b_{1\downarrow}b^\dagger_{2\downarrow}b_{2\downarrow} + V_{34}b^\dagger_{1\downarrow}b_{1\downarrow}b^\dagger_{2\uparrow}b_{2\downarrow} + \text{h.c.}\Big). \quad
\end{aligned} \tag{A.13}$$

Now we insert the substitution of spin operators

$$
\begin{aligned}
H_{\text{mag}} = {} & V_{11}\left(\frac{1}{2}+S_1^z\right)\left(\frac{1}{2}+S_2^z\right) + V_{22}\left(\frac{1}{2}+S_1^z\right)\left(\frac{1}{2}-S_2^z\right) \\
& + V_{33}\left(\frac{1}{2}-S_1^z\right)\left(\frac{1}{2}+S_2^z\right) + V_{44}\left(\frac{1}{2}-S_1^z\right)\left(\frac{1}{2}-S_2^z\right) \\
& + \left(V_{12}\left(\frac{1}{2}+S_1^z\right)S_2^+ + V_{13}S_1^+\left(\frac{1}{2}+S_2^z\right) + V_{14}S_1^+S_2^+\right. \\
& \left. + V_{23}S_1^+S_2^- + V_{24}S_1^+\left(\frac{1}{2}-S_2^z\right) + V_{34}\left(\frac{1}{2}-S_1^z\right)S_2^+ + \text{h.c.}\right).
\end{aligned}
$$

$$(\text{A.14})$$

Next we begin the process of arranging these terms

$$
\begin{aligned}
H_{\text{mag}} = {} & \frac{1}{4}\left(V_{11}+V_{22}+V_{33}+V_{44}\right) + \left(V_{11}-V_{22}-V_{33}+V_{44}\right)S_1^zS_2^z \\
& + \frac{1}{2}\left(V_{11}+V_{22}-V_{33}-V_{44}\right)S_1^z + \frac{1}{2}\left(V_{11}-V_{22}+V_{33}-V_{44}\right)S_2^z \\
& + \frac{1}{2}\big[(V_{13}+V_{24})S_1^+ + (V_{12}+V_{34})S_2^+ \\
& \quad + (V_{13}+V_{24})^*S_1^- + (V_{12}+V_{34})^*S_2^-\big] \\
& + (V_{12}-V_{34})S_1^zS_2^+ + (V_{13}-V_{24})S_1^+S_2^z \\
& + (V_{12}-V_{34})^*S_1^zS_2^- + (V_{13}-V_{24})^*S_1^-S_2^z \\
& + V_{14}S_1^+S_2^+ + V_{23}S_1^+S_2^- + V_{14}^*S_1^-S_2^- + V_{23}^*S_1^-S_2^+.
\end{aligned}
$$

$$(\text{A.15})$$

The spin raising and lowering operators are convenient for many purposes, but here we revert back to the cartesian components through $S^\pm = S^x \pm iS^y$, and, throwing out the overall constant term, we write

$$
\begin{aligned}
H_{\text{mag}} = {} & (V_{11}-V_{22}-V_{33}+V_{44})\,S_1^zS_2^z \\
& + \frac{1}{2}\left(V_{11}+V_{22}-V_{33}-V_{44}\right)S_1^z + \frac{1}{2}\left(V_{11}-V_{22}+V_{33}-V_{44}\right)S_2^z \\
& + \Re\left(V_{13}+V_{24}\right)S_1^x + \Re\left(V_{12}+V_{34}\right)S_2^x \\
& - \Im\left(V_{13}+V_{24}\right)S_1^y - \Im\left(V_{12}+V_{34}\right)S_2^y \\
& + 2\Re\left(V_{12}-V_{34}\right)S_1^zS_2^x + 2\Re\left(V_{13}-V_{24}\right)S_1^xS_2^z
\end{aligned}
$$

$$-2\Im\left(V_{12}+V_{34}\right)S_1^z S_2^y - 2\Im\left(V_{13}-V_{24}\right)S_1^y S_2^z$$
$$+2\Re\left(V_{23}+V_{14}\right)S_1^x S_2^x + 2\Re\left(V_{23}-V_{14}\right)S_1^y S_2^y$$
$$-2\Im\left(V_{14}+V_{23}\right)S_1^y S_2^x - 2\Im\left(V_{14}-V_{23}\right)S_1^x S_2^y. \tag{A.16}$$

We recognize that this can be written in a much more compact form

$$H_{\text{mag}} = \sum_{ab} S_1^a J_{ab} S_2^b + b_1 \cdot S_1 + b_2 \cdot S_2 \tag{A.17}$$

with

$$b_1 = \Re\left(V_{13}+V_{24}\right)\hat{x} - \Im\left(V_{13}+V_{24}\right)\hat{y} + \frac{1}{2}\left(V_{11}+V_{22}-V_{33}-V_{44}\right)\hat{z}, \tag{A.18}$$

$$b_2 = \Re\left(V_{12}+V_{34}\right)\hat{x} - \Im\left(V_{12}+V_{34}\right)\hat{y} + \frac{1}{2}\left(V_{11}-V_{22}+V_{33}-V_{44}\right)\hat{z}, \tag{A.19}$$

and the exchange tensor given by

$$J = \begin{pmatrix} 2\Re\left(V_{23}+V_{14}\right) & -2\Im\left(V_{14}-V_{23}\right) & 2\Re\left(V_{13}-V_{24}\right) \\ -2\Im\left(V_{14}+V_{23}\right) & 2\Re\left(V_{23}-V_{14}\right) & -2\Im\left(V_{13}-V_{24}\right) \\ 2\Re\left(V_{12}-V_{34}\right) & -2\Im\left(V_{12}+V_{34}\right) & \left(V_{11}-V_{22}-V_{33}+V_{44}\right) \end{pmatrix}. \tag{A.20}$$

Finally, we may decompose J into its symmetric $J_S = (J + J^T)/2$ and antisymmetric $J_A = (J - J^T)/2$ parts, the latter of which is entirely responsible for the Dzyaloshinski-Moriya interaction in Eq. (28).

Having built the above framework in some generality, one may now insert a particular model to calculate the matrix elements $(H_{\text{mag}})_{\alpha\beta}$, and finally generate the exchange matrix and b vectors. Thus, inserting

$$h^{+\hat{x}} = -t \begin{pmatrix} \cos\alpha & \sin\alpha \\ -\sin\alpha & \cos\alpha \end{pmatrix}, \quad h^{+\hat{y}} = -t \begin{pmatrix} \cos\alpha & i\sin\alpha \\ i\sin\alpha & \cos\alpha \end{pmatrix}, \tag{A.21}$$

as well as $U_{\uparrow\uparrow} = U_{\downarrow\downarrow} = U$, $U_{\uparrow\downarrow} = U_{\downarrow\uparrow} = \lambda U$, we can quickly obtain that the matrix elements conspire such that $b_1 = b_2 = 0$ along either bond. This should be expected as the underlying model was time-reversal symmetric, so we could have thrown these terms out by hand. The exchange tensors are, respectively,

$$J^{+\hat{x}} = \frac{4t^2}{gU} \begin{pmatrix} -\cos(2\theta) & 0 & g\sin(2\theta) \\ 0 & -1 & 0 \\ -g\sin(2\theta) & 0 & (1-2g)\cos(2\theta) \end{pmatrix}, \qquad (A.22)$$

$$J^{+\hat{y}} = \frac{4t^2}{gU} \begin{pmatrix} -1 & 0 & 0 \\ 0 & -\cos(2\theta) & -g\sin(2\theta) \\ 0 & g\sin(2\theta) & (1-2g)\cos(2\theta) \end{pmatrix}, \qquad (A.23)$$

which match precisely the expressions of Eq. (28) and the paragraph that follows it.

References

1. L. Pitaevskii and S. Stringari, *Bose-Einstein Condensation*. International Series of Monographs on Physics, Clarendon Press (2003). ISBN 9780198507192.
2. C. J. Pethick and H. Smith, *Bose–Einstein Condensation in Dilute Gases*. Cambridge University Press (2008).
3. M. Z. Hasan and C. L. Kane, Colloquium: topological insulators, *Reviews of Modern Physics*. **82**(4), 3045 (2010).
4. M. Z. Hasan and J. E. Moore, Three-Dimensional Topological Insulators, *Annual Review of Condensed Matter Physics*. **2**(1), 55–78 (Mar., 2011).
5. X. L. Qi and S.-C. Zhang, Topological insulators and superconductors, *Reviews of Modern Physics*. **83**(4), 1057 (2011).
6. A. Vishwanath and T. Senthil, Physics of Three-Dimensional Bosonic Topological Insulators: Surface-Deconfined Criticality and Quantized Magnetoelectric Effect, *Physical Review X*. **3**(1), 011016 (Feb., 2013).
7. X. Chen, Z.-C. Gu, Z.-X. Liu, and X.-G. Wen, Symmetry protected topological orders and the group cohomology of their symmetry group, *Physical Review B*. **87**(15), 155114 (Apr., 2013).
8. T. Senthil and M. Levin, Integer Quantum Hall Effect for Bosons, *Physical Review Letters*. **110**(4), 046801 (Jan., 2013).
9. T. Esslinger, Fermi-Hubbard Physics with Atoms in an Optical Lattice, *Annual Review of Condensed Matter Physics*. **1**(1), 129–152 (Aug., 2010).
10. P. Windpassinger and K. Sengstock, Engineering novel optical lattices, *Reports on Progress in Physics*. **76**(8), 086401 (July, 2013).
11. G. Jotzu, M. Messer, R. Desbuquois, M. Lebrat, T. Uehlinger, D. Greif, and T. Esslinger, Experimental realization of the topological Haldane model with ultracold fermions, *Nature* **515**(7526), 237–240 (November, 2014).
12. M. Aidelsburger, M. Lohse, C. Schweizer, M. Atala, J. T. Barriero, S. Nascimbène, N. R. Cooper, I. Bloch and N. Goldman, Measuring the Chern number of Hofstadter bands with ultracold bosonic atoms, *Nature Physics* **11**(2), 162–166 (Dec., 2014).

13. J. Dalibard, F. Gerbier, G. Juzeliūnas, and P. Öhberg, Colloquium: Artificial gauge potentials for neutral atoms, *Reviews of Modern Physics.* **83**(4), 1523–1543 (Nov., 2011).

14. N. Goldman, G. Juzeliunas, P. Ohberg, and I. B. Spielman, Light-induced gauge fields for ultracold atoms, *ArXiv e-prints* (Aug., 2013).

15. V. Galitski and I. B. Spielman, Spin-orbit coupling in quantum gases, *Nature.* **494** (7435), 49–54 (Apr., 2014).

16. H. Zhai, Degenerate Quantum Gases with Spin-Orbit Coupling, *ArXiv e-prints* (Mar., 2014).

17. Y. J. Lin, K. Jiménez-García, and I. B. Spielman, Spin-orbit-coupled Bose-Einstein condensates, *Nature.* **470**(7336), 83–86 (Apr., 2012).

18. L. Cheuk, A. Sommer, Z. Hadzibabic, T. Yefsah, W. Bakr, and M. W. Zwierlein, Spin-Injection Spectroscopy of a Spin-Orbit Coupled Fermi Gas, *Physical Review Letters.* **109**(9), 095302 (Aug., 2012).

19. P. Wang, Z.-Q. Yu, Z. Fu, J. Miao, L. Huang, S. Chai, H. Zhai, and J. Zhang, Spin-Orbit Coupled Degenerate Fermi Gases, *Physical Review Letters.* **109**(9), 095301 (Aug., 2012).

20. J.-Y. Zhang, S.-C. Ji, Z. Chen, L. Zhang, Z.-D. Du, B. Yan, G.-S. Pan, B. Zhao, Y.-J. Deng, H. Zhai, S. Chen, and J.-W. Pan, Collective Dipole Oscillations of a Spin-Orbit Coupled Bose-Einstein Condensate, *Physical Review Letters.* **109**(11), 115301 (Sept., 2012).

21. C. Hamner, Y. Zhang, M. A. Khamehchi, M. J. Davis, and P. Engels, Spin-orbit coupled Bose-Einstein condensates in a one-dimensional optical lattice, *ArXiv e-prints* (May, 2014).

22. J. Struck, C. Oelschlaeger, R. Le Targat, P. Soltan-Panahi, A. Eckardt, M. Lewenstein, P. Windpassinger, and K. Sengstock, Quantum Simulation of Frustrated Classical Magnetism in Triangular Optical Lattices on line supporting material, *Science.* **333** (6045), 996–999 (Aug., 2011).

23. J. Struck, C. Olschlager, M. Weinberg, P. Hauke, J. Simonet, A. Eckardt, M. Lewenstein, K. Sengstock, and P. Windpassinger, Tunable Gauge Potential for Neutral and Spinless Particles in Driven Optical Lattices, *Physical Review Letters.* **108**(22), 225304 (May, 2012).

24. J. Struck, M. Weinberg, C. Ölschläger, C. Oelschlaeger, P. Windpassinger, J. Simonet, K. Sengstock, R. Hoeppner, P. Hauke, A. Eckardt, M. Lewenstein, and L. Mathey, Engineering Ising-XY spin-models in a triangular lattice using tunable artificial gauge fields, *Nature Physics.* **9**(11), 738–743 (Nov., 2013).

25. W. Zheng and H. Zhai, Floquet topological states in shaking optical lattices, *Phys. Rev. A.* **89**, 061603 (Jun, 2014). doi: 10.1103/PhysRevA.89.061603. URL http://link.aps.org/doi/10.1103/PhysRevA.89.061603.

26. S.-L. Zhang and Q. Zhou, Shaping topological properties of the band structures in a shaken optical lattice, *Phys. Rev. A.* **90**, 051601 (Nov, 2014). doi: 10.1103/PhysRevA.90.051601. URL http://link.aps.org/doi/10.1103/PhysRevA.90.051601.

27. N. Goldman and J. Dalibard, Periodically driven quantum systems: Effective Hamiltonians and engineered gauge fields, *Phys. Rev. X.* **4**, 031027 (Aug, 2014). doi:

10.1103/PhysRevX.4.031027. URL http://link.aps.org/doi/10.1103/PhysRevX.4.031027.

28. J. Struck, J. Simonet, and K. Sengstock, Spin-orbit coupling in periodically driven optical lattices, *Phys. Rev. A*. **90**, 031601 (Sep, 2014). doi: 10.1103/PhysRevA.90.031601. URL http://link.aps.org/doi/10.1103/PhysRevA.90.031601.

29. R. A. Williams, L. J. LeBlanc, K. Jiménez-García, M. C. Beeler, A. R. Perry, W. D. Phillips, and I. B. Spielman, Synthetic Partial Waves in Ultracold Atomic Collisions, *Science*. **335**(6066), 314–317 (Jan., 2012).

30. M. C. Beeler, R. A. Williams, K. Jiménez-García, L. J. LeBlanc, A. R. Perry, and I. B. Spielman, The spin Hall effect in a quantum gas, *Nature*. pp. 1–6 (June, 2013).

31. L. J. LeBlanc, M. C. Beeler, K. Jiménez-García, A. R. Perry, S. Sugawa, R. A. Williams, and I. B. Spielman, Direct observation of zitterbewegung in a Bose–Einstein condensate, *New Journal of Physics*. **15**(7), 073011 (July, 2013).

32. S.-C. Ji, J.-Y. Zhang, L. Zhang, Z.-D. Du, W. Zheng, Y.-J. Deng, H. Zhai, S. Chen, and J.-W. Pan, Experimental determination of the finite-temperature phase diagram of a spin–orbit coupled Bose gas, *Nature Physics*. **10**(4), 314–320 (Mar., 2014).

33. S.-C. Ji, L. Zhang, X.-T. Xu, Z. Wu, Y. Deng, S. Chen, and J.-W. Pan, Softening of Roton and Phonon Modes in a Bose-Einstein Condensate with Spin-Orbit Coupling, *ArXiv e-prints* (Aug., 2014).

34. Z. Fu, L. Huang, Z. Meng, P. Wang, L. Zhang, S. Zhang, H. Zhai, P. Zhang, and J. Zhang, Production of Feshbach molecules induced by spin–orbit coupling in Fermi gases, *Nature Physics*. **10**(12), 1–6 (Dec., 2013).

35. A. J. Olson, S.-J. Wang, R. J. Niffenegger, C.-H. Li, C. H. Greene, and Y. P. Chen, Tunable landau-zener transitions in a spin-orbit-coupled bose-einstein condensate, *Phys. Rev. A*. **90**, 013616 (Jul, 2014). doi: 10.1103/PhysRevA.90.013616. URL http://link.aps.org/doi/10.1103/PhysRevA.90.013616.

36. M. A. Khamehchi, Y. Zhang, C. Hamner, T. Busch, and P. Engels, Long-range interactions and roton minimum softening in a spin-orbit coupled Bose-Einstein condensate, *ArXiv e-prints* (Sept., 2014).

37. J. Zhang, H. Hu, X.-J. Liu, and H. Pu, Fermi gases with synthetic spin-orbit coupling, *Annual Review of Cold Atoms and Molecules*. **2**, 81 (2014).

38. C. J. Kennedy, G. A. Siviloglou, H. Miyake, W. C. Burton, and W. Ketterle, Spin-Orbit Coupling and Quantum Spin Hall Effect for Neutral Atoms without Spin Flips, *Physical Review Letters*. **111**(22), 225301 (Nov., 2013).

39. K. Jiménez-García, L. J. LeBlanc, R. A. Williams, M. C. Beeler, A. R. Perry, and I. B. Spielman, Peierls Substitution in an Engineered Lattice Potential, *Physical Review Letters*. **108**(22), 225303 (May, 2012).

40. J. Higbie and D. Stamper-Kurn, Periodically Dressed Bose-Einstein Condensate: A Superfluid with an Anisotropic and Variable Critical Velocity, *Physical Review Letters*. **88**(9), 090401 (Feb., 2002).

41. I. Spielman, Raman processes and effective gauge potentials, *Physical Review A*. **79** (6), 063613 (June, 2009).

42. T.-L. Ho and S. Zhang, Bose-Einstein Condensates with Spin-Orbit Interaction, *Physical Review Letters*. **107**(15), 150403 (Oct., 2011).

43. Y. J. Lin, R. L. Compton, K. Jiménez-García, J. V. Porto, and I. B. Spielman, Synthetic magnetic fields for ultracold neutral atoms, *Nature.* **462**(7273), 628–632 (Dec., 2009).
44. T. Graß, K. Saha, K. Sengupta, and M. Lewenstein, Quantum phase transition of ultracold bosons in the presence of a non-Abelian synthetic gauge field, *Physical Review A.* **84**(5), 053632 (Nov., 2011).
45. W. Cole, S. Zhang, A. Paramekanti, and N. Trivedi, Bose-Hubbard Models with Synthetic Spin-Orbit Coupling: Mott Insulators, Spin Textures, and Superfluidity, *Physical Review Letters.* **109**(8), 085302 (Aug., 2012).
46. J. Radić, A. Di Ciolo, K. Sun, and V. Galitski, Exotic Quantum Spin Models in Spin-Orbit-Coupled Mott Insulators, *Physical Review Letters.* **109**(8), 085303 (Aug., 2012).
47. A. R. Kolovsky, Creating artificial magnetic fields for cold atoms by photon-assisted tunneling, *Epl.* **93**(2), 20003 (Feb., 2011).
48. C. E. Creffield and F. Sols, Comment on "Creating artificial magnetic fields for cold atoms by photon-assisted tunneling" by Kolovsky A. R., *Epl.* **101**(4), 40001 (Feb., 2013).
49. D. Jaksch and P. Zoller, Creation of effective magnetic fields in optical lattices: the Hofstadter butterfly for cold neutral atoms, *New Journal of Physics.* **5**, 56 (2003).
50. F. Gerbier and J. Dalibard, Gauge fields for ultracold atoms in optical superlattices, *New Journal of Physics.* **12**(3), 033007 (Mar., 2010).
51. M. Aidelsburger, M. Atala, S. Nascimbène, S. Trotzky, C. Y-A, and I. Bloch, Experimental Realization of Strong Effective Magnetic Fields in an Optical Lattice, *Physical Review Letters.* **107**(25), 255301 (Dec., 2011).
52. M. Aidelsburger, M. Atala, S. Nascimbène, S. Trotzky, C. Y-A, and I. Bloch, Experimental realization of strong effective magnetic fields in optical superlattice potentials, *Applied Physics B* (May, 2013).
53. M. Aidelsburger, M. Atala, M. Lohse, J. T. Barreiro, B. Paredes, and I. Bloch, Realization of the Hofstadter Hamiltonian with Ultracold Atoms in Optical Lattices, *Physical Review Letters.* **111**(18), 185301 (Oct., 2013).
54. H. Miyake, G. A. Siviloglou, C. J. Kennedy, W. C. Burton, and W. Ketterle, Realizing the Harper Hamiltonian with Laser-Assisted Tunneling in Optical Lattices, *Physical Review Letters.* **111**(18), 185302 (Oct., 2013).
55. A. Eckardt, P. Hauke, P. Soltan-Panahi, C. Becker, K. Sengstock, and M. Lewenstein, Frustrated quantum antiferromagnetism with ultracold bosons in a triangular lattice, *Epl.* **89**(1), 10010 (Jan., 2010).
56. C. V. Parker, L. C. Ha, and C. Chin, Direct observation of effective ferromagnetic domains of cold atoms in a shaken optical lattice, *Nature Physics* (2013).
57. N. R. Cooper, Optical flux lattices for ultracold atomic gases, *Phys. Rev. Lett.* **106**, 175301 (Apr, 2011). doi: 10.1103/PhysRevLett.106.175301. URL http://link.aps.org/doi/10.1103/PhysRevLett.106.175301.
58. B. Bernevig and T. Hughes, *Topological Insulators and Topological Superconductors.* Princeton University Press (2013). ISBN 9780691151755.
59. Z. Cai, X. Zhou, and C. Wu, Magnetic phases of bosons with synthetic spin-orbit coupling in optical lattices, *Physical Review A.* **85**(6), 061605 (June, 2012).
60. M. Gong, Y. Qian, V. W. Scarola, and C. Zhang, Dzyaloshinskii-Moriya Interaction and Spiral Order in Spin-orbit Coupled Optical Lattices, *ArXiv e-prints* (May, 2012).

61. D. Cocks, P. Orth, S. Rachel, M. Buchhold, K. Le Hur, and W. Hofstetter, Time-Reversal-Invariant Hofstadter-Hubbard Model with Ultracold Fermions, *Physical Review Letters.* **109**(20), 205303 (Nov., 2012).

62. P. P. Orth, D. Cocks, S. Rachel, M. Buchhold, K. L. Hur, and W. Hofstetter, Correlated topological phases and exotic magnetism with ultracold fermions, *Journal of Physics B: Atomic, Molecular and Optical Physics.* **46**(13), 134004 (2013). URL http://stacks.iop.org/0953-4075/46/i=13/a=134004.

63. T. Moriya, Anisotropic Superexchange Interaction and Weak Ferromagnetism, *Physical Review.* **120**(1), 91–98 (1960).

64. I. Dzyaloshinsky, A Thermodynamic Theory of Weak Ferromagnetism of Antiferromagnetics, *Journal of Physics And Chemistry of Solids.* **4**(4), 241–255 (1958).

65. A. Kitaev, Anyons in an exactly solved model and beyond, *Annals of Physics.* **321**(1), 2–111 (2006).

66. G. Jackeli and G. Khaliullin, Mott Insulators in the Strong Spin-Orbit Coupling Limit: From Heisenberg to a Quantum Compass and Kitaev Models, *Physical Review Letters.* **102**(1), 017205 (Jan., 2009).

67. S. Mandal, K. Saha, and K. Sengupta, Superfluid-insulator transition of two-species bosons with spin-orbit coupling, *Phys. Rev. B.* **86**, 155101 (Oct. 2012). doi: 10.1103/PhysRevB.86.155101. URL http://link.aps.org/doi/10.1103/PhysRevB.86.155101.

68. Y. Qian, M. Gong, V. W. Scarola, and C. Zhang, Spin-Orbit Driven Transitions Between Mott Insulators and Finite Momentum Superfluids of Bosons in Optical Lattices, *ArXiv e-prints* (Dec., 2013).

69. L. He, A. Ji, and W. Hofstetter, Bose-Bose Mixtures with Synthetic Spin-Orbit Coupling in Optical Lattices, *ArXiv e-prints* (Apr. 2014).

70. J. Zhao, S. Hu, J. Chang, P. Zhang, and X. Wang, Ferromagnetism in a two-component bose-hubbard model with synthetic spin-orbit coupling, *Phys. Rev. A.* **89**, 043611 (Apr, 2014). doi: 10.1103/PhysRevA.89.043611. URL http://link.aps.org/doi/10.1103/PhysRevA.89.043611.

71. J. Zhao, S. Hu, J. Chang, F. Zheng, P. Zhang, and X. Wang, The evolution of magnetic structure driven by a synthetic spin-orbit coupling in two-component Bose-Hubbard model, *ArXiv e-prints* (Mar., 2014).

72. Z. Xu, W. S. Cole, and S. Zhang, Mott-superfluid transition for spin-orbit-coupled bosons in one-dimensional optical lattices, *Physical Review A.* **89**(5), 051604 (May, 2014).

73. M. Piraud, Z. Cai, I. P. McCulloch, and U. Schollwöck, Quantum magnetism of bosons with synthetic gauge fields in one-dimensional optical lattices: A density-matrix renormalization-group study, *Phys. Rev. A.* **89**, 063618 (Jun. 2014). doi: 10.1103/PhysRevA.89.063618. URL http://link.aps.org/doi/10.1103/PhysRevA.89.063618.

74. S. Peotta, L. Mazza, E. Vicari, M. Polini, R. Fazio and D. Rossini, The XYZ chain with Dzyaloshinsky-Moriya interactions: from spin-orbit-coupled lattice bosons to interacting Kitaev chains, *Journal of Statistical Mechanics: Theory and Experiment* **2014**(9), P09005, http://stacks.iop.org/1742-5468/i=9/a=P09005, (2014).

75. J. Perk and H. Capel, Antisymmetric exchange, canting and spiral structure, *Physics Letters A.* **58**(2), 115–117 (1976).
76. R. Jafari, M. Kargarian, A. Langari, and M. Siahatgar, Phase diagram and entanglement of the Ising model with Dzyaloshinskii-Moriya interaction, *Physical Review B.* **78**(21), 214414 (Dec., 2008).
77. A. Y. Kitaev, Unpaired Majorana fermions in quantum wires, *Physics-Uspekhi.* **44**, 131 (2001).
78. S. Ryu, A. P. Schnyder, A. Furusaki, and A. W. W. Ludwig, Topological insulators and superconductors: tenfold way and dimensional hierarchy, *New Journal of Physics.* **12** (6), 065010 (June, 2010).
79. C. Wang, C. Gao, C.-M. Jian, and H. Zhai, Spin-Orbit Coupled Spinor Bose-Einstein Condensates, *Physical Review Letters.* **105**(16) (Oct. 2010).
80. T. Ozawa and G. Baym, Stability of Ultracold Atomic Bose Condensates with Rashba Spin-Orbit Coupling against Quantum and Thermal Fluctuations, *Physical Review Letters.* **109**(2), 025301 (July, 2012).
81. T. Ozawa and G. Baym, Ground-state phases of ultracold bosons with Rashba-Dresselhaus spin-orbit coupling, *Physical Review A.* **85**(1) (Jan., 2012).
82. H. Hu, B. Ramachandhran, H. Pu, and X.-J. Liu, Spin-Orbit Coupled Weakly Interacting Bose-Einstein Condensates in Harmonic Traps, *Physical Review Letters.* **108**(1), 010402 (Jan., 2012).
83. R. M. Wilson, B. M. Anderson, and C. W. Clark, Meron ground state of rashba spin-orbit-coupled dipolar bosons, *Phys. Rev. Lett.* **111**, 185303 (Oct. 2013). doi: 10.1103/PhysRevLett.111.185303. URL http://link.aps.org/doi/10.1103/PhysRevLett.111.185303.
84. S. Gopalakrishnan, I. Martin, and E. A. Demler, Quantum quasicrystals of spin-orbit-coupled dipolar bosons, *Phys. Rev. Lett.* **111**, 185304 (Oct. 2013). doi: 10.1103/PhysRevLett.111.185304. URL http://link.aps.org/doi/10.1103/PhysRevLett.111.185304.
85. T. Graß, K. Saha, K. Sengupta, and M. Lewenstein, Quantum phase transition of ultracold bosons in the presence of a non-abelian synthetic gauge field, *Phys. Rev. A.* **84**, 053632 (Nov, 2011). doi: 10.1103/PhysRevA.84.053632. URL http://link.aps.org/doi/10.1103/PhysRevA.84.053632.
86. C. H. Wong and R. A. Duine, Topological Transport in Spin-Orbit Coupled Bosonic Mott Insulators, *Physical Review Letters.* **110**(11), 115301 (Mar., 2013).
87. M. Killi and A. Paramekanti, Use of quantum quenches to probe the equilibrium current patterns of ultracold atoms in an optical lattice, *Phys. Rev. A.* **85**, 061606 (Jun. 2012). doi: 10.1103/PhysRevA.85.061606. URL http://link.aps.org/doi/10.1103/PhysRevA.85.061606.
88. D. L. Campbell, G. Juzeliūnas, and I. B. Spielman, Realistic rashba and dresselhaus spin-orbit coupling for neutral atoms, *Phys. Rev. A.* **84**, 025602 (Aug, 2011). doi: 10.1103/PhysRevA.84.025602. URL http://link.aps.org/doi/10.1103/PhysRevA.84.025602.
89. J. D. Sau, R. Sensarma, S. Powell, I. B. Spielman, and S. Das Sarma, Chiral rashba spin textures in ultracold fermi gases, *Phys. Rev. B.* **83**, 140510 (Apr, 2011). doi: 10.1103/PhysRevB.83.140510. URL http://link.aps.org/doi/10.1103/PhysRevB.83.140510.

90. Z. F. Xu and L. You, Dynamical generation of arbitrary spin-orbit couplings for neutral atoms, *Phys. Rev. A*. **85**, 043605 (Apr, 2012). doi: 10.1103/PhysRevA.85.043605. URL http://link.aps.org/doi/10.1103/PhysRevA.85.043605.

91. B. M. Anderson, I. B. Spielman, and G. Juzeliūnas, Magnetically generated spin-orbit coupling for ultracold atoms, *Phys. Rev. Lett.* **111**, 125301 (Sep, 2013). doi: 10.1103/PhysRevLett.111.125301. URL http://link.aps.org/doi/10.1103/PhysRevLett.111.125301.

92. Ciarán, Hickey and Arun Paramekanti, Thermal Phase transitions of strongly correlated Bosons with spin-obit coupling, *Phys. Rev. Lett.* **113**(26), 265302 (2014).

93. T. Fukuhara, Y. Takasu, M. Kumakura, and Y. Takahashi, Degenerate Fermi Gases of Ytterbium, *Physical Review Letters*. **98**(3), 030401 (Jan., 2007).

94. S. Taie, R. Yamazaki, S. Sugawa, and Y. Takahashi, An SU(6) Mott insulator of an atomic Fermi gas realized by large-spin Pomeranchuk cooling, *Nature Physics*. **8**(11), 825–830 (Sept., 2012).

95. F. Scazza, C. Hofrichter, M. Höfer, P. C. De Groot, I. Bloch, and S. Folling, Observation of two-orbital spin-exchange interactions with ultracold SU(N)-symmetric fermions, *Nature Physics*. **10**(10), 779–784 (Aug., 2014).

96. G. Cappellini, M. Mancini, G. Pagano, P. Lombardi, L. Livi, M. Siciliani de Cumis, P. Cancio, M. Pizzocaro, D. Calonico, F. Levi, C. Sias, J. Catani, M. Inguscio, and L. Fallani, Direct Observation of Coherent Interorbital Spin-Exchange Dynamics, *Physical Review Letters*. **113**(12), 120402 (Sept., 2014).

97. X. Zhang, M. Bishof, S. L. Bromley, C. V. Kraus, M. S. Safronova, P. Zoller, A. M. Rey, and J. Ye, Spectroscopic observation of SU(N)-symmetric interactions in Sr orbital magnetism, *Science*. **345**(6203), 1467–1473 (Sept., 2014).

98. M. Lu, N. Q. Burdick, and B. L. Lev, Quantum degenerate dipolar fermi gas, *Phys. Rev. Lett.* **108**, 215301 (May, 2012). doi: 10.1103/PhysRevLett.108.215301. URL http://link.aps.org/doi/10.1103/PhysRevLett.108.215301.

99. K. Aikawa, A. Frisch, M. Mark, S. Baier, A. Rietzler, R. Grimm, and F. Ferlaino, Bose-einstein condensation of erbium, *Phys. Rev. Lett.* **108**, 210401 (May, 2012). doi: 10.1103/PhysRevLett.108.210401. URL http://link.aps.org/doi/10.1103/PhysRevLett.108.210401.

100. X. Cui, B. Lian, T.-L. Ho, B. L. Lev, and H. Zhai, Synthetic gauge field with highly magnetic lanthanide atoms, *Phys. Rev. A*. **88**, 011601 (Jul, 2013). doi: 10.1103/PhysRevA.88.011601. URL http://link.aps.org/doi/10.1103/PhysRevA.88.011601.

101. T. Neupert, L. Santos, C. Chamon, and C. Mudry, Fractional quantum hall states at zero magnetic field, *Phys. Rev. Lett.* **106**, 236804 (Jun. 2011). doi: 10.1103/PhysRevLett.106.236804. URL http://link.aps.org/doi/10.1103/PhysRevLett.106.236804.

102. M. S. Scheurer, S. Rachel, and P. P. Orth, Dimensional crossover and cold-atom realization of topological Mott insulators, *ArXiv e-prints* (June, 2014).

103. M. A. Naides, R. W. Turner, R. A. Lai, J. M. DiSciacca, and B. L. Lev, Trapping ultracold gases near cryogenic materials with rapid reconfigurability, *Applied Physics Letters*. 103(25):251112 (2013). doi: http://dx.doi.org/10.1063/1.4852017. URL http://scitation.aip.org/content/aip/journal/apl/103/25/10.1063/1.4852017.

CHAPTER 4

MICROSCOPY OF MANY-BODY STATES
IN OPTICAL LATTICES

Christian Gross and Immanuel Bloch*

Max-Planck-Institut für Quantenoptik
85748 Garching, Germany
*christian.gross@mpq.mpg.de

Ultracold atoms in optical lattices have proven to provide an extremely clean and controlled setting to explore quantum many-body phases of matter. Now, imaging of atoms in such lattice structures has reached the level of single-atom sensitive detection combined with the highest resolution down to the level of individual lattice sites. It has opened up fundamentally new opportunities for the characterization and the control of quantum many-body systems. Here we give a brief overview of this field and explore the opportunities offered for future research.

1. Introduction

Over the past years, ultracold atoms in optical lattices have emerged as versatile new system to explore the physics of quantum many-body systems. On the one hand they can be helpful in gaining a better understanding of known phases of matter and their dynamical behavior; on the other hand they allow one to realize completely novel quantum systems that have not been studied before in nature.[1–3] Commonly, the approach of exploring quantum many-body systems in such a way is referred to as "quantum simulations". Examples of some of the first strongly-interacting many-body phases that have been realized both in lattices and in the continuum include the quantum

---

*Ludwigs-Maximilians-Universität, Fakultät für Physik, 80799 München, Germany

phase transition from a Superfluid to a Mott insulator,[4–6] the achievement of a Tonks-Girardeau gas[7,8] and the realization of the BEC-BCS crossover in Fermi gas mixture[9] using Feshbach resonances.[10]

In almost all of these experiments, detection was limited to time-of-flight imaging or more refined derived techniques that mainly characterized the momentum distribution of the quantum gas.[2] However, quantum optics experiments on single or few atoms or ions had shown how powerful the detection and control of individual quantum particles can be. For several years, researchers in the field have therefore aspired to employ such single particle detection methods for the analysis of ultracold quantum gases. Only recently it has become possible to realize such imaging techniques, marking a milestone for the characterization and manipulation of ultracold quantum gases.[11–15] In our discussion, we will focus on one the most successful one among these techniques based on high-resolution fluorescence imaging. Despite being a rather new technique, such quantum gas microscopy has already proven to be an enabling technology for probing and manipulating quantum many-body systems. For the first time controllable and strongly interacting many-body systems, as realized with ultracold atoms, could be observed on a local scale.[14,15] The power of the technique became even more apparent with the advent of local hyperfine state specific addressing in optical lattices.[16] Together with the local detection this provides a complete toolbox for the manipulation of one- and two-dimensional lattice gases on the scale of a few hundred Nanometers. Our review will concentrate on the experiments employing ultracold bosonic atoms in optical lattices, showing applications in correlation measurements,[17–19] quantum magnetism[20–22] and quantum critical behaviour close to a quantum phase transition.[23]

2. Site Resolved Imaging

One of the standard imaging techniques in ultracold quantum gases — absorption imaging — cannot be easily extended to the regime of single atom sensitivity. This is mainly due to the limited absorption a laser beam experiences when interacting with a single atom. For typical experimental conditions, the absorption signal is always smaller than the accompanying photon shot noise. While high resolution images of down to $1\,\mu$m resolution have been successfully used to record *in-situ* absorption images of trapped quantum gases,[24] they have not reached the single-atom sensitive detection regime. Fluorescence imaging can however overcome this limited signal-to-noise and therefore provides a viable route for combining high-resolution imaging with single-atom sensitivity. By using laser induced fluorescence in an optical molasses configuration and by trapping the atoms in a very deep potential, several hundred thousand photons can be scattered from a single atom, of which a few thousand are ultimately detected. An excellent signal-to-noise in the detection of a single atom can therefore be achieved.

This idea was first pioneered for the case of optical lattices by the group of D. Weiss, who loaded atoms from a magneto-optical trap into a three-dimensional lattice with a lattice constant of $6\,\mu$m.[11] However, for typical condensed matter oriented experiments, such large spaced lattices are of limited use, due to their almost vanishing tunnel coupling between neighboring potential wells. Extending fluorescence imaging to a regime where the resolution can be comparable to typical sub-micron lattice spacings, thus requires large numerical apertures (NA) microscope objectives, as the smallest resolvable distances in classical optics are determined by $\sigma = \lambda/(2\text{NA})$.

In recent works, Bakr *et al.*[13, 14] and Sherson *et al.*[15] have demonstrated such high-resolution imaging and applied it to image the transition of a superfluid to a Mott insulator in 2D. In the experiments, 2D Bose-Einstein condensates were first created in tightly confining potential planes. Subsequently, the depth of a two-dimensional simple-cubic type lattice was increased, leaving the system either in a superfluid or Mott insulating regime. The lattice depths were then suddenly increased to very deep values of approximately 300μK, essentially freezing out the density distribution of the atoms in the lattice. A near-resonant optical molasses was then used to induce fluorescence of the atoms in the deep lattice and also provide laser cooling, such that atoms remained on lattice sites while fluorescing. High resolution microscope objectives with numerical apertures of NA $\approx 0.7 - 0.8$ were used to record the fluorescence and image the atomic density distribution on CCD cameras (see Fig. 1). A limitation of the detection method is that so called inelastic light-induced collisions occurring during the illumination period only allow one to record the parity of the on-site atom number. Whenever pairs of atoms are present on a single lattice site, both atoms are rapidly lost within the first few hundred microseconds of

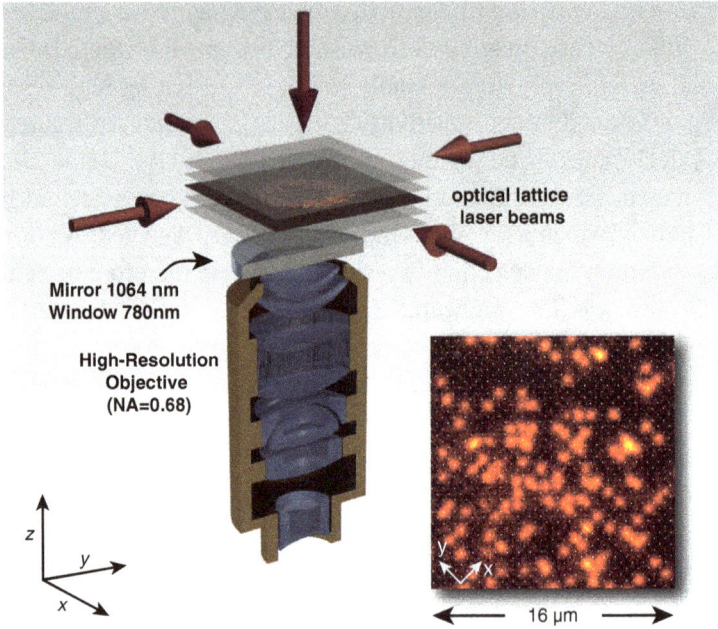

Fig. 1.   Schematic setup for high resolution fluorescence imaging of a 2D quantum gas. Two-dimensional bosonic quantum gases are prepared in a single 2D plane of an optical standing wave along the z-direction, which is created by retro-reflecting a laser beam ($\lambda = 1064\,\mathrm{nm}$) on the coated vacuum window. Additional lattice beams along the x- and y-directions are used to bring the system into the strongly correlated regime of a Mott insulator. The atoms are detected using fluorescence imaging via a high resolution microscope objective. Fluorescence of the atoms was induced by illuminating the quantum gas with an optical molasses that simultaneously laser cools the atoms. The inset shows a section from a fluorescence picture of a dilute thermal cloud (points mark the lattice sites). Adapted from Sherson *et al.*[15]

illumination, due to a large energy release caused by radiative escape and fine-structure changing collisions.[25]

In both experiments, high resolution imaging has allowed one to reconstruct the atom distribution (modulo 2) on the lattice down to a single-site level. Results for the case of a Bose-Einstein condensate and Mott insulators of such a particle number reconstruction are displayed in Fig. 2. The fidelity of the imaging process is currently limited to approximately 99% by atom loss during the illumination due to background gas collisions.

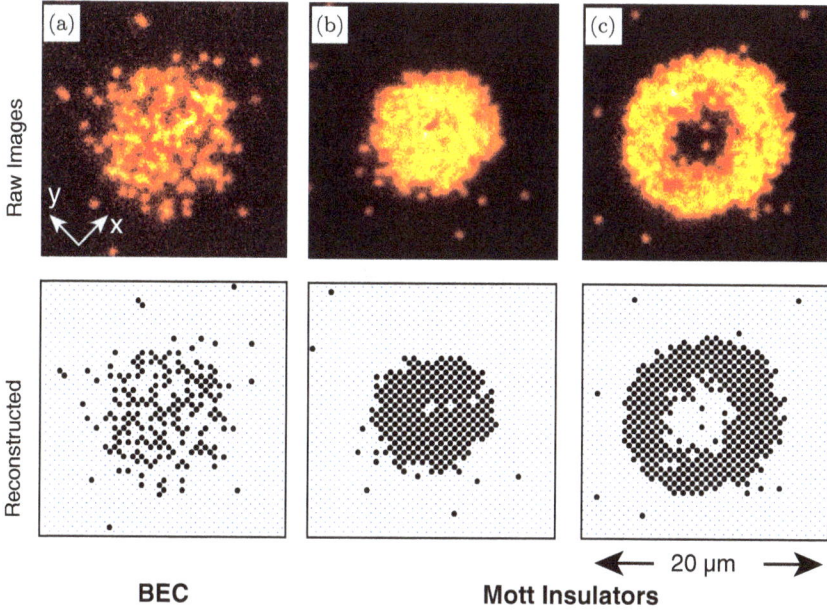

Fig. 2. High resolution fluorescence images of a weakly interacting Bose-Einstein condensate and Mott insulators. (a) Bose-Einstein condensate exhibiting large particle number fluctuations and (b,c) wedding cake structure of $n = 1$ and $n = 2$ Mott insulators. Using a numerical algorithm, the corresponding atom distribution on the lattice can be reconstructed. The reconstructed images can be seen in the row below (small points mark lattice sites, large points mark position of a single atom). Figure adapted from Sherson et al..[15]

3. Thermometry at the Limit of Individual Thermal Excitations

Within a tight-binding approximation and for interaction energies that do not exceed the vibrational level splitting on a single lattice site, the behaviour of interacting bosons on a lattice can be described by the Bose-Hubbard Hamiltonian:[4–6]

$$\hat{H} = -J \sum_{\langle \mathbf{R}, \mathbf{R}' \rangle} \hat{a}_{\mathbf{R}'}^{\dagger} \hat{a}_{\mathbf{R}} + \frac{1}{2} U \sum_{\mathbf{R}} \hat{n}_{\mathbf{R}} (\hat{n}_{\mathbf{R}} - 1), \tag{1}$$

where $\hat{a}_{\mathbf{R}}^{\dagger} (\hat{a}_{\mathbf{R}})$ denote the bosonic creation (annihilation) operators on site $\mathbf{R}$, $\hat{n}_{\mathbf{R}} = \hat{a}_{\mathbf{R}}^{\dagger} \hat{a}_{\mathbf{R}}$, J characterizes the tunnel coupling between neighbouring lattice sites and U is the on-site interaction energy of two atoms on a given lattice site. For the following discussion, we will assume repulsive interactions $U > 0$.

Deep in the Mott-insulating regime the strongly interacting bosonic quantum gas becomes essentially classical. In this so called *atomic-limit of the Bose-Hubbard model* the individual wells are disconnected, that is, the tunneling is $J = 0$ and the ratio of interaction to tunneling U/J diverges. Hence, the grand canonical partition function of the trapped quantum gas $Z^{(0)}$ can be written as a product of on-site partition functions $Z^{(0)} = \prod_{\mathbf{R}} Z_{\mathbf{R}}^{(0)}$, where the on-site partition function is given by

$$Z^{(0)} = \sum_n e^{-\beta(E_n - \mu(\mathbf{R})n)}. \tag{2}$$

The local chemical potential at lattice site $\mathbf{R}$ is denoted by $\mu(\mathbf{R})$ and the eigenenergy of n atoms on this lattice site is given by the standard single-band Bose-Hubbard interaction term $E_n = 1/2Un(n-1)$. In particular we can use the above to calculate the on-site probability of finding n atoms per lattice site as

$$P_{\mathbf{R}}(n) = \frac{e^{-\beta(E_n - \mu(\mathbf{R})n)}}{Z^{(0)}}. \tag{3}$$

Thus, the thermodynamics is determined only by the ratio of $U/(k_B T)$ and the local chemical potential. In this limit the problem becomes analytically tractable and simple to analyze.

As a simple application of our result, let us calculate the density profile and its fluctuations for a two-dimensional radially symmetric trapping potential. All sites with the same distance r from the trapping center exhibit the same chemical potential $\mu(r)$. The average density at this radial distance is thus given by $\bar{n} = \frac{1}{Z(r)} \sum_n n e^{-\beta(E_n - \mu(r))}$. In order to evaluate this, we would need to sum over all possible occupation states in our on-site partition function. In practice, we may truncate our sum around occupation numbers $n_{\max} = \text{ceil}(\mu/U) + 1$ for temperatures $k_B T \ll U$, as thermal fluctuations become exponentially suppressed in this regime. This corresponds to the so called *particle-hole* approximation. In this regime individual thermal excitations, corresponding either to a missing or and extra atom on a single site are directly detected by the local parity sensitive imaging.

Taking into the light assisted collisions into account we find for the detected average density:

$$\bar{n}_{\text{det}} = \frac{1}{Z(r)} \sum_n \text{mod}_2(n) e^{-\beta(E_n - \mu(r))}.$$ (4)

The parity projection during the imaging process assures that the experimentally detected atoms number per site is either 0 or 1. Thus, the second moment of the measured onsite atom number distribution is equal to its mean $\overline{n_{\text{det}}^2} = \bar{n}_{\text{det}}$. Within the particle-hole approximation the physical atom number per site can only fluctuate by ±1 around its average value, such that its variance σ^2 can be measured despite parity projection $\sigma^2 = \bar{n}_{\text{det}} - \bar{n}_{\text{det}}^2$. Both the average density profile and its fluctuations are functions of three parameters μ/U, $k_B T/U$ and the trapping frequency ω of the overall harmonic confinement. While the trap frequency can be independently measured, the chemical potential and temperature of the quantum gas can be extracted via a fit to azimuthally averaged radial density profiles of single images of the quantum gas. This is shown in Fig. 3 for the two images of an $n = 1$ and $n = 2$ Mott insulator in the core of the gas.

In the atomic limit, these fit-functions thus allow an efficient determination of temperature and chemical potential of the quantum gas. The radial density and fluctuation profiles can be converted to density and fluctuation profiles vs chemical potential by using again the local density approximation $\mu_{\text{loc}}(r) = \mu - 1/2m\omega^2 r^2$. We see that the data for the two distinct measurements of the $n = 1$ and $n = 2$ Mott insulators fall on top of each other when plotting in this way, underlining the fact that radial profiles correspond to cuts through the phase diagram (see inset in Fig. 3(c)) of the Bose-Hubbard model. Residual small differences between the two curves can be attributed to the slightly different temperatures of the atom clouds. Interestingly, as in the case of fermionic atoms, one notes that fluctuations are concentrated to the border of the Mott insulating regions, where the system is superfluid for $T = 0$. In the local density approximation picture, the energy gap is minimal here and, thus, it is thermodynamically easiest to introduce fluctuations in this spatial region. Note that this thermometry method can be extended to work also in the vicinity of the superfluid to Mott insulator transition. In this case the local quantum fluctuations can be

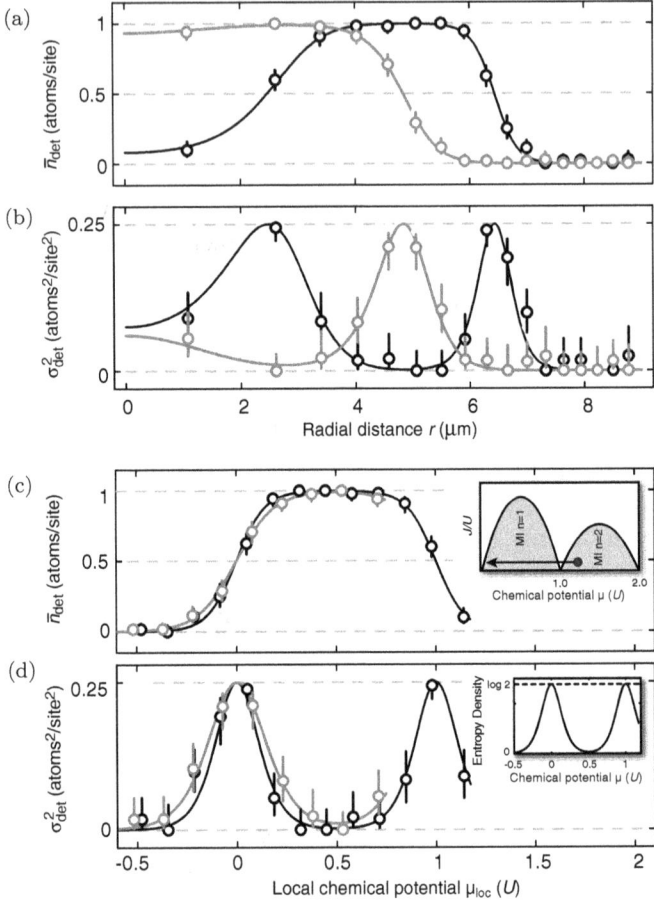

Fig. 3. Radial atom density and variance profiles. Radial profiles were obtained from the reconstructed images by azimuthal averaging. (a), (b), grey and black points correspond to the $n = 1$ and $n = 2$ MI images of Fig. 2(d),(e). For the two curves, the fits yielded temperatures $T = 0.090(5)U/k_B$ and $T = 0.074(5)U/k_B$, chemical potentials $\mu = 0.73(3)U$ and $\mu = 1.17(1)U$, and radii $r_0 = 5.7(1)$ μm and $r_0 = 5.95(4)$ μm respectively. From the fitted values of T, μ and r_0, we determined the atom numbers of the system to $N = 300(20)$ and $N = 610(20)$. (c), (d), The same data plotted versus the local chemical potential using local-density approximation. The inset of (c) is a Bose-Hubbard phase diagram ($T = 0$) showing the transition between the characteristic MI lobes and the superfluid region. The line starting at the maximum chemical potential μ shows the part of the phase diagram existing simultaneously at different radii in the trap due to the external harmonic confinement. The inset of (d) is the entropy density calculated for the displayed $n = 2$ MI. From Sherson et al.[15]

suppressed prior to the imaging by a properly timed lattice ramp that ends deep in the Mott insulator regime.[23]

4. Single-Site Resolved Addressing of Individual Atoms

Being able to spatially resolve single lattice sites also allows to manipulate atoms with single-site resolution. A laser beam can be sent in reverse through the high-resolution objective and, hence, is focused onto the atoms. Thereby the high-resolution objective is used twice — for imaging and for local addressing. In typical cases, the resulting spot size of the laser beam will still be on the order of a lattice spacing and for most applications too large in order to reliably address atoms on single lattice sites. One possibility to increase the spatial resolution is to make use of a resonance imaging technique: the focused laser is tuned to such a wavelength that it creates a differential energy shift between two internal hyperfine ground states of an atom. Then global microwave radiation will be resonant only at the position of the focused beam and thus, can be used to control the spin state of the atom.[16,26] The spatial resolution for the addressing of single atoms can thereby be increased up to a limit given by (often magnetic field driven) fluctuations of the energy splitting between the two hyperfine states. For typical parameters this corresponds to an increase by almost an order of magnitude down to $\simeq 50\,\mathrm{nm}$, well below the optical diffraction limit.

In the experiment, such addressing was demonstrated in a 2D Mott insulator with unity occupation per lattice site.[16] In order to prepare an arbitrary pattern of spins in the array, the laser beam was moved to a specific site and a Landau-Zener microwave sweep was applied in order to flip the spin of the atom located at the lattice site. The laser beam was then moved to the next lattice site and the procedure was repeated. In order to detect the resulting spin pattern, unflipped atoms were removed by applying a resonant laser beam that rapidly expelled these atoms from the trap.[16] The remaining spin-flipped atoms were then detected using standard high-resolution fluorescence imaging, as described above. The resulting atomic patterns can be seen in Fig. 4, showing that almost arbitrary atomic orderings can be produced in this way. The described scheme can be enhanced to allow for simultaneous addressing of multiple lattice sites using an intensity

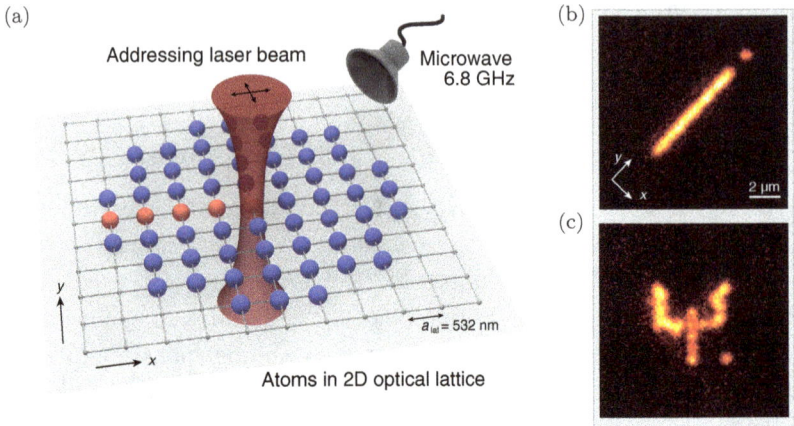

Fig. 4. High-resolution addressing of single atoms. (a), Atoms in a Mott insulator with unity filling arranged on a square lattice with period $a_{lat} = 532$ nm were addressed using an off-resonant laser beam. The beam was focused onto individual lattice sites by a high-aperture microscope objective (not shown) and could be moved in the $x-y$ plane with an accuracy of better than $0.1 a_{lat}$. (b,c) Fluorescence images of spin-flipped atoms following the addressing procedure. From Weitenberg *et al.*.[16]

shaped laser beam instead of a focused Gaussian beam. Such a beam can be prepared in the lab using spatial light modulators.[20]

In order to demonstrate that the addressing does not affect the motional state of the atoms on the lattice site, the tunneling of particles was investigated after an addressing sequence. Using the addressing sequence described above, a line of atoms in y-direction was prepared from a Mott insulator in a deep lattice. Thereafter, the lattice depth along the x-direction was lowered in order to initiate tunneling of the particles along this direction. After a variable evolution time, the position of the atoms was measured (see Fig. 5). By repeating the experiment several times, the probability of finding the atom at a certain lattice site for a specific evolution time could be determined and compared to the probability distribution predicted by the Schrödinger equation for the quantum evolution of a single particle tunneling on a lattice. Excellent agreement was found between the experimental data and the theoretical prediction, indicating that most atoms indeed were still in the lower energy band of the lattice despite the addressing. Atoms in higher energy bands typically exhibit an order of magnitude larger tunnel coupling, allowing them to travel much further

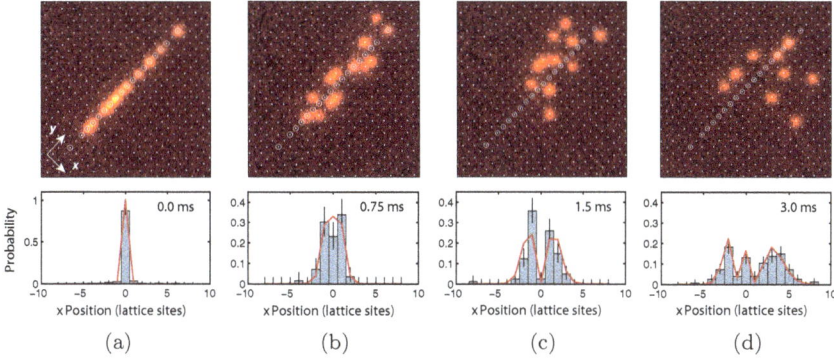

Fig. 5. Tunneling of single particles on a lattice. (a), Atoms were prepared in a single line along the y-direction before the lattice along the x axis was lowered, allowing the atoms to tunnel in this direction (b)–(d). The top row shows snapshots of the atomic distribution after different hold times. White circles indicate the lattice sites at which the atoms were prepared (not all sites initially contained an atom). The bottom row shows the respective position distribution obtained from an average over $10-20$ of such pictures, the error bars give the 1σ statistical uncertainty. The red curve corresponds to the prediction by theory. From Weitenberg *et al.*.[16]

given the same evolution time. However, in the experiment a negligible fraction of atoms was detected at such positions in the experiment.

High resolution imaging and addressing can be very useful for preparing almost arbitrary initial configurations of the many-body system that can e.g., be used to investigate a specific non-equilibrium evolution. It can also be highly beneficial for quantum information applications, where e.g., in the case of a one-way quantum computer,[27] it is essential to measure the spin state of an atom at a specified lattice site.

5. Quantum Gas Microscopy — An Enabling Technology

Combining the techniques described above, quantum gas microscopy has proven to be an enabling technology for probing and controlling quantum many-body systems. The imaging method allows for the measurement of local counting statistics of the atomic parity for strongly correlated many-body states. For example, correlation functions — not necessary restricted to two-point correlators — can be extracted from the data.[19,28] Based on those, the *quantum melting* of one- and two-dimensional Mott insulators through a proliferation of correlated particle-hole pairs has been directly observed. Furthermore, non-local multi-point correlators have been extracted to analyze the emerging string order, a hidden order parameter for Mott-insulating states at zero temperature.[17] In the context of topologically ordered phases of matter, non-local order parameters play a crucial role to characterize the complex entanglement order present in these states.[29–31] So far, it was believed that non-local order is merely a theoretical concept, not accessible to experiments. Quantum gases microscopy now makes probing such highly non-trivial order a reality for experiments. The local parity sensitive detection is also ideally suited to study low lying excitations of the strongly correlated system close to the Mott-insulating phase. Here, the excitations in the system can be converted into particle-hole excitations by a sweep of the lattice depth such that the final state is deep in the Mott-insulating regime. In this regime, the detection scheme is sensitive to single quasiparticles (i.e., these particle-hole excitations) such that bolometric measurements with highest sensitivity are possible. Such measurements enabled the detection of a mode softening around the particle-hole symmetric critical point in two-dimensions, which could be attributed to a Higgs-like excitation on the superfluid side of the transition.[23]

Next to equilibrium physics, also dynamical properties of strongly correlated systems in optical lattices can be studied. This is especially remarkable, since it allows for real-time tracking of the dynamics in the system. A controlled quench of the lattice height of a one-dimensional lattice gas into the Mott-insulating state excites the energetically low-lying particle-hole excitations homogeneously within the system. These excitations manifest themselves in characteristic correlations based on entangled quasiparticles that spread out across the system with a fixed velocity. The light-cone like spreading of correlations, first predicted by Lieb-Robinson,[32] could thereby be revealed for the first time experimentally.[18]

The hyperfine state selective microscopic detection and manipulation technique is ideally suited to study bosonic quantum magnetism in optical lattices. Heisenberg type magnetic couplings can be implemented by using two internal hyperfine degrees of freedom, on which a pseudo-spin $1/2$ is defined. For such systems, the anisotropy in the spin couplings can in principle be controlled either by Feshbach resonances or by state dependent hoppings.[33–35] In the case of Rubidium and spin independent lattices the resulting Heisenberg Hamiltonian is — up to a few percent — symmetric in the spin coupling. It takes the simple form

$$\hat{H} = -J_{\text{ex}} \sum_i \hat{\mathbf{S}}_i \cdot \hat{\mathbf{S}}_i , \qquad (5)$$

where J_{ex} is the superexchange coupling $J_{\text{ex}} = 4J^2/U$. This superexchange coupling describes spin exchange processes obtained from second order perturbation terms in the Mott insulating phase at unity filling. This is commonly referred to as the *strong-coupling limit*.[33,34] In first experiments, direct observation of superexchange couplings[36] and detection of singlet-triplet spin correlations in double wells[37,38] were achieved and more complex plaquette resonating valence bond states[39,40] were observed. Now, using quantum gas microscopes, the detection and control possibilities for quantum magnetism have also been dramatically enhanced.

When characterizing the spin-spin exchange couplings, one already finds that in one-dimensional systems the corresponding energy scale is small $J_{\text{ex}} = h \times \mathcal{O}(10\,\text{Hz})$. In higher dimensions an even smaller ratio J/U is required to reach the Mott insulating phase resulting in an even more reduced exchange coupling.[41–44] These tiny energy scales pose a major open challenge to observe characteristic magnetic quantum correlations in thermal equilibrium, as temperature is typically larger than such exchange couplings. However, in a spin-polarized Mott insulator, entropy is not distributed uniformly throughout the system, but is rather confined to narrow regions at the boundary of the system (see Sec. 3). The core of such a fully polarized Mott insulator can therefore be regarded to be at almost zero temperature, forming an ideal initial state for the observation of coherent quantum magnetic phenomena. Especially, in combination with high fidelity local addressing, this allows for the deterministic preparation of precisely controlled initial spin distributions, whose ensuing quantum evolution can be readily tracked. Using such a technique, the coherent dynamics of a single magnetic quasiparticle, a magnon, could be observed in Heisenberg spin chains.[20] These measurements were carried out in the subspace of a single spin impurity, such that the next neighbor spin-interaction term $\propto S_i^z S_j^z$ does not play any role for the dynamical evolution.

Given this ultimate control over the initial local magnetization, complexity can be added step wise to the problem. The simplest setting in which the magnetic interaction, i.e., the $\hat{S}_i^z \hat{S}_{i+1}^z$ coupling, becomes important is the case of two spin impurities on the Heisenberg chain. This scenario can be readily studied by flipping two adjacent spins in the initially fully polarized chain. Such a state has overlap both with free magnon as well as bound magnon states and one therefore expects to observe both propagation phenomena in the subsequent dynamical evolution of the initial state. The emergence of the low energy bound states in the excitation spectrum is probably the most striking microscopic effect of ferromagnetism[45,46] and in fact can be seen as the most elementary magnetic soliton. These bound states have recently been directly observed and characterized by site resolved correlation measurements.[21] Their signature, a high probability of finding the two impurity spins on adjacent sites even after a long time, can be seen in Fig. 6.

Studies of quantum magnetism is not limited to the symmetric Heisenberg scenario described above. Quantum Ising models, that is, models with classical Ising coupling in addition to transverse single spin couplings by external fields, can be achieved by a different, spatial encoding of the pseudo

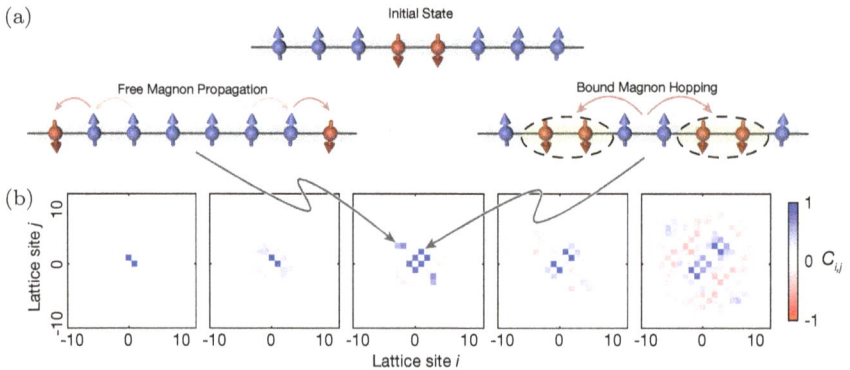

Fig. 6. Magnon dynamics. **(a)**, Initially two adjacent spins are flipped in an otherwise fully polarized Heisenberg chain. This state has roughly equal overlap with *free* magnon and *bound* magnon states. During the subsequent evolution the free and bound magnon contributions develop distinct features that are revealed in correlation measurements shown in **(b)**. The plots show the correlation (color scale) between two sites for 0 ms, 40 ms, 60 ms, 80 ms and 120 ms from left to right. The bound magnons tend to stick together leading to a high nearest neighbor correlation signal on the "lower left to upper right" diagonal for all times. On the contrary, the free magnons show anti-bunching behavior, maximizing their distance. This shows up as a signal in the "upper left to lower right" diagonal. Adapted from Fukuhara *et al.*[21]

spin that is defined on the bonds of a tilted lattice, i.e., in between two sites.[47] The latter technique maps an empty site next to a doubly occupied site to one of the two spin states and two singly occupied states next to each other to the other one. Thus, parity resolved local detection is an ideal tool to study Ising spin chains using this mapping as realized in M. Greiner's group.[22]

6. Outlook

The novel techniques that we have outlined above to image and control individual atoms mark a milestone in the experimental control over quantum many-body systems. In fact we believe that we have only scratched the surface of the wide range of applications that will emerge in the future. The possibility to reveal hidden order parameters of topological phases of matter in higher dimensions,[28,29] the ability to measure the full counting statistics in a many-body setting or the possibility to directly measure entanglement entropies[48,49] will open new avenues for our understanding of correlated quantum phases of matter. Currently, work is progressing in several groups to realize such imaging and control also for fermionic quantum gases. There, it could be directly used to characterize long-ranged magnetic correlations[50] in the fermionic Hubbard model and to excite individual magnetic quasiparticles in the system and observe their propagation. Single atom imaging has now also found useful applications in the field of Rydberg gases, where the imaging of crystalline structures of an ensemble of Rydberg atoms has only been possible due to these new detection capabilities.[51,52] Extending the control towards individual Rydberg atoms might enable one to realize novel multi-particle interactions in a Rydberg quantum simulator that could be used to implement exotic spin models such as those underlying Kitaev's toric code,[53] spin-ice models[54] or even non-Abelian lattice gauge theories.[55] High-resolution imaging will certainly also find other applications, e.g., in realizing the confinement and providing the resolution needed to probe edge state physics in the recently realized topologically band structures in optical lattices.[56–61] Next to imaging the local occupation in such states, it could also be employed to directly probe the particle currents along individual bonds in the lattice[59,62,63] thereby providing fundamentally complementary information to measurements in the occupation basis. While technically challenging to implement, we believe that the imaging and control techniques outlined here offer so many advantages and potential that they will in fact become standard techniques on any cold atom or molecule experiment in the future.

References

1. D. Jaksch and P. Zoller, The cold atoms Hubbard toolbox, *Ann. Phys.* **315**, 52–79 (2005).
2. I. Bloch, J. Dalibard, and W. Zwerger, Many-body physics with ultracold gases, *Rev. Mod. Phys.* **80**, 885–964 (2008).
3. M. Lewenstein, A. Sanpera, V. Ahufinger, B. Damski, A. S. De, and U. Sen, Ultracold atomic gases in optical lattices: mimicking condensed matter physics and beyond, *Adv. Phys.* **56**, 243–379 (2007).
4. M. P. A. Fisher, P. B. Weichman, G. Grinstein, and D. S. Fisher, Boson localization and the superfluid-insulator transition, *Phys. Rev. B.* **40**, 546–570 (1989).
5. D. Jaksch, C. Bruder, J. Cirac, C. Gardiner, and P. Zoller, Cold bosonic atoms in optical lattices, *Phys. Rev. Lett.* **81**, 3108–3111 (1998).
6. M. Greiner, O. Mandel, T. Esslinger, T. W. Hänsch, and I. Bloch, Quantum phase transition from a superfluid to a Mott insulator in a gas of ultracold atoms, *Nature.* **415**, 39–44 (2002).
7. B. Paredes, A. Widera, V. Murg, O. Mandel, S. Fölling, J. I. Cirac, G. V. Shlyapnikov, T. W. Hänsch, and I. Bloch, Tonks-Girardeau gas of ultracold atoms in an optical lattice, *Nature.* **429**, 277–281 (2004).
8. T. Kinoshita, T. Wenger, and D. Weiss, Observation of a one-dimensional Tonks-Girardeau gas, *Science.* **305**, 1125–1128 (2004).
9. *The BCS-BEC Crossover and the Unitary Fermi Gas.* vol. 836, *Lecture Notes in Physics*, Springer (2012).
10. C. Chin, R. Grimm, P. Julienne, and E. Tiesinga, Feshbach resonances in ultracold gases, *Rev. Mod. Phys.* **82**, 1225–1286 (2010).
11. K. D. Nelson, X. Li, and D. S. Weiss, Imaging single atoms in a three-dimensional array, *Nat. Phys.* **3**, 556–560 (2007).
12. T. Gericke, P. Würtz, D. Reitz, T. Langen, and H. Ott, High-resolution scanning electron microscopy of an ultracold quantum gas, *Nat. Phys.* **4**, 949–953 (2008).
13. W. S. Bakr, J. I. Gillen, A. Peng, S. Fölling, and M. Greiner, A quantum gas microscope for detecting single atoms in a Hubbard-regime optical lattice, *Nature.* **462**, 74–77 (2009).
14. W. S. Bakr, A. Peng, M. E. Tai, R. Ma, J. Simon, J. I. Gillen, S. Fölling, L. Pollet, and M. Greiner, Probing the superfluid-to-Mott insulator transition at the single-atom level., *Science.* **329**, 547–550 (2010).
15. J. F. Sherson, C. Weitenberg, M. Endres, M. Cheneau, I. Bloch, and S. Kuhr, Single-atom-resolved fluorescence imaging of an atomic Mott insulator., *Nature.* **467**, 68–72 (2010).
16. C. Weitenberg, M. Endres, J. F. Sherson, M. Cheneau, P. Schauß, T. Fukuhara, I. Bloch, and S. Kuhr, Single-spin addressing in an atomic Mott insulator., *Nature.* **471**, 319–324 (2011).

17. M. Endres, M. Cheneau, T. Fukuhara, C. Weitenberg, P. Schauß, C. Gross, L. Mazza, M. Bañuls, L. Pollet, I. Bloch, and S. Kuhr, Observation of correlated particle-hole pairs and string order in low-dimensional Mott insulators, *Science*. **334**, 200–203 (2011).
18. M. Cheneau, P. Barmettler, D. Poletti, M. Endres, P. Schauß, T. Fukuhara, C. Gross, I. Bloch, C. Kollath, and S. Kuhr, Light-cone-like spreading of correlations in a quantum many-body system, *Nature*. **481**, 484–487 (2012).
19. M. Endres, M. Cheneau, T. Fukuhara, C. Weitenberg, P. Schauß, C. Gross, L. Mazza, M. C. Bañuls, L. Pollet, I. Bloch, and S. Kuhr, Single-site-and single-atom-resolved measurement of correlation functions, *Appl. Phys. B*. **113**, 27–39 (2013).
20. T. Fukuhara, A. Kantian, M. Endres, M. Cheneau, P. Schauß, S. Hild, D. Bellem, U. Schollwöck, T. Giamarchi, C. Gross, I. Bloch, and S. Kuhr, Quantum dynamics of a mobile spin impurity, *Nat. Phys.* **9**, 235–241 (2013).
21. T. Fukuhara, P. Schauß, M. Endres, S. Hild, M. Cheneau, I. Bloch, and C. Gross, Microscopic observation of magnon bound states and their dynamics, *Nature*. **502**, 76–79 (2013).
22. J. Simon, W. S. Bakr, R. Ma, M. E. Tai, P. M. Preiss, and M. Greiner, Quantum simulation of antiferromagnetic spin chains in an optical lattice., *Nature*. **472**, 307–312 (2011).
23. M. Endres, T. Fukuhara, D. Pekker, M. Cheneau, P. Schauß, C. Gross, E. Demler, S. Kuhr, and I. Bloch, The 'Higgs' amplitude mode at the two-dimensional superfluid/Mott insulator transition, *Nature*. **487**, 454–458 (2012).
24. N. Gemelke, X. Zhang, C.-L. Hung, and C. Chin, In situ observation of incompressible Mott-insulating domains in ultracold atomic gases, *Nature*. **460**, 995–998 (2009).
25. M. T. M. DePue, C. McCormick, S. L. Winoto, S. Oliver, and D. D. S. Weiss, Unity occupation of sites in a 3d optical lattice, *Phys. Rev. Lett.* **82**, 2262–2265 (1999).
26. D. Weiss, J. Vala, A. Thapliyal, S. Myrgren, U. Vazirani, and K. Whaley, Another way to approach zero entropy for a finite system of atoms, *Phys. Rev. A*. **70**, 040302(R) (2004).
27. R. Raussendorf and H. J. Briegel, A one-way quantum computer, *Phys. Rev. Lett.* **86**, 5188–5191 (2001).
28. S. P. Rath, W. Simeth, M. Endres, and W. Zwerger, Non-local order in Mott insulators, duality and Wilson loops, *Ann. Phys.* **334**, 256–271 (2013).
29. X. Wen, *Quantum field theory of many-body systems*. Oxford Graduate Texts, Oxford University Press, Oxford (2004).
30. E. G. Dalla Torre, E. Berg, and E. Altman, Hidden order in 1d Bose insulators, *Phys. Rev. Lett.* **97**, 260401 (2006).
31. F. Anfuso and A. Rosch, String order and adiabatic continuity of Haldane chains and band insulators, *Phys. Rev. B*. **75**, 144420 (2007).
32. E. H. Lieb and D. W. Robinson, The finite group velocity of quantum spin systems, *Commun. Math. Phys.* **28**, 251–257 (1972).
33. A. B. Kuklov and B. V. Svistunov, Counterflow superfluidity of two-species ultracold atoms in a commensurate optical lattice, *Phys. Rev. Lett.* **90**, 100401 (2003).
34. L.-M. Duan, E. Demler, and M. Lukin, Controlling spin exchange interactions of ultracold atoms in optical lattices, *Phys. Rev. Lett.* **91**, 090402 (2003).
35. E. Altman, W. Hofstetter, E. Demler, and M. D. Lukin, Phase diagram of two-component bosons on an optical lattice, *New J. Phys.* **5**, 113 (2003).

36. S. Trotzky, P. Cheinet, S. Fölling, M. Feld, U. Schnorrberger, A. M. Rey, A. Polkovnikov, E. A. Demler, M. D. Lukin, and I. Bloch, Time-resolved observation and control of superexchange interactions with ultracold atoms in optical lattices, *Science*. **319**, 295–299 (2008).
37. S. Trotzky, Y.-A. Chen, U. Schnorrberger, P. Cheinet, and I. Bloch, Controlling and detecting spin correlations of ultracold atoms in optical lattices, *Phys. Rev. Lett.* **105**, 265303 (2010).
38. D. Greif, T. Uehlinger, G. Jotzu, L. Tarruell, and T. Esslinger, Short-range quantum magnetism of ultracold fermions in an optical lattice, *Science*. **340**, 1307–1310 (2013).
39. B. Paredes and I. Bloch, Minimum instances of topological matter in an optical plaquette, *Phys. Rev. A.* **77**, 023603 (2008).
40. S. Nascimbène, Y.-A. Chen, M. Atala, M. Aidelsburger, S. Trotzky, B. Paredes, and I. Bloch, Experimental realization of plaquette resonating valence-bond states with ultracold atoms in optical superlattices, *Phys. Rev. Lett.* **108**, 205301 (2012).
41. V. A. Kashurnikov and B. V. Svistunov, Exact diagonalization plus renormalization-group theory: Accurate method for a one-dimensional superfluid-insulator-transition study, *Phys. Rev. B.* **53**, 11776–11778 (1996).
42. T. D. Kühner, S. R. White, and H. Monien, One-dimensional Bose-Hubbard model with nearest-neighbor interaction, *Phys. Rev. B.* **61**, 12474–12489 (2000).
43. B. Capogrosso-Sansone, N. V. Prokof'ev, and B. V. Svistunov, Phase diagram and thermodynamics of the three-dimensional Bose-Hubbard model, *Phys. Rev. B.* **75**, 134302 (2007).
44. B. Capogrosso-Sansone, S. G. Söyler, N. Prokof'ev, and B. Svistunov, Monte Carlo study of the two-dimensional Bose-Hubbard model, *Phys. Rev. A.* **77**, 015602 (2008).
45. H. Bethe, Zur Theorie der Metalle, *Z. Phys.* **71**, 205–226 (1931).
46. M. Wortis, Bound states of two spin waves in the Heisenberg ferromagnet, *Phys. Rev.* **132**, 85 (1963).
47. S. Sachdev, K. Sengupta, and S. M. Girvin, Mott insulators in strong electric fields, *Phys. Rev. B.* **66**, 075128 (2002).
48. A. Daley, H. Pichler, J. Schachenmayer, and P. Zoller, Measuring entanglement growth in quench dynamics of bosons in an optical lattice, *Phys. Rev. Lett.* **109**, 020505 (2012).
49. H. Pichler, L. Bonnes, A. J. Daley, A. M. Läuchli, and P. Zoller, Thermal versus entanglement entropy: a measurement protocol for fermionic atoms with a quantum gas microscope, *New J. Phys.* **15**, 063003 (2013).
50. P. Lee, N. Nagaosa, and X.-G. Wen, Doping a Mott insulator: Physics of high-temperature superconductivity, *Rev. Mod. Phys.* **78**, 17–85 (2006).
51. P. Schauß, M. Cheneau, M. Endres, T. Fukuhara, S. Hild, A. Omran, T. Pohl, C. Gross, S. Kuhr, and I. Bloch, Observation of spatially ordered structures in a two-dimensional Rydberg gas, *Nature*. **490**, 87–91 (2012).
52. P. Schauß, J. Zeiher, T. Fukuhara, S. Hild, M. Cheneau, T. Macrì, T. Pohl, I. Bloch, and C. Gross, Dynamical crystallization in a low-dimensional Rydberg gas, *arXiv:1404.0980* (2014).
53. H. Weimer and H. P. Büchler, Two-stage melting in systems of strongly interacting Rydberg atoms, *Phys. Rev. Lett.* **105**, 230403 (2010).

54. A. W. Glaetzle, M. Dalmonte, R. Nath, I. Rousochatzakis, R. Moessner, and P. Zoller, Quantum Spin Ice and dimer models with Rydberg atoms, *arXiv:1404.5326* (2014).

55. L. Tagliacozzo, A. Celi, P. Orland, M. W. Mitchell, and M. Lewenstein, Simulation of non-abelian gauge theories with optical lattices., *Nat. Commun.* **4**, 2615 (2013).

56. M. Atala, M. Aidelsburger, J. T. Barreiro, D. Abanin, T. Kitagawa, E. Demler, and I. Bloch, Direct measurement of the Zak phase in topological bloch bands, *Nat. Phys.* **9**, 795–800 (2013).

57. M. Aidelsburger, M. Atala, M. Lohse, J. T. Barreiro, B. Paredes, and I. Bloch, Realization of the Hofstadter Hamiltonian with ultracold atoms in optical lattices, *Phys. Rev. Lett.* **111**, 185301 (2013).

58. H. Miyake, G. A. Siviloglou, C. J. Kennedy, W. C. Burton, and W. Ketterle, Realizing the Harper Hamiltonian with laser-assisted tunneling in optical lattices, *Phys. Rev. Lett.* **111**, 185302 (2013).

59. M. Atala, M. Aidelsburger, M. Lohse, J. T. Barreiro, B. Paredes, and I. Bloch, Observation of chiral currents with ultracold atoms in bosonic ladders, *Nat. Phys.* **10**, 588–593 (2014).

60. G. Jotzu, M. Messer, R. Desbuquois, M. Lebrat, T. Uehlinger, D. Greif, and T. Esslinger, Experimental realisation of the topological Haldane model, *arXiv:1406.7874* (2014).

61. M. Aidelsburger, M. Lohse, C. Schweizer, M. Atala, J. T. Barreiro, S. Nascimbène, N. R. Cooper, I. Bloch, and N. Goldman, Revealing the topology of Hofstadter bands with ultracold bosonic atoms, *arXiv:1407.4205* (2014).

62. S. Trotzky, Y.-A. Chen, A. Flesch, I. P. McCulloch, U. Schollwöck, J. Eisert, and I. Bloch, Probing the relaxation towards equilibrium in an isolated strongly correlated one-dimensional Bose gas, *Nat. Phys.* **8**, 325–330 (2012).

63. S. Keßler and F. Marquardt, Single-site-resolved measurement of the current statistics in optical lattices, *Phys. Rev. A.* **89**, 061601 (2014).

CHAPTER 5

SPIN-ORBIT-COUPLED BOSE-EINSTEIN CONDENSATES

Yun Li[*,†] and Giovanni I. Martone[*] and Sandro Stringari[*]

*Dipartimento di Fisica, Università di Trento and INO-CNR BEC Center,
I-38123 Povo, Italy
†Centre for Quantum Technologies, National University of Singapore,
3 Science Drive 2, 117542, Singapore

The recent realization of synthetic spin-orbit coupling represents an outstanding achievement in the physics of ultracold quantum gases. In this review we explore the properties of a spin-orbit-coupled Bose-Einstein condensate with equal Rashba and Dresselhaus strengths. This system presents a rich phase diagram, which exhibits a tricritical point separating a zero-momentum phase, a spin-polarized plane-wave phase, and a stripe phase. In the stripe phase translational invariance is spontaneously broken, in analogy with supersolids. Spin-orbit coupling also strongly affects the dynamics of the system. In particular, the excitation spectrum exhibits intriguing features, including the suppression of the sound velocity, the emergence of a roton minimum in the plane-wave phase, and the appearance of a double gapless band structure in the stripe phase. Finally, we discuss a combined procedure to make the stripes visible and stable, thus allowing for a direct experimental detection.

1. Introduction

A large variety of exotic phenomena in solid-state systems can take place when their constituent electrons are coupled to an external gauge field, or in the presence of strong spin-orbit coupling. For example, magnetic fields influencing the motion of the electrons are at the base of the well-known quantum Hall effect,[1] whereas spin-orbit coupling, i.e., the coupling between an electron's spin and its momentum, is crucial for topological insulators,[2,3] Majorana fermions,[4] spintronic devices,[5] etc. Ultracold atomic gases are good candidates to investigate these interesting quantum phenomena. In this respect, the main difficulty arises from the fact that atoms are neutral particles, and consequently they cannot be coupled to a gauge field. In addition, they do not exhibit any coupling between their spin and their center-of-mass motion.

In the last few years there have been several proposals to realize artificial gauge fields for quantum gases, thus overcoming the problem of their neutrality.[6] One of these schemes relies on the notion of geometric phase,[7] which emerges when the motion of a particle with some internal level structure is slow enough, so that the particle follows adiabatically one of these levels. In such conditions, the particle experiences an effective vector potential. In ultracold atomic gases, several methods to implement these ideas exploit the space-dependent coupling of the atoms with a properly designed configuration of laser beams; the synthetic gauge field arises when the system follows adiabatically one of the local eigenstates of the light-atom interaction Hamiltonian (dressed states).[8–11] Other approaches are also possible, such as the periodic shaking of an optical lattice with special frequencies, which couples different Bloch bands.[12]

Since 2009, several experiments have been successful in realizing ultracold atomic gases coupled to artificial gauge fields.[13–17] For instance, in the experiment of Ref. 14 a space-dependent atom-light coupling was employed to simulate an effective magnetic field exerting a Lorentz-like force on neutral bosons; this procedure has been used to generate quantized vortices in Bose-Einstein condensates (BECs).

Another interesting situation occurs when the local dressed states are degenerate, giving rise to spin-orbit-coupled configurations. In particular, by using a suitable arrangement of Raman lasers, the authors of Ref. 18 managed to engineer a one-dimensional spin-orbit coupling, characterized by equal Rashba[19] and Dresselhaus[20] strengths, on a neutral atomic BEC. The same scheme has been subsequently extended to realize spin-orbit-coupled Fermi gases.[21,22]

These first experimental achievements have stimulated a growing interest in this field of research, resulting in a wide number of papers devoted to artificial gauge fields and, more specifically, to spin-orbit-coupled quantum gases, both from the theoretical and the experimental side. In this review we will focus on the properties of Bose-Einstein condensates with the kind of spin-orbit coupling first realized by the NIST team.[18] Readers who are interested in a broader overview about spin-orbit-coupled quantum gases and, more generally, about artificial gauge fields on neutral atoms, can refer to some recent reviews[6,23–25] and references therein.

This paper is organized as follows. In Sec. 2 we illustrate the quantum phase diagram of the system. The dynamic behavior of the gas in the two uniform phases is studied in Sec. 3. Section 4 deals with the collective excitations in the presence of harmonic trapping. Section 5 is entirely devoted to the phase exhibiting density modulations in the form of stripes: we discuss both the ground state and the excitation spectrum, and we illustrate a procedure allowing for the direct observation of the stripes. Finally, in Sec. 6 we report some brief concluding remarks.

2. Ground-state Phase Diagram

2.1. *Single-particle picture*

The experimental setup employed in Ref. 18 to realize spin-orbit coupling consists of a ^{87}Rb Bose-Einstein condensate in the $F = 1$ hyperfine manifold, with a bias magnetic field providing a nonlinear Zeeman splitting between the three levels of the manifold. The BEC is coupled to the field of two Raman lasers having orthogonal linear polarizations, frequencies ω_L and $\omega_L + \Delta\omega_L$, and wave vector difference $\mathbf{k}_0 = k_0\hat{\mathbf{e}}_x$, with $\hat{\mathbf{e}}_x$ the unit vector along the x direction. The laser field induces transitions between the three states characterized by a Rabi frequency Ω fixed by the intensity of the lasers. This Raman process is illustrated schematically in Fig. 1. The frequency splitting ω_Z between the states $|F = 1, m_F = 0\rangle$ and $|F = 1, m_F = -1\rangle$ is chosen to be very close to the frequency difference $\Delta\omega_L$ between the two lasers, while the separation $\omega_Z - \omega_q$ between $|F = 1, m_F = 0\rangle$ and $|F = 1, m_F = 1\rangle$ contains a large additional shift from Raman resonance due to the quadratic Zeeman effect. This implies that the state $|m_F = 1\rangle$ can be neglected, and we are left with an effective spin-1/2 system, with the two spin states given by $|\uparrow\rangle = |m_F = 0\rangle$ and $|\downarrow\rangle = |m_F = -1\rangle$. The single-particle Hamiltonian of this system takes the form (we set $\hbar = m = 1$)

$$h_0 = \frac{\mathbf{p}^2}{2} + \frac{\Omega}{2}\sigma_x \cos(2k_0 x - \Delta\omega_L t) + \frac{\Omega}{2}\sigma_y \sin(2k_0 x - \Delta\omega_L t) - \frac{\omega_Z}{2}\sigma_z,$$

$$(1)$$

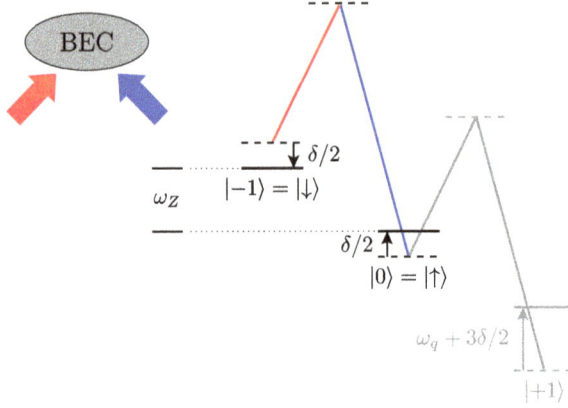

Fig. 1. Level diagram. Two Raman lasers with orthogonal linear polarizations couple the two states $|\uparrow\rangle = |m_F = 0\rangle$ and $|\downarrow\rangle = |m_F = -1\rangle$ of the $F = 1$ hyperfine manifold of ^{87}Rb, which differ in energy by a Zeeman splitting ω_Z. The lasers have frequency difference $\Delta\omega_L = \omega_Z + \delta$, where δ is a small detuning from the Raman resonance. The state $|m_F = 1\rangle$ can be neglected since it has a much larger detuning, due to the quadratic Zeeman shift ω_q.

where σ_k with $k = x, y, z$ denotes the usual 2×2 Pauli matrices. Hamiltonian (1) is not translationally invariant but exhibits a screwlike symmetry, being invariant with respect to helicoidal translations of the form $e^{id(p_x - k_0\sigma_z)}$, consisting of a combination of a rigid translation by distance d and a spin rotation by angle $-dk_0$ around the z axis.

Let us now apply the unitary transformation $e^{i\Theta\sigma_z/2}$, corresponding to a position and time-dependent rotation in spin space by the angle $\Theta = 2k_0 x - \Delta\omega_L t$, to the wave function obeying the Schrödinger equation. As a consequence of the transformation, the single-particle Hamiltonian (1) is transformed into the translationally invariant and time-independent form

$$h_0^{\text{SO}} = \frac{1}{2}\left[(p_x - k_0\sigma_z)^2 + p_\perp^2\right] + \frac{\Omega}{2}\sigma_x + \frac{\delta}{2}\sigma_z. \tag{2}$$

The spin-orbit nature acquired by the Hamiltonian results from the noncommutation of the kinetic energy and the position-dependent rotation, while the renormalization of the effective magnetic field $\delta = \Delta\omega_L - \omega_Z$ results from the additional time dependence exhibited by the wave function in the rotating frame. The new Hamiltonian is characterized by equal contributions of Rashba[19] and Dresselhaus[20] couplings. It has the peculiar property of violating both parity and time-reversal symmetry. It is worth pointing out that the operator $\mathbf{p}$ entering Eq. (2) is the canonical momentum $-i\nabla$, with the physical velocity being given by $\mathbf{v}_\pm = \mathbf{p} \mp k_0\hat{\mathbf{e}}_x$ for the spin-up and spin-down particles. In terms of $\mathbf{p}$ the eigenvalues of Eq. (2) are given by

$$\varepsilon_\pm(\mathbf{p}) = \frac{p_x^2 + p_\perp^2}{2} + E_r \pm \sqrt{\left(k_0 p_x - \frac{\delta}{2}\right)^2 + \frac{\Omega^2}{4}}, \qquad (3)$$

where $E_r = k_0^2/2$ is the recoil energy. The double-branch structure exhibited by the dispersion (3) reflects the spinor nature of the system.

We now focus on the case $\delta = 0$ and $\Omega \geq 0$. In Fig. 2 we plot the dispersion (3) as a function of p_x, for different values of Ω. The lower branch $\varepsilon_-(\mathbf{p})$ exhibits, for $\Omega < 2k_0^2$, two degenerate minima at momenta $\mathbf{p} = \pm k_1\hat{\mathbf{e}}_x$ with $k_1 = k_0\sqrt{1 - \Omega^2/4k_0^4}$, both capable to host Bose-Einstein condensation. At larger values of Ω the spectrum has instead a single minimum at $\mathbf{p} = 0$. The effective mass of particles moving along x, fixed by the relation $m/m^* = \mathrm{d}^2\varepsilon/\mathrm{d}p_x^2$, also shows a nontrivial Ω dependence. Near the minimum one finds

$$\frac{m}{m^*} = 1 - \left(\frac{\Omega}{2k_0^2}\right)^2 \quad \text{for } \Omega < 2k_0^2, \qquad (4)$$

$$\frac{m}{m^*} = 1 - \frac{2k_0^2}{\Omega} \quad \text{for } \Omega > 2k_0^2. \qquad (5)$$

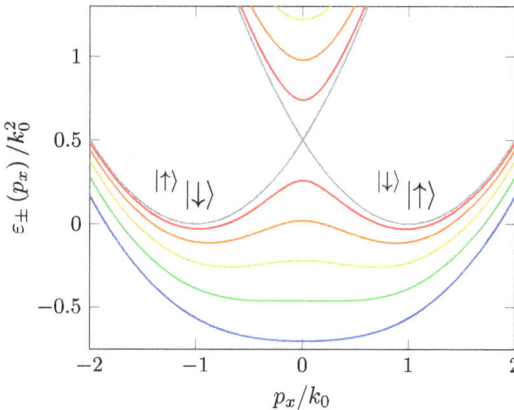

Fig. 2. Single-particle dispersion (3) at $\delta = 0$. Eigenenergies calculated for Raman coupling ranging from $\Omega = 0$ (grey) to $\Omega = 2.4\,k_0^2$ (blue). The two minima in the lower branch disappear at $\Omega = 2k_0^2$.

Thus, the effective mass exhibits a divergent behavior at $\Omega = 2k_0^2$, where the double-well structure disappears and the spectrum has a p_x^4 dispersion near the minimum.

Before concluding the present section, it is worth mentioning that a single-particle dispersion similar to (3) can also be achieved by trapping the atoms in a shaken optical lattice, as recently realized experimentally.[17] In such systems, different Bloch bands coupled through lattice shaking bear several analogies with the spin states involved in the Raman process described above.[26]

2.2. Many-body ground state

We shall now illustrate how the peculiar features of the single-particle dispersion (3) are at the origin of new interesting phases in the many-body ground state of the BEC. For a gas of N particles enclosed in a volume V, in the presence of two-body interactions, the many-body Hamiltonian takes the form

$$H = \sum_j h_0^{SO}(j) + \sum_{\sigma, \sigma'} \frac{1}{2} \int d\mathbf{r} \, g_{\sigma\sigma'} \rho_\sigma(\mathbf{r})\rho_{\sigma'}(\mathbf{r}), \tag{6}$$

where h_0^{SO} is given by (2), $j = 1, \ldots, N$ is the particle index, and σ, σ' are the spin indices $(\uparrow, \downarrow = \pm)$ characterizing the two spin states. The spin-up and spin-down density operators entering Eq. (6) are defined by $\rho_\pm(\mathbf{r}) = (1/2) \sum_j \left(1 \pm \sigma_{z,j}\right) \delta(\mathbf{r} - \mathbf{r}_j)$, while $g_{\sigma\sigma'} = 4\pi a_{\sigma\sigma'}$ are the relevant coupling constants in the different spin channels, with $a_{\sigma\sigma'}$ the corresponding s-wave scattering lengths.

To investigate the ground state of the system we resort to the Gross-Pitaevskii mean-field approach, and we write the energy functional associated to Hamiltonian (6) as

$$E = \int d\mathbf{r} \, \Psi^\dagger(\mathbf{r}) h_0^{SO} \Psi(\mathbf{r})$$

$$+ \int d\mathbf{r} \left[\frac{g_{\uparrow\uparrow}}{2} |\psi_\uparrow(\mathbf{r})|^4 + \frac{g_{\downarrow\downarrow}}{2} |\psi_\downarrow(\mathbf{r})|^4 + g_{\uparrow\downarrow} |\psi_\uparrow(\mathbf{r})|^2 |\psi_\downarrow(\mathbf{r})|^2 \right], \tag{7}$$

where $\Psi = \left(\psi_\uparrow \ \psi_\downarrow\right)^T$ is the two-component condensate wave function. For simplicity, in this review we will assume $\delta = 0$ and equal intraspecies interactions $g_{\uparrow\uparrow} = g_{\downarrow\downarrow} \equiv g$, unless otherwise specified; the effect of asymmetry of the coupling constants will be briefly discussed at the end of the present section. The ground-state wave function can be determined through a variational procedure based on the ansatz[27]

$$\Psi = \sqrt{\bar{n}} \left[C_+ \begin{pmatrix} \cos\theta \\ -\sin\theta \end{pmatrix} e^{ik_1 x} + C_- \begin{pmatrix} \sin\theta \\ -\cos\theta \end{pmatrix} e^{-ik_1 x} \right] \tag{8}$$

where $\bar{n} = N/V$ is the average density, and k_1 represents the canonical momentum where Bose-Einstein condensation takes place. For a given value of $\bar{n}$ and Ω, the variational parameters are C_+, C_-, k_1 and θ. Their values are determined by minimizing the energy (7) with the normalization constraint $\int d\mathbf{r} \, \Psi^* \Psi = N$, i.e., $|C_+|^2 + |C_-|^2 = 1$. Minimization with respect to k_1 yields the general relation $2\theta = \arccos(k_1/k_0)$ fixed by the single-particle Hamiltonian (2). Once the other variational parameters are determined, it is possible to calculate key physical quantities like, for example, the momentum distribution accounted for by the parameter k_1, the total density $n(\mathbf{r}) = \Psi^* \Psi$, the longitudinal ($s_z(\mathbf{r})$) and transverse ($s_x(\mathbf{r})$, $s_y(\mathbf{r})$) spin densities, given by

$$s_z(\mathbf{r}) = \Psi^\dagger \sigma_z \Psi = \bar{n} \left(|C_+|^2 - |C_-|^2 \right) \frac{k_1}{k_0}, \tag{9}$$

$$s_x(\mathbf{r}) = \Psi^\dagger \sigma_x \Psi = -\bar{n} \left[\frac{\sqrt{k_0^2 - k_1^2}}{k_0} + 2|C_+ C_-| \cos(2k_1 x + \phi) \right], \tag{10}$$

$$s_y(\mathbf{r}) = \Psi^\dagger \sigma_y \Psi = \bar{n} |C_+ C_-| \frac{2k_1}{k_0} \sin(2k_1 x + \phi), \tag{11}$$

with ϕ the relative phase between C_+ and C_-, and the corresponding spin polarizations $\langle \sigma_k \rangle = N^{-1} \int d\mathbf{r} \, s_k$ with $k = x, y, z$. Before going on, we notice that results (10) and (11) hold in the spin-rotated frame where the Hamiltonian takes the form (6). Since the operators σ_x and σ_y do not commute with σ_z, in the original laboratory frame the average value of these operators exhibits an additional oscillatory behavior analogous to the one characterizing the laser potential of Eq. (1).

The ansatz (8) exactly describes the ground state of the single-particle Hamiltonian h_0^{SO} (ideal Bose gas), reproducing all the features presented in Sec. 2.1, including the values of the canonical momentum k_1. In this case the energy is independent of $C_\pm$, reflecting the degeneracy of the ground state.

The same ansatz is well suited also for discussing the role of interactions, which crucially affect the explicit values of C_+, C_- and k_1. By inserting (8) into (7), one finds that the energy per particle $\varepsilon = E/N$ takes the form

$$\varepsilon = \frac{k_0^2}{2} - \frac{\Omega}{2k_0}\sqrt{k_0^2 - k_1^2} - F(\beta)\frac{k_1^2}{2k_0^2} + G_1(1 + 2\beta), \qquad (12)$$

where we have defined the quantities $\beta = |C_+|^2|C_-|^2 \in [0, 1/4]$, $G_1 = \bar{n}(g + g_{\uparrow\downarrow})/4$, $G_2 = \bar{n}(g - g_{\uparrow\downarrow})/4$ and the function $F(\beta) = (k_0^2 - 2G_2) + 4(G_1 + 2G_2)\beta$. By minimizing (12) with respect to β and k_1 we obtain the mean-field ground state of the system. Depending on the values of the relevant parameters k_0, Ω, g, $g_{\uparrow\downarrow}$ and $\bar{n}$, the minimum can occur either at $k_1 = 0$ or at $k_1 \neq 0$ and β equal to one of the limiting values 0 and 1/4. Therefore, the ground state is compatible with three distinct quantum phases; the corresponding phase diagram is shown in Fig. 3.

(I) Stripe phase. For small values of the Raman coupling Ω and $g > g_{\uparrow\downarrow}$, the ground state is a linear combination of the two plane-wave states $e^{\pm ik_1 x}$ with equal weights ($|C_+| = |C_-| = 1/\sqrt{2}$), yielding a vanishing longitudinal spin polarization (see Eq. (9)). The most striking feature of this phase is the appearance of density modulations in the form of stripes

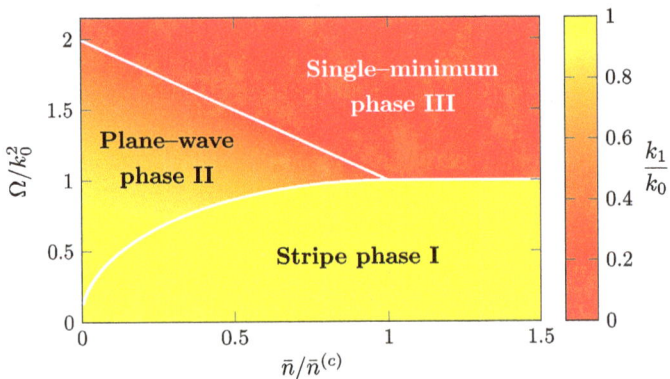

Fig. 3. Phase diagram of a spin-orbit-coupled BEC. The color represents the value of k_1/k_0. The white solid lines identify the phase transitions (I–II), (II–III) and (I–III). The diagram corresponds to a configuration with $\gamma = (g - g_{\uparrow\downarrow})/(g + g_{\uparrow\downarrow}) = 0.0012$ consistent with the value of Ref. 18.

according to the law

$$n(\mathbf{r}) = \bar{n} \left[1 + \frac{\Omega}{2 \left(k_0^2 + G_1 \right)} \cos \left(2k_1 x + \phi \right) \right]. \tag{13}$$

The periodicity of the fringes π / k_1 is determined by the wave vector

$$k_1 = k_0 \sqrt{1 - \frac{\Omega^2}{4 \left(k_0^2 + G_1 \right)^2}} \tag{14}$$

and differs from the one of the laser potential, equal to π / k_0 (see Eq. (1)). These modulations have a deeply different nature with respect to those exhibited by the density profile in the presence of usual optical lattices. Indeed, they appear as the result of a spontaneous breaking mechanism of translational invariance, with the actual position of the fringes being given by the value of the phase ϕ. Because of the coexistence of BEC and crystalline order, the stripe phase shares important analogies with supersolids.[28] It also shares similarities with the spatial structure of smectic liquid crystals. The contrast in $n(\mathbf{r})$ is given by

$$\frac{n_{\max} - n_{\min}}{n_{\max} + n_{\min}} = \frac{\Omega}{2(k_0^2 + G_1)} \tag{15}$$

and vanishes as $\Omega \to 0$ as a consequence of the orthogonality of the two spin states entering Eq. (8) (in this limit $\theta \to 0$ and $k_1 \to k_0$). It is also worth mentioning that the ansatz, Eq. (8), for the stripe phase provides only a first approximation which ignores higher-order harmonics caused by the nonlinear interaction terms in the Hamiltonian.

(II) Plane-wave phase. For larger values of the Raman coupling, the system enters a new phase, the so-called plane-wave phase (also called the spin-polarized or de-mixed phase), where Bose-Einstein condensation takes place in a single plane-wave state with momentum $\mathbf{p} = k_1 \hat{\mathbf{e}}_x$ ($C_- = 0$), lying on the x direction (in the following we choose $k_1 > 0$). In this phase, the density is uniform and the spin polarization is given by

$$\langle \sigma_z \rangle = \frac{k_1}{k_0} \tag{16}$$

with

$$k_1 = k_0 \sqrt{1 - \frac{\Omega^2}{4\left(k_0^2 - 2G_2\right)^2}}. \tag{17}$$

An energetically equivalent configuration is obtained by considering the BEC in the single-particle state with $\mathbf{p} = -k_1\hat{\mathbf{e}}_x$ ($C_+ = 0$). The choice between the two configurations is determined by a mechanism of spontaneous symmetry breaking, typical of ferromagnetic configurations.

(III) Single-minimum phase. At even larger values of Ω, the system enters the single-minimum phase (also called zero-momentum phase), where the condensate has zero momentum ($k_1 = 0$), the density is uniform, and the average spin polarization $\langle\sigma_z\rangle$ identically vanishes, while $\langle\sigma_x\rangle = -1$. Contrary to what one would naively expect, also the single-minimum phase exhibits nontrivial properties, as we will see in Secs. 3 and 4.

The chemical potential in the three phases can be calculated from the energy per particle (12), and takes the form

$$\mu^{(\mathrm{I})} = 2G_1 - \frac{k_0^2\Omega^2}{8\left(k_0^2 + G_1\right)^2}, \tag{18}$$

$$\mu^{(\mathrm{II})} = 2\left(G_1 + G_2\right) - \frac{k_0^2\Omega^2}{8\left(k_0^2 - 2G_2\right)^2}, \tag{19}$$

$$\mu^{(\mathrm{III})} = 2G_1 + \frac{k_0^2 - \Omega}{2}. \tag{20}$$

The critical values of the Rabi frequencies Ω characterizing the phase transitions can be identified by imposing that the chemical potential Eqs. (18)–(20) and the pressure $P = n\mu(n) - \int \mu(n)\,dn$ be equal in the two phases at equilibrium. The transition between the stripe and the plane-wave phases has a first-order nature and is characterized by different values of the densities of the two phases. The density differences are, however, extremely small and are not visible in Fig. 3. The transition between the plane-wave and the single-minimum phases has instead a second-order nature and is characterized by a jump in the compressibility $n^{-1}(\partial\mu/\partial n)^{-1}$ if $G_2 \neq 0$ and by a divergent behavior of the magnetic polarizability (see Sec. 2.3). In the low density (or weak coupling) limit, i.e., G_1, $G_2 \ll k_0^2$, the critical value of the Raman coupling $\Omega^{(I-II)}$ characterizing the transition between phases I and II is given by the density-independent expression[27,29]

$$\Omega^{(I-II)} = 2k_0^2\sqrt{\frac{2\gamma}{1+2\gamma}}, \tag{21}$$

with $\gamma = G_2/G_1$. The transition between phases II and III instead takes place at the higher value[27]

$$\Omega^{(II-III)} = 2\left(k_0^2 - 2G_2\right), \tag{22}$$

provided that the condition $\bar{n} < \bar{n}^{(c)}$ is satisfied, with $\bar{n}^{(c)} = k_0^2/(2g\gamma)$ being the value of the density at the tricritical point shown in Fig. 3, where the three phases connect each other. For higher densities one has instead a first-order transition directly between phases I and III. We also remark that, if $g < g_{\uparrow\downarrow}$, only phases II and III are available, the stripe phase being always energetically unfavorable.

The previous results can be extended to the case $\delta \neq 0$ and $g_{\uparrow\uparrow} \neq g_{\downarrow\downarrow}$. In general one can introduce three interaction parameters: $G_1 = \bar{n}(g_{\uparrow\uparrow} + g_{\downarrow\downarrow} + 2g_{\uparrow\downarrow})/8$, $G_2 = \bar{n}(g_{\uparrow\uparrow} + g_{\downarrow\downarrow} - 2g_{\uparrow\downarrow})/8$ and $G_3 = \bar{n}(g_{\uparrow\uparrow} - g_{\downarrow\downarrow})/4$. In the case of the states $|\uparrow\rangle = |F = 1, m_F = 0\rangle$ and $|\downarrow\rangle = |F = 1, m_F = -1\rangle$ of ^{87}Rb the values of the scattering lengths are $a_{\uparrow\uparrow} = 101.41\,a_B$ and $a_{\downarrow\downarrow} = a_{\uparrow\downarrow} = 100.94\,a_B$, where a_B is the Bohr radius. This corresponds to $0 < G_2 = G_3/2 \ll G_1$. However, since the differences among the scattering lengths are very small, by properly choosing the detuning δ, this effect can be well compensated. For example, using first order perturbation theory, one finds that the correction to the energy per particle is given (see Eq. (12)), in the low density (weak coupling) limit, by

$$\varepsilon^{(1)} = \left(G_3 + \frac{\delta}{2}\right)\frac{k_1}{k_0}(|C_+|^2 - |C_-|^2). \tag{23}$$

By choosing $\delta = -2G_3$ the correction (23) vanishes, thus ensuring that the properties of the ground state of the system and the transition frequencies are not affected by the inclusion of the new terms in the Hamiltonian. If the weak coupling condition is not satisfied, the value of δ ensuring exact compensation should depend on Ω.

The emergence of a double minimum in the single-particle spectrum and the Ω dependence of the value of k_1 was experimentally observed by

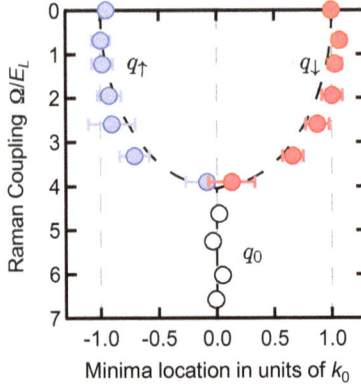

Fig. 4. Measured values of the canonical momentum versus Ω at $\delta = 0$. The data points correspond to the minima of the dispersion $\varepsilon_-(\mathbf{q})$ given in Eq. (3). The Raman coupling is expressed in units of the recoil energy $E_L = k_0^2/2$. Reprinted with permission from Macmillan Publishers Ltd: Lin *et al.*, Nature **471**, 83–86, © 2011.

Lin *et al.* by measuring the velocity of the expanding cloud after the release of the trap[18] (see Fig. 4). The double-minimum structure vanishes at the predicted value (22) of the Raman coupling giving the transition between the plane-wave and the single-minimum phases. In the same experiment, at a lower value of Ω, they identified another transition between a mixed phase, characterized by two different canonical momentum components overlapping in space, and a de-mixed phase, where the two components coexist but are spatially separated. The critical Raman coupling at which the latter transition has been observed is in good agreement with the prediction $\Omega^{(I-II)} = 0.19\, E_r$ for the transition frequency between the stripe and the plane-wave phases, obtained from Eq. (21) with the ^{87}Rb value $\gamma = 0.0012$. However, it has not been possible to observe directly the density modulations because of the smallness of their contrast and periodicity (see Sec. 5).

Finally, we mention that the critical density $\bar{n}^{(c)}$ is very large in the experimental conditions of Ref. 18, thus preventing the access to the regime where the first-order transition between the stripe and the single-minimum phases takes place. A strong reduction of the value of $\bar{n}^{(c)}$ could be achieved, for example, by considering configurations where the interspecies coupling strength $g_{\uparrow\downarrow}$ is significantly smaller than the intraspecies ones $g_{\uparrow\uparrow}$, $g_{\downarrow\downarrow}$, as discussed in Sec. 5.3.

2.3. *Magnetic polarizability and compressibility*

As already pointed out, the transition between the plane-wave and the single-minimum phases is characterized by a divergent behavior of the magnetic polarizability χ_M. This quantity is defined as the linear response $\chi_M = (\langle h|\sigma_z|h\rangle - \langle h = 0|\sigma_z|h = 0\rangle)/h$ to a static perturbation of the form $-h\sigma_z$, and can be calculated by generalizing the ground-state condensate wave function (8) to include the presence of a small magnetic field h. In the plane-wave and the single-minimum phases, the magnetic polarizability takes the simple form[30]

$$\chi_M^{(II)} = \frac{\Omega^2}{\left(k_0^2 - 2G_2\right)\left[4\left(k_0^2 - 2G_2\right)^2 - \Omega^2\right]}, \tag{24}$$

$$\chi_M^{(III)} = \frac{2}{\Omega - 2\left(k_0^2 - 2G_2\right)}, \tag{25}$$

and exhibits a divergent behavior at the transition between the two phases. Indeed, when approaching the transition (22) from above or below, the values of χ_M differ by a factor 2, revealing the second-order nature of the phase transition [31, §144]. It is worth pointing out that, if $G_2 = 0$, the calculation of χ_M reduces to the ideal gas value, which is found to be related to the effective mass (4) and (5) by the simple relation

$$\frac{m^*}{m} = 1 + k_0^2 \chi_M. \tag{26}$$

The divergent behavior of the magnetic polarizability near the second-order phase transition was experimentally confirmed by Zhang *et al.* through the study of the center-of-mass oscillation[32] (see also the discussion in Sec. 4). Concerning the stripe phase, the calculation of χ_M yields a complicated expression which, in the weak coupling limit G_1, $G_2 \ll k_0^2$, reduces to the simplified form

$$\chi_M^{(I)} = \frac{\Omega^2 - 4k_0^4}{(G_1 + 2G_2)\Omega^2 - 8G_2k_0^4}. \tag{27}$$

Notice that $\chi_M^{(I)}$ diverges at the critical frequency providing the transition to the plane-wave phase (see Eq. (21)). However, Eq. (27) is valid only in the weak coupling limit, and the inclusion of higher-order terms makes the value of χ_M finite at the transition.

The thermodynamic compressibility $\kappa_T = n^{-1}(\partial\mu/\partial n)^{-1}$ in all the phases can be calculated from the expressions of the chemical potential (see Eqs. (18)–(20)),

$$1/\kappa_T^{(I)} = 2G_1 + \frac{G_1 k_0^2 \Omega^2}{4\left(k_0^2 + G_1\right)^3}, \tag{28}$$

$$1/\kappa_T^{(II)} = 2\left(G_1 + G_2\right) - \frac{G_2 k_0^2 \Omega^2}{2\left(k_0^2 - 2G_2\right)^3}, \tag{29}$$

$$1/\kappa_T^{(III)} = 2G_1. \tag{30}$$

For an interacting Bose gas, the compressibility has always a finite value (see Eqs. (28)–(30)). It is discontinuous at the first-order transition between the stripe and the plane-wave phases; furthermore, if $G_2 \neq 0$, it exhibits a jump also at the second-order transition between the plane-wave and the single-minimum phases. However, as we will show in the next section, the sound velocity is continuous across the latter transition.

3. Dynamic Properties of the Uniform Phases

Spin-orbit coupling affects in a deep way also the dynamic behavior of a BEC, giving rise to exotic features in the excitation spectrum, such as the emergence of a rotonic structure when one approaches the transition from the plane-wave to the stripe phase,[33,34] the suppression of the sound velocity near the transition between the plane-wave and the single-minimum phases,[34] a double gapless band structure in the stripe phase,[35] etc. In the present section and in the next one, we focus on the dynamic behavior of a spin-orbit-coupled BEC in the uniform phases II and III. The properties of the stripe phase will be discussed in Sec. 5.

3.1. Dynamic density response function. Excitation spectrum

In order to investigate the dynamic properties of a spin-orbit-coupled BEC, it is useful to evaluate its dynamic density response function. This can be done by adding a time-dependent perturbation $V_\lambda = -\lambda e^{i(\mathbf{q}\cdot\mathbf{r}-\omega t)} + \text{H.c.}$ to the single-particle Hamiltonian (2). The direction of the wave vector

$\mathbf{q}$ is characterized by the polar angle $\alpha \in [0, \pi]$ with respect to the x axis. The density response function is then calculated through the usual definition $\chi(\mathbf{q}, \omega) = \lim_{\lambda \to 0} \delta n_{\mathbf{q}}/(\lambda e^{-i\omega t})$, where $\delta n_{\mathbf{q}}$ are the fluctuations of the $\mathbf{q}$ component of the density induced by the external perturbation.[a] In the following we derive $\chi(\mathbf{q}, \omega)$ by solving the time-dependent Gross-Pitaevskii equation

$$i\frac{\partial \Psi}{\partial t} = \left[h_0^{SO} + V_\lambda + \frac{1}{2}\left(g + g_{\uparrow\downarrow}\right)\left(\Psi^\dagger \Psi\right) + \frac{1}{2}\left(g - g_{\uparrow\downarrow}\right)\left(\Psi^\dagger \sigma_z \Psi\right)\sigma_z \right]\Psi,$$

(31)

where h_0^{SO} is the single-particle Hamiltonian (2) with $\delta = 0$. Since we are focusing on phases II and III, where the ground-state density is uniform, the spinor wave function Ψ in Eq. (31) can be written in the simple form

$$\Psi = e^{-i\mu t}\left[\sqrt{\bar{n}}\begin{pmatrix} \cos\theta \\ -\sin\theta \end{pmatrix}e^{ik_1 x} + \begin{pmatrix} u_\uparrow(\mathbf{r}) \\ u_\downarrow(\mathbf{r}) \end{pmatrix}e^{-i\omega t} + \begin{pmatrix} v_\uparrow^*(\mathbf{r}) \\ v_\downarrow^*(\mathbf{r}) \end{pmatrix}e^{i\omega t} \right].$$

(32)

The terms depending on the Bogoliubov amplitudes u and v provide the deviations in the order parameter with respect to equilibrium, caused by the external perturbation. In the linear (small λ) limit we find the result (near the poles one should replace ω with $\omega + i0$)

$$\chi(\mathbf{q}, \omega) = \frac{-Nq^2\left[\omega^2 - 4k_1 q\cos\alpha\,\omega + a(q, \alpha)\right]}{\omega^4 - 4k_1 q\cos\alpha\,\omega^3 + b_2(q, \alpha)\omega^2 \atop +k_1 q\cos\alpha\,b_1(q, \alpha)\omega + b_0(q, \alpha)},$$

(33)

where the coefficients a and b_i are even functions of $q \equiv |\mathbf{q}|$ and $\cos\alpha$, implying that $b_i(q, \alpha) = b_i(q, \pi \pm \alpha)$ (the same for a), and their actual values depend on whether one is in phase II or III (see App. A.1). In the plane-wave phase, the odd terms in ω entering the response function reflect the lack of parity and time-reversal symmetry of the ground-state wave function; in the single-minimum phase, however, one has $k_1 = 0$ and thus the symmetry is restored.

---

[a]The spin-density response function can be calculated with an analogous procedure by adding a perturbation $\sigma_z V_\lambda$ to (2).

Notice that the response function (33) reduces to the usual Bogoliubov form $\chi(\mathbf{q}, \omega) = -Nq^2/[\omega^2 - q^2(2G_1 + q^2/4)]$ when $G_2 = 0$ and $\Omega = 0$, characterizing the response of a BEC gas in the absence of spin-orbit coupling. It is also worth pointing out that, since V_λ commutes with the unitary transformation yielding the Hamiltonian in the spin-rotated frame (see Sec. 2.1), the expression for $\chi(\mathbf{q}, \omega)$ is the same as in the original laboratory frame, and thus all the results based on the calculation in the spin-rotated frame are relevant for actual experiments.

The frequencies of the elementary excitations are given by the poles of the response function χ, i.e., by the zeros of

$$\omega^4 - 4k_1 q \cos\alpha\, \omega^3 + b_2\, \omega^2 + k_1 q \cos\alpha\, b_1\, \omega + b_0 = 0. \qquad (34)$$

The solutions of this equation provide two separated branches, as shown in Fig. 5(a) and (b) for phase II and phase III respectively. The lower one is gapless and exhibits a phonon dispersion at small q, while the upper one is gapped as a consequence of the Raman coupling. For example, in phase III the gap between the two branches is given, at $\mathbf{q} = 0$, by $\Delta = \sqrt{\Omega(\Omega + 4G_2)}$. Differently from the single-minimum phase, the excitation spectrum in the plane-wave phase is not symmetric under inversion of q_x into $-q_x$, as a consequence of the symmetry-breaking terms appearing in Eq. (33). For negative values of q_x, the lower branch in phase II exhibits a very peculiar feature, resulting in the emergence of a roton minimum, which becomes more and more pronounced as one approaches the transition to the stripe phase. The occurrence of the rotonic structure in spin-orbit-coupled BECs shares interesting analogies with the case of dipolar gases in quasi-2D configurations[36] and of condensates with soft-core, finite-range interactions.[37,38] The physical origin of the roton minimum is quite clear. In phase II the ground state is twofold degenerate, and it is very favorable for atoms to be transferred from the BEC state with momentum $\mathbf{p} = k_1\hat{\mathbf{e}}_x$ to the empty state at $\mathbf{p} = -k_1\hat{\mathbf{e}}_x$. The excitation spectrum has been recently measured using Bragg spectroscopy techniques, confirming the occurrence of a characteristic rotonic structure[39,40] (see also Ref. 41 for the case of shaken optical lattices).[b]

---

[b]In the experiments of Refs. 39–41 the excitation spectrum has been measured on top of the BEC state with momentum $\mathbf{p} = -k_1\hat{\mathbf{e}}_x$, for which the roton minimum, differently from the case discussed above, appears at positive values of q_x.

(a)

(b)

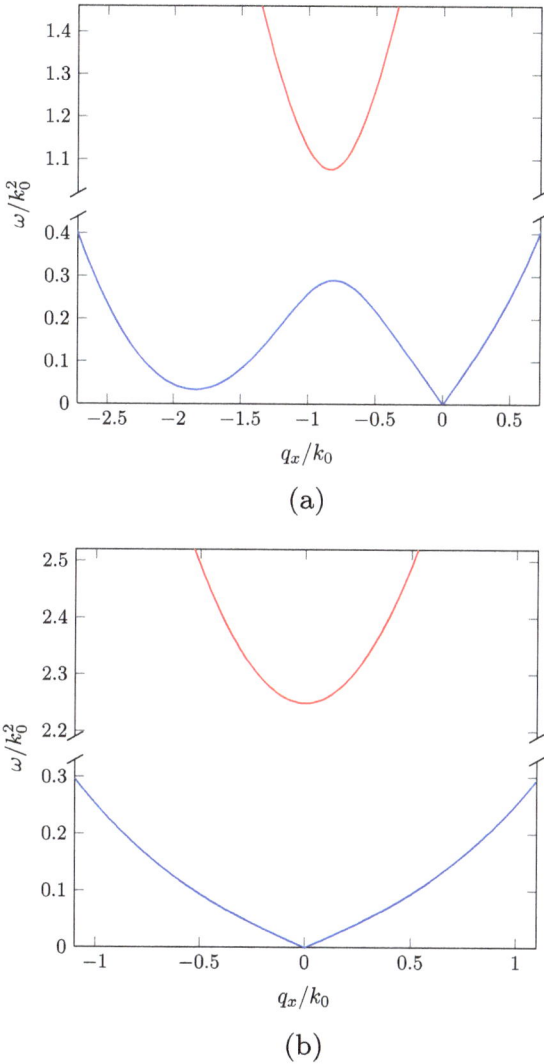

Fig. 5. Excitation spectrum (a) in phase II ($\Omega/k_0^2 = 0.85$) and (b) in phase III ($\Omega/k_0^2 = 2.25$) as a function of q_x ($q_y = q_z = 0$), calculated in the experimental conditions of Ref. 32. The blue and red lines represent the lower and upper branches, respectively. In phase II the spectrum is not symmetric and exhibits a roton minimum for negative q_x, whose energy becomes smaller and smaller as one approaches the transition to the stripe phase at $\Omega/k_0^2 = 0.095$. The other parameters: $G_1/k_0^2 = 0.12$, $\gamma = G_2/G_1 = 10^{-3}$.

3.2. Static response function and static structure factor

The static response function $\chi(\mathbf{q}) \equiv \chi(\mathbf{q}, \omega = 0)/N$ can be derived directly from Eq. (33). Its $q = 0$ value $\mathcal{K} \equiv \chi(q = 0)$ is given by

$$\mathcal{K}_{\mathrm{II}}^{-1} = 2G_1 + \frac{2G_2 k_1^2 \left(k_1^2 \cos^2 \alpha + k_0^2 \sin^2 \alpha - 2G_2\right)}{k_1^2 \left(k_0^2 \cos^2 \alpha - 2G_2\right) + k_0^4 \sin^2 \alpha}, \tag{35}$$

$$\mathcal{K}_{\mathrm{III}}^{-1} = 2G_1 \tag{36}$$

in the plane-wave phase II and the single-minimum phase III, respectively. The anisotropy of $\mathcal{K}$ in phase II caused by the spin interaction term G_2 is revealed by the last term of Eq. (35) which depends on the polar angle α. It is also worth pointing out that $\mathcal{K}_{\mathrm{II}}$ coincides with the thermodynamic compressibility $\kappa_T^{(\mathrm{II})}$ (see Eq. (29)) only along the x direction, i.e., when $\sin \alpha = 0$. In this case, $\mathcal{K}$ also exhibits a jump at the transition between phases II and III. This marks a difference with respect to the behavior of the frequencies of the elementary excitations, fixed by Eq. (34), which are always continuous functions of Ω at the transition for all values of $\mathbf{q}$.

Far from the phonon regime, the occurrence of the roton minimum is reflected in an enhancement in the static response function $\chi(q_x)$ close to the roton momentum, as shown in Fig. 6, representing a typical tendency of the system towards crystallization. When the roton frequency vanishes, $\chi(q_x)$ exhibits a divergent behavior. A simple analytic expression for the corresponding value of the Raman coupling Ω is obtained in the weak

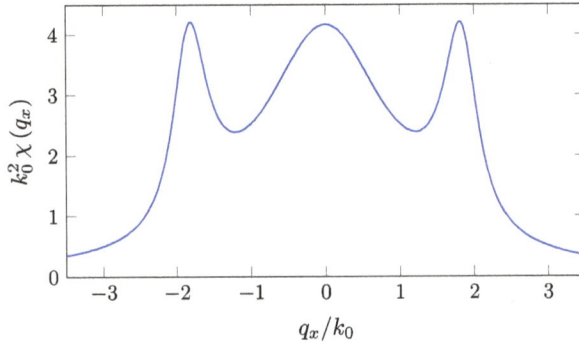

Fig. 6. Static response in phase II as a function of q_x ($q_y = q_z = 0$). The curve is symmetric and exhibits a typical peak near the roton momentum. The parameters are $\Omega/k_0^2 = 0.85$, $G_1/k_0^2 = 0.12$ and $\gamma = G_2/G_1 = 10^{-3}$.

coupling limit $G_1, G_2 \ll k_0^2$, where we find that the critical value exactly coincides with the value (see Eq. (21)) characterizing the transition between the plane-wave and the stripe phases. For larger values of the coupling constants G_1 and G_2, the critical value takes place for values of the Raman coupling smaller than the value at the transition, exhibiting the typical spinoidal behavior of first-order liquid-crystal phase transitions.

The dynamic structure factor at $T = 0$ can be calculated from the response function (33) through the relation $S(\mathbf{q}, \omega) = \pi^{-1}\mathrm{Im}\,\chi(\mathbf{q}, \omega)$ for $\omega \geq 0$ and $S(\mathbf{q}, \omega) = 0$ for negative ω. In the plane-wave phase, the condition $\mathrm{Im}\,\chi(\mathbf{q}, \omega) = -\mathrm{Im}\,\chi(-\mathbf{q}, -\omega)$, characterizing the imaginary part of the response function, is still satisfied, but the symmetry relation $\mathrm{Im}\,\chi(\mathbf{q}, \omega) = \mathrm{Im}\,\chi(-\mathbf{q}, \omega)$ is not ensured, and consequently one finds $S(\mathbf{q}, \omega) \neq S(-\mathbf{q}, \omega)$. This affects several well-known equalities involving sum rules, which have to be formulated in a more general way to account for the breaking of inversion symmetry in the plane-wave phase. An example is the f-sum rule $\int d\omega\,[S(\mathbf{q}, \omega) + S(-\mathbf{q}, \omega)]\,\omega = Nq^2$, which is exactly satisfied, as one can deduce from the correct large ω behavior of the density response function: $\chi(\mathbf{q}, \omega)_{\omega\to\infty} = -Nq^2/\omega^2$.[42] On the other hand, the inversion symmetry of the static structure factor $S(\mathbf{q}) = \int_0^\infty d\omega\,S(\mathbf{q}, \omega)/N$ is always ensured, since it is a general feature following from the completeness relation and the commutation relation involving the density operators: $S(\mathbf{q}) - S(-\mathbf{q}) = \langle\,[\rho_\mathbf{q}, \rho_{-\mathbf{q}}]\,\rangle = 0$.

It is worth pointing out that, despite the strong enhancement exhibited by the static response function $\chi(q_x)$, the static structure factor $S(q_x)$ does not exhibit any peaked structure near the roton point. This is different from what happens, for example, in superfluid helium.[c] In Fig. 7 we show the static structure factor $S(q_x)$ together with the contribution to the integral $S(q_x) = \int d\omega\,S(q_x, \omega)/N$ arising from the lower branch of the elementary excitations. The figure shows that the lower-branch contribution is not symmetric for exchange of q_x into $-q_x$, even if the total $S(q_x)$ is symmetric, as we have showed previously. Remarkably, the strength carried by the

---

[c]At finite temperature T one instead expects the static structure factor to be significantly peaked near the roton minimum, provided the roton energy is small compared to T, as a consequence of the thermal excitations of rotons, similarly to what is predicted for quasi-2D dipolar gases.[43]

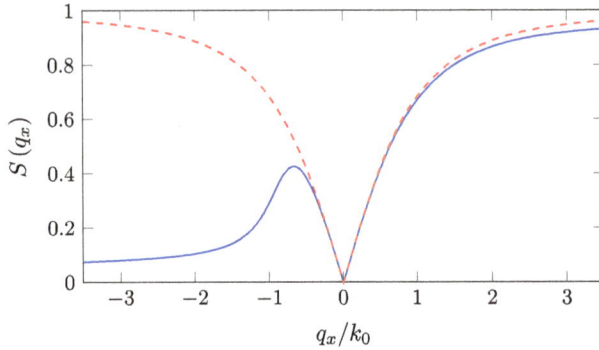

Fig. 7. Contribution of the lower branch to the static structure factor in phase II as a function of q_x (blue solid line), compared with the total $S(q_x)$ (red dashed line). The parameters are $\Omega/k_0^2 = 0.85$, $G_1/k_0^2 = 0.12$ and $\gamma = G_2/G_1 = 10^{-3}$.

lower branch is significantly peaked for intermediate values of q_x between the phonon and the roton regimes, in the so-called maxon region, where the lower branch of the excitation spectrum exhibits a maximum.

3.3. *Velocity and density vs spin nature of the sound mode*

The low-frequency excitations at small q, i.e., the sound waves, can be easily obtained by setting $\omega = cq$, where c is the sound velocity, and keeping the leading terms proportional to q^2 in Eq. (34). This allows us to obtain the sound velocity in the plane-wave and the single-minimum phases,

$$c_{\text{II}} = \frac{1}{k_0^4 - 2G_2k_1^2}\left\{ G_2k_1 \left(k_0^2 - k_1^2\right)\cos\alpha \right.$$

$$\left. + \sqrt{\begin{array}{c} 2\left[G_1k_0^4 + G_2k_1^2\left(k_0^2 - 2G_1 - 2G_2\right)\right] \\ \left[k_0^4 - 2G_2k_1^2 - k_0^2\left(k_0^2 - k_1^2\right)\cos^2\alpha\right] \end{array}} \right\}, \qquad (37)$$

$$c_{\text{III}} = \sqrt{2G_1\left(1 - \frac{2k_0^2\cos^2\alpha}{\Omega + 4G_2}\right)}. \qquad (38)$$

Approaching the transition between the two phases, both sound velocities exhibit a strong reduction along the x direction ($\cos\alpha = \pm1$), caused by the spin-orbit coupling. This suppression can be understood in terms of the increase of the effective mass associated with the single-particle dispersion (see Eq. (3)). At the transition, where the velocity of sound modes propagating along the x direction vanishes, the elementary excitations exhibit a different q^2 dependence. On the other hand, the sound velocities along the other directions ($\alpha \neq 0, \alpha \neq \pi$) remain finite at the transition. The sound velocity in phase II shows a further interesting feature caused by the lack of parity symmetry. The asymmetry effect in c_{II} is due to the presence of the first term in the numerator of Eq. (37), therefore the symmetry will be recovered if $G_2 = 0$ or $\alpha = \pi/2$ (corresponding to phonons propagating along the directions orthogonal to the x axis).

The role played by the spin degree of freedom in the propagation of the sound can be better understood by relating the sound velocity to the magnetic polarizability χ_M (see Eqs. (24) and (25)) and the $q = 0$ static response $\mathcal{K}$ (see Eqs. (35) and (36)). One finds the result

$$c(\alpha)c(\alpha + \pi) = \frac{1 + k_0^2 \, \chi_M \sin^2 \alpha}{\mathcal{K}\left(1 + k_0^2 \, \chi_M\right)}, \qquad (39)$$

holding in both phases II and III. The above equation generalizes the usual relation $c^2 = 1/\mathcal{K} = n(\partial \mu / \partial n)$ between the sound velocity and the compressibility holding in usual superfluids. It explicitly shows that, along the x direction, where $\sin \alpha = 0$, the sound velocity c vanishes at the transition because of the divergent behavior of the magnetic polarizability. The sound velocity along the x axis as a function of Ω is shown in Fig. 8 for a

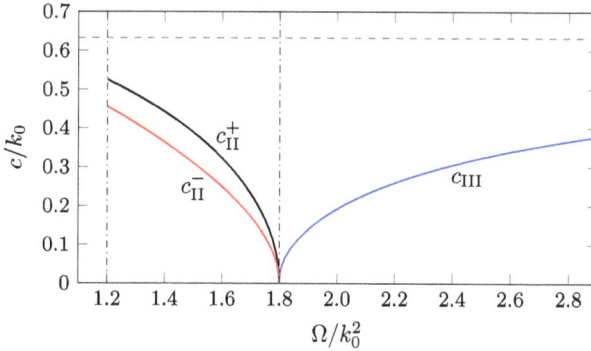

Fig. 8. Sound velocity as a function of the Raman coupling for the following choice of parameters: $G_1/k_0^2 = 0.2$, $G_2/k_0^2 = 0.05$. The two sound velocities in phase II correspond to phonons propagating in the direction parallel (c_{II}^+) and antiparallel (c_{II}^-) to k_1. The horizontal dashed line corresponds to the value $\sqrt{2G_1} = 0.63\,k_0$ of the sound velocity in the absence of spin-orbit and Raman coupling. The vertical dash-dotted lines indicate the critical values of Ω at which the I-II and II-III phase transitions take place.

configuration with relatively large G_2, emphasizing the difference between $c_{\mathrm{II}}^+ = c_{\mathrm{II}}(\alpha = 0)$ and $c_{\mathrm{II}}^- = c_{\mathrm{II}}(\alpha = \pi)$. The suppression effect exhibited by the sound velocity near the II-III phase transition is particularly remarkable in the single-minimum phase III where BEC takes place in the $\mathbf{p} = 0$ state and the compressibility of the gas is unaffected by spin-orbit coupling. It explicitly reveals the mixed density and spin nature of the sound waves, with the spin nature becoming more and more important as one approaches the phase transition where χ_M diverges.

The combined density and spin nature of sound waves is also nicely revealed by the relative amplitudes of the density δn and spin-density δs oscillations in the $q \to 0$ limit, characterizing the propagation of sound. In terms of the magnetic polarizability χ_M we find

$$
\left(\frac{\delta s}{\delta n}\right)_{\mathrm{II}} = \frac{\sqrt{1 + \left(k_0^2 - 2G_2\right)\chi_M}}{1 + k_0^2\chi_M}
$$

$$
+ \frac{k_0\,\chi_M\cos\alpha}{1 + k_0^2\,\chi_M}\sqrt{\frac{2\left[G_2 + G_1\left(1 + k_0^2\,\chi_M\right)\right]}{1 + k_0^2\,\chi_M\sin^2\alpha}}, \tag{40}
$$

$$
\left(\frac{\delta s}{\delta n}\right)_{\mathrm{III}} = \frac{2k_0\,\chi_M\cos\alpha\,\sqrt{G_1}}{\sqrt{2\left(1 + k_0^2\,\chi_M\right)\left(1 + k_0^2\,\chi_M\sin^2\alpha\right)}} \tag{41}
$$

in the plane-wave and the single-minimum phases respectively. The above equations show that, near the transition between phases II and III, the amplitude of the spin-density fluctuations δs of the sound waves propagating along the x direction ($\sin\alpha = 0$) is strongly enhanced with respect to the density fluctuations δn, as a consequence of the divergent behavior of the magnetic polarizability. In particular, very close to the phase transition the relative amplitude is given by

$$
\frac{\delta s}{\delta n} \sim \sqrt{2G_1\chi_M} \tag{42}
$$

in both phases II and III. This suggests that an effective way to excite these phonon modes near the transition is through a coupling with the spin degree of freedom as recently achieved in two-photon Bragg experiments on Fermi gases.[44] For sound waves propagating in the direction orthogonal to x the situation is instead different. In particular in phase III sound waves are purely density oscillations ($\delta s = 0$).

It is finally interesting to understand the role played by the sound waves in terms of sum rules. The phonon mode exhausts the compressibility sum rule $\int d\omega \, [S(\mathbf{q}, \omega) + S(-\mathbf{q}, \omega)]/\omega$ at small q, as one can easily prove from Eq. (33). However, different from ordinary superfluid, it gives only a small contribution to the f-sum rule as one approaches the second-order transition. This contribution becomes vanishingly small at the transition for wave vectors $\mathbf{q}$ oriented along the x direction. Also, the static structure factor $S(\mathbf{q})$ is strongly quenched compared to usual BECs. This results in an enhancement of the quantum fluctuations of the order parameter, as predicted by the uncertainty principle inequality.[45,46] This effect is, however, small because the sound velocity vanishes only along the x direction.[27]

4. Collective Excitations in the Trap

In this section we discuss the collective excitations for a harmonically trapped BEC with spin-orbit coupling. First one should notice that, in typical experimental conditions, the spin-orbit coupling strength, usually quantified by the recoil energy $E_r = k_0^2/2$, is much larger than the trapping frequencies. As a consequence, one expects that the three phases occurring in uniform matter due to the spin-orbit coupling survive also in the presence of harmonic trapping. This can be verified by solving numerically the Gross-Pitaevskii equation

$$i \frac{\partial \Psi}{\partial t} = \left[h_0^{\mathrm{SO}} + V_{\mathrm{ext}}(\mathbf{r}) + \frac{1}{2}(g + g_{\uparrow\downarrow})(\Psi^\dagger \Psi) + \frac{1}{2}(g - g_{\uparrow\downarrow})(\Psi^\dagger \sigma_z \Psi)\sigma_z \right] \Psi$$

(43)

for the condensate wave function, with $V_{\mathrm{ext}}(\mathbf{r}) = (\omega_x^2 x^2 + \omega_y^2 y^2 + \omega_z^2 z^2)/2$ representing the external trapping potential. Figure 9 gives an example of the momentum distribution and the spin polarization of a trapped spin-orbit-coupled BEC as a function of the Raman coupling. For simplicity, we have considered harmonic trapping only along the x direction. One can see that the three phases discussed in the bulk case show up also here. It is worth mentioning that in the low density limit, where the interaction energy is much smaller than the recoil energy, the value of Ω/k_0^2 at the transitions

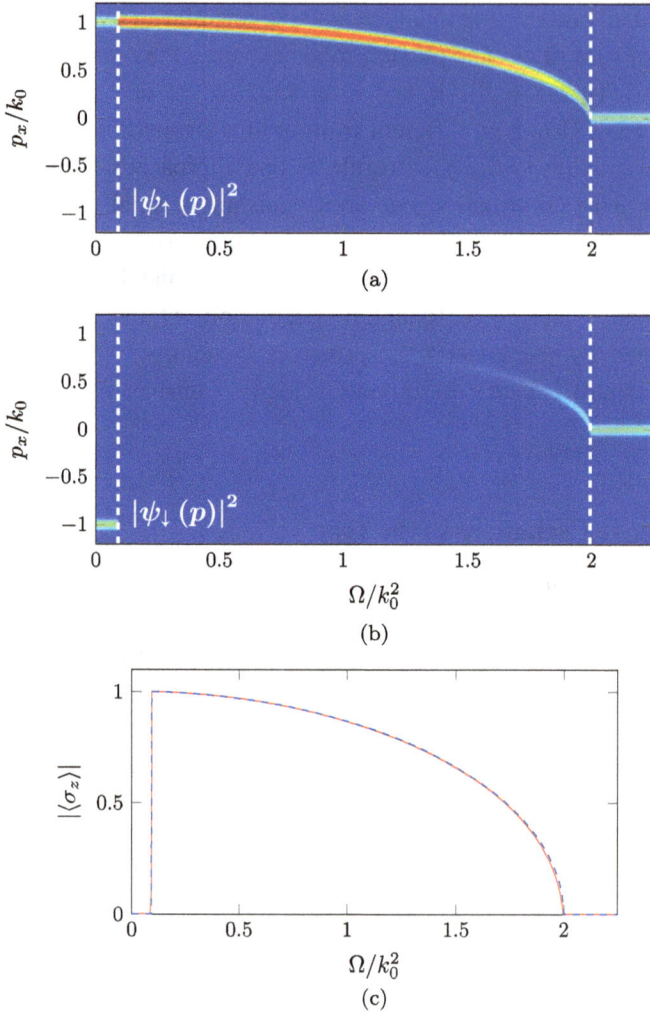

Fig. 9. (a–b). Momentum distribution for the two spin components as a function of Ω. The white dashed lines indicate the transition frequencies calculated using (21) and (22). (c). Spin polarization $|\langle\sigma_z\rangle| = |N_\uparrow - N_\downarrow|/N$ as a function of Ω in the trapped case (red solid line) and in the uniform case using the density in the center of the trap (blue dashed line). The parameters are chosen as follows: $\omega_x = 2\pi \times 40\,\text{Hz}$, $\omega_x/k_0^2 = 0.01$, $\delta = 0$, $g_{\uparrow\uparrow} = g_{\downarrow\downarrow} = 4\pi \times 101.20\,a_B$, $g_{\uparrow\downarrow} = 4\pi \times 100.94\,a_B$, where a_B is the Bohr radius. The density in the center of the trap corresponds to $n \simeq 1.9 \times 10^{13}\,\text{cm}^{-3}$.

(21) and (22) is almost density-independent, therefore even in the presence of a trap they can be well identified using the results obtained in the bulk.

4.1. *Dipole mode: a sum-rule approach*

Among the various excitations exhibited by a trapped spin-orbit-coupled gas, the dipole mode deserves a special attention. It corresponds to the oscillation of the center-of-mass of the system, and can be easily excited experimentally.[47] For a conventional trapped gas without spin-orbit coupling, the oscillation along a certain direction, for example the x axis, is excited by the dipole operator $X = \sum_j x_j$, and its frequency is equal to the frequency ω_x of the harmonic trap. In the presence of spin-orbit-coupling, the behavior of the dipole oscillation can be studied using the formalism of sum rules.[30] A major advantage of this method is that it can reduce the calculation of the dynamical properties of the many-body system to the knowledge of a few key parameters relative to the ground state.

The starting point of our analysis is represented by the definition of the k-th moment of the dynamic structure factor for a general operator F, given at zero temperature by[d]

$$m_k(F) = \sum_n (E_n - E_0)^k \, |\langle 0| F |n\rangle|^2 . \qquad (44)$$

Here $|0\rangle$ and $|n\rangle$ are, respectively, the ground state and the n-th excited state of the many-body Hamiltonian (6), now including the external trapping potential in the single-particle contributions

$$h_0^{SO}(j) = \frac{1}{2}\left[(p_{x,j} - k_0\sigma_{z,j})^2 + p_{\perp,j}^2\right] + \frac{\Omega}{2}\sigma_{x,j} + \frac{\delta}{2}\sigma_{z,j} + V_{\text{ext}}(\mathbf{r}_j), \qquad (45)$$

and E_0, E_n are the corresponding energies. The quantity $|\langle 0| F |n\rangle|^2$ is called the strength of the operator F relative to the state $|n\rangle$.

Some moments can be easily calculated by employing the closure relation and the commutation rules involving the Hamiltonian of the system. In the case of the dipole operator $F = X$ one finds, for example, that the energy-weighted moment takes the well-known model-independent value

---

[d]At finite temperature, the moments m_k should include the proper Boltzmann factors.[42]

(also called f-sum rule)

$$m_1(X) = \frac{1}{2}\langle 0| [X, [H, X]] |0\rangle = \frac{N}{2} \qquad (46)$$

with N the total number of atoms. Notice that this sum rule is not affected by the spin terms in the Hamiltonian, despite the fact that the commutator of H with X explicitly depends on the spin-orbit coupling:

$$[H, X] = -i(P_x - k_0 \Sigma_z), \qquad (47)$$

where $P_x = \sum_j p_{x,j}$ is the total momentum of the gas along the x direction, and $\Sigma_z = \sum_j \sigma_{z,j}$ is the total spin operator along z. Equation (47) actually reflects the fact that the equation of continuity (and hence in our case the dynamic behavior of the center-of-mass coordinate) is deeply influenced by the coupling with the spin variable.

Another important sum rule is the inverse energy-weighted sum rule (also called dipole polarizability). In the presence of harmonic trapping, this sum rule can be calculated in a straightforward way using the commutation relation

$$[H, P_x] = i\omega_x^2 X \qquad (48)$$

and the closure relation. One finds

$$m_{-1}(X) = \frac{m_1(P_x)}{\omega_x^4} = \frac{N}{2\omega_x^2}. \qquad (49)$$

Both sum rules (46) and (49) are insensitive to the presence of the spin terms in the single-particle Hamiltonian (45), as well as to the two-body interaction. This does not mean, however, that the dipole dynamics is not affected by the spin-orbit coupling. This effect is accounted for by another sum rule, particularly sensitive to the low-energy region of the excitation spectrum: the inverse cubic energy-weighted sum rule, for which we find the exact result

$$m_{-3}(X) = \frac{m_{-1}(P_x)}{\omega_x^4} = \frac{N}{2\omega_x^2}(1 + k_0^2 \chi_M), \qquad (50)$$

where χ_M corresponds to the magnetic polarizability already defined in Sec. 2.3, and given in terms of sum rules by $\chi_M = 2m_{-1}(\Sigma_z)/N$. It is worth mentioning that the above results for the sum rules $m_1(X)$, $m_{-1}(X)$

and $m_{-3}(X)$ hold exactly for the Hamiltonian (6), including the interaction terms. Their validity is not restricted to the mean-field approximation and is ensured for both Bose and Fermi statistics, at zero as well as at finite temperature. In particular the sum rule $m_{-3}(X)$, being sensitive to the magnetic polarizability, is expected to exhibit a nontrivial temperature dependence across the BEC transition.

Equation (50) exploits the crucial role played by the spin-orbit coupling proportional to k_0. The effect is particularly important when the magnetic polarizability takes a large value. A large increase of χ_M is associated with the occurrence of a dipole soft mode as can be inferred by taking the ratio between the inverse and cubic inverse energy-weighted sum rules $m_{-1}(X)$ and $m_{-3}(X)$, yielding the rigorous upper bound

$$\omega_D^2 = \frac{m_{-1}(X)}{m_{-3}(X)} = \frac{\omega_x^2}{1 + k_0^2 \chi_M} \tag{51}$$

to the lowest dipole excitation energy. The value of χ_M for a trapped BEC can be calculated in the same way as in uniform matter (see Sec. 2.3), with the difference that the condensate wave function is now provided by the solution of Eq. (43) rather than by the ansatz (8). Figure 10(a) shows the behavior of the magnetic polarizability in the plane-wave and the single-minimum phases as a function of the Raman coupling Ω, calculated by numerically solving Eq. (43) in the presence of harmonic trapping along the x direction (red dashed lines), and by the relations (24) and (25) in uniform matter using the density in the center of the trap (blue solid lines). The choice of the parameters corresponds to the experimental conditions of Ref. 32. The black squares represent the magnetic polarizability extracted from the measurement of the oscillation amplitudes of some relevant quantities[32] (see the discussion in Sec. 4.2). Figure 10(b) shows the frequency of the dipole oscillation predicted from Eq. (51) using the same values of χ_M presented in (a). It reveals important deviations from the trap frequency ω_x caused by the spin-orbit coupling. The circles are the experimental results of Ref. 32. Far from the transition point at $\Omega \simeq 2k_0^2$ the theoretical curves agree very well with the experimental data, while near the transition nonlinear effects play a major role, as discussed in Ref. 32. The lack of data points in the region below the transition is due to the occurrence of a dynamic instability, which makes the observation of the dipole oscillation very difficult.[48]

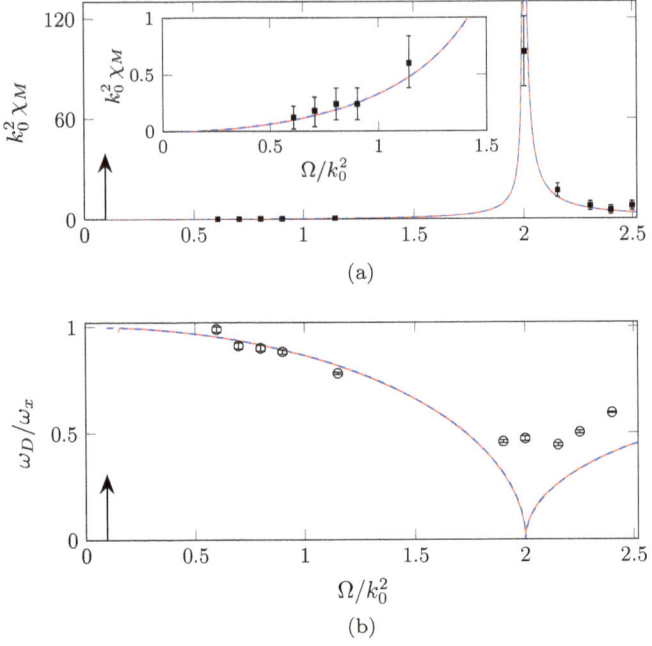

Fig. 10. (a). Magnetic polarizability χ_M as a function of Ω calculated in a trap (red solid lines) and in uniform matter using the density in the center of the trap (blue dashed lines). (b). Dipole frequency predicted by (51), using the values of χ_M shown above, represented by the red solid lines and the blue dashed lines respectively. The parameters are $k_0^2 = 2\pi \times 4.42\,\text{kHz}$, $\omega_x = 2\pi \times 45\,\text{Hz}$, the scattering lengths $a_{\uparrow\uparrow} = a_{\downarrow\downarrow} = 101.20\,a_B$, $a_{\uparrow\downarrow} = 100.94\,a_B$, where a_B is Bohr radius, and the atomic mass of ^{87}Rb. The density in the center of the trap is $n \simeq 1.37 \times 10^{14}\,\text{cm}^{-3}$. The black squares and circles in the figures are the experimental data of Ref. 32. The black arrows indicate the transition between phases I and II.

4.2. *Dipole mode and oscillation amplitudes*

The combined spin-orbit nature of the lowest dipole mode is also nicely revealed by the relative amplitudes of the oscillating values of the center-of-mass position (A_X), the momentum (A_{P_x}) and the spin polarization (A_{Σ_z}). These amplitudes can be calculated in the present approach by writing the many-body oscillating wave function as

$$|\Psi(t)\rangle = e^{i\alpha(t)\delta P_x} e^{\beta(t)G}|0\rangle, \tag{52}$$

where $\delta P_x = P_x - \langle P_x \rangle_0$ plays the role of the excitation operator, $\langle\ \rangle_0$ denoting the expectation value on the ground state $|0\rangle$, while G represents the restoring force defined by the commutation relation $[H, G] = \delta P_x$, and α, β are time-dependent parameters. The equations governing the time evolution of these parameters can be obtained through a variational Lagrange procedure; at the lowest order in α and β they read

$$\dot{\alpha}(t) = -\beta(t), \tag{53}$$

$$\dot{\beta}(t) = \frac{\omega_x^2}{1 + k_0^2 \chi_M} \alpha(t). \tag{54}$$

The time dependence of the relevant quantities $\langle X \rangle$, $\langle P_x \rangle$ and $\langle \Sigma_z \rangle$ can be obtained by solving Eqs. (53) and (54). The relations between the spin, the center-of-mass and the momentum oscillation amplitudes eventually take the useful form

$$A_{\Sigma_z} = A_X \frac{\omega_x k_0 \chi_M}{\sqrt{1 + k_0^2 \chi_M}}, \tag{55}$$

$$\frac{A_{P_x}}{k_0} = A_{\Sigma_z} \frac{1 + k_0^2 \chi_M}{k_0^2 \chi_M}. \tag{56}$$

The connection between the momentum and spin amplitudes has been already pointed out in Ref. 32 (see Fig. 4 therein). It provides a practical way to determine experimentally the magnetic polarizability χ_M. Near the transition point between the plane-wave and the single-minimum phase the ratio A_{Σ_z}/A_X between the spin and the center-of-mass amplitudes diverges like $\sqrt{\chi_M}$, in analogy with the behavior exhibited by the ratio between the spin and the density amplitudes in the propagation of sound (see Eq. (42)).

It is worth pointing out that the results (55)–(56), as well as the upper bound of the excitation frequency (51), are expected to be accurate when the Raman coupling Ω is larger than the trapping frequency ω_x. Instead, in the opposite limit $\Omega \ll \omega_x$, the lowest mode is mainly a spin oscillation, which does not exhibit any significant coupling to the center-of-mass motion. The corresponding frequency can be estimated with a sum-rule approach by considering the ratio $m_1(\Sigma_z)/m_{-1}(\Sigma_z)$ of the moments of the spin operator Σ_z. The excitation frequency calculated in this way is found to vanish linearly with Ω. Finally, in the intermediate regime between the two limits discussed above, one can define $F = P_x + \eta k_0 \Sigma_z$ and use the ansatz $\delta F = F - \langle F \rangle_0$ for the operator exciting the dipole oscillation, where the value of the variational parameter η is found by minimizing the estimate $m_1(F)/m_{-1}(F)$ for the excitation frequency. The corresponding oscillation amplitudes of the relevant physical quantities can be calculated by a procedure analogous to the one discussed above.[30]

4.3. Hydrodynamic formalism

A useful approach to describe the phonon regime in the excitation spectrum of a superfluid is provided by hydrodynamic theory. For a spinor BEC this theory can be derived by writing the spin-up and spin-down components of the order parameter in terms of their modulus and phase.[33,34] In the resulting equations the quantum pressure terms can be safely neglected in the phonon regime, characterized by long wavelengths and low frequencies. Furthermore, since the phonon frequencies are much smaller than the gap between the two branches of the excitation spectrum, which is of the order of Ω, the relative phase of the two spin components is locked ($\phi_\uparrow = \phi_\downarrow$). As a consequence, the relevant hydrodynamic equations reduce to the equations for the change in the total density δn and in the phase $\delta\phi = \delta\phi_\uparrow = \delta\phi_\downarrow$. Assuming for simplicity $g_{\uparrow\downarrow} = g$, these two equations assume the simple form

$$\frac{\partial}{\partial t}\delta n + \nabla_\perp \cdot (n\,\nabla_\perp\delta\phi) + \frac{m}{m^*}\partial_x\,(n\,\partial_x\delta\phi) = 0 \qquad (57)$$

and

$$\frac{\partial}{\partial t}\delta\phi + \delta\mu = 0, \qquad (58)$$

with m/m^* given by Eqs. (4) and (5) and $\delta\mu = g\delta n$. Notice that, due to the assumption $g_{\uparrow\downarrow} = g$, the above hydrodynamic picture can describe the dynamics only in the plane-wave and the single-minimum phases (the investigation of the phonon modes in the stripe phase requires a more sophisticated calculation, see Sec. 5). Remarkably, the equation of continuity (57) is crucially affected by the spin-orbit coupling. This follows from the fact that the current is not simply given by the canonical momentum operator, as happens in usual superfluids, but contains an additional spin contribution, accounted for, in Eq. (57), through the effective mass term. The current density operator should actually satisfy the continuity equation $[H, \rho(\mathbf{r})] = i\nabla\cdot\mathbf{j}$, where $\rho(\mathbf{r}) = \sum_j \delta(\mathbf{r}-\mathbf{r}_j)$ is the total density operator.

By explicitly carrying out the commutator one identifies the current as $\mathbf{j}(\mathbf{r}) = \mathbf{p}(\mathbf{r}) - k_0 \sigma_z(\mathbf{r}) \hat{\mathbf{e}}_x$, where $\mathbf{p}(\mathbf{r}) = \sum_j [\mathbf{p}_j \, \delta(\mathbf{r} - \mathbf{r}_j) + \text{H.c.}]/2$ and $\sigma_z(\mathbf{r}) = \sum_j \sigma_{z,j} \, \delta(\mathbf{r} - \mathbf{r}_j)$ are the momentum and spin-density operators, respectively. This expression for the current explicitly reveals the presence of a gauge field associated to the vector potential $\mathbf{A} = k_0 \sigma_z \hat{\mathbf{e}}_x$. It is worth noticing that at equilibrium the momentum and spin-dependent terms exactly compensate each other, yielding $\langle \mathbf{j}(\mathbf{r}) \rangle = 0$. The presence of the spin term in the current also reflects the violation of Galilean invariance in the spin-orbit Hamiltonian.[48]

Combining (57) and (58) one finds the following equation for the density:

$$\frac{\partial^2}{\partial t^2} \delta n = g \left[\nabla_\perp \cdot (n \, \nabla_\perp \delta n) + \frac{m}{m^*} \partial_x (n \, \partial_x \delta n) \right]. \tag{59}$$

In uniform matter, characterized by a constant density $n = \bar{n}$, Eq. (59) yields the relation $c^2 = g\bar{n}/m^*$ for the sound velocity along the x direction, consistent with the results (37) and (38) for $g_{\uparrow\downarrow} = g$. In the presence of harmonic trapping, where the equilibrium density profile is given by an inverted parabola, the solutions of the hydrodynamic equations (59) coincide with those one finds for usual BECs, with the simple replacement of the trap frequency ω_x with $\omega_x \sqrt{m/m^*}$. This gives the result $\omega_D = \omega_x \sqrt{m/m^*}$ for the dipole frequency, which is consistent with the estimate (51) based on a sum-rule approach, once the relation (26) between the effective mass and the magnetic polarizability (holding for $G_2 = 0$) is taken into account. Equation (59) also shows that, for any other hydrodynamic mode involving a motion of the gas along the x axis, a similar effect of strong reduction of the frequency close to the second-order transition should be expected. This is the case, for example, of the scissors mode for deformed traps in the x-y or x-z plane, where the collective frequency takes the form $\sqrt{(m/m^*)\omega_x^2 + \omega_y^2}$ and $\sqrt{(m/m^*)\omega_x^2 + \omega_z^2}$ respectively.

5. Static and Dynamic Properties of the Stripe Phase

The stripe phase is doubtlessly the most intriguing phase appearing in the phase diagram of Sec. 2. It has been the object of several recent theoretical investigations.[29,35,49–57] As we already pointed out, the stripe phase is characterized by the spontaneous breaking of two continuous symmetries. The breaking of gauge symmetry yields superfluidity, while the breaking of translational invariance is responsible for the occurrence of a crystalline structure. The simultaneous presence of these two broken symmetries is typical of supersolids.[28,58–60] As we shall see, it is at the origin of the appearance of two gapless excitations as well as of a band structure in the excitation spectrum.[35]

Some important properties of the ground state and the dynamics of the stripe phase in uniform matter will be discussed in Secs. 5.1 and 5.2. Many relevant quantities that we will consider, such as the contrast of the density modulations (15), will turn out to depend crucially on the value of the Raman coupling Ω. Therefore, in order to enhance the effects of the presence of the stripes one needs to use relatively large values of Ω. On the other hand, the stripe phase is favored only in a range of low values of the Raman coupling lying below the transition frequency $\Omega^{(I-II)}$. In the following we will consider configurations with relatively large values of the parameter G_2 which, as can be seen from Eq. (21), allows to obtain a significant increase of the critical value of Ω. This is not, however, the situation in current experiments with ^{87}Rb atoms,[18,32] where G_2 is instead extremely small. In Sec. 5.3 we will illustrate a procedure to increase the value of G_2 with available experimental techniques.

5.1. *Ground state and excitation spectrum*

In Sec. 2.2 the ground state in the stripe phase has been described by means of an approximated wave function, based on the ansatz (8), which takes into account only first-order harmonic terms. The exact wave function includes also higher-order harmonics, whose appearance is a consequence of the nonlinearity of the Gross-Pitaevskii theory. It can be written in the form

$$\begin{pmatrix} \psi_{0\uparrow} \\ \psi_{0\downarrow} \end{pmatrix} = \sqrt{\bar{n}} \sum_{\bar{K}} \begin{pmatrix} a_{-k_1+\bar{K}} \\ -b_{-k_1+\bar{K}} \end{pmatrix} e^{i(\bar{K}-k_1)x}, \tag{60}$$

where $k_1 = \pi/d$ is related to the period d of the stripes, $\bar{K} = 2nk_1$, with $n = 0, \pm 1, \ldots$, are the reciprocal lattice vectors, while $a_{-k_1+\bar{K}}$ and $b_{-k_1+\bar{K}}$ are expansion coefficients to be determined, together with the value of k_1,

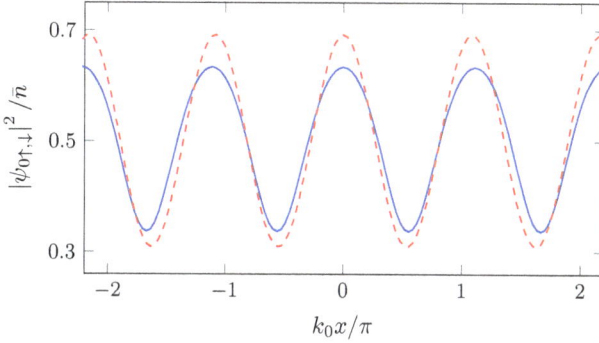

Fig. 11. Density profile in the stripe phase along the x direction, calculated within the first-order harmonic approximation (8) (red dashed line) and from Eq. (60) including the higher-order harmonics (blue solid line). The parameters are $\Omega/k_0^2 = 1.0$, $G_1/k_0^2 = 0.3$, and $G_2/k_0^2 = 0.08$, yielding the transition frequency $\Omega^{(I-II)}/k_0^2 \simeq 1.3$.

by a procedure of minimization of the mean-field energy functional (7). The energy minimization gives rise to the presence of terms with opposite phase ($e^{\pm i k_1 x}$, $e^{\pm 3 i k_1 x}$, ...), responsible for the density modulations and characterized by the symmetry condition $a_{-k_1+\bar{K}} = b^*_{k_1-\bar{K}}$, causing the vanishing of the spin polarization $\langle \sigma_z \rangle$. Figure 11 shows an example of density profile in the stripe phase, calculated for a configuration with relatively large values of G_2 and Ω/k_0^2 in order to emphasize the contrast in the density modulations.

As in the case of the uniform phases, also in the stripe phase we can evaluate the elementary excitations by the standard Bogoliubov approach, writing the deviations of the order parameter with respect to equilibrium as

$$\Psi = e^{-i\mu t}\left[\begin{pmatrix} \psi_{0\uparrow} \\ \psi_{0\downarrow} \end{pmatrix} + \begin{pmatrix} u_\uparrow(\mathbf{r}) \\ u_\downarrow(\mathbf{r}) \end{pmatrix} e^{-i\omega t} + \begin{pmatrix} v_\uparrow^*(\mathbf{r}) \\ v_\downarrow^*(\mathbf{r}) \end{pmatrix} e^{i\omega t}\right] \tag{61}$$

and solving the corresponding linearized time-dependent Gross-Pitaevskii equations. The equations are conveniently solved by expanding $u_{\uparrow,\downarrow}(\mathbf{r})$ and $v_{\uparrow,\downarrow}(\mathbf{r})$ in the Bloch form in terms of the reciprocal lattice vectors:

$$u_{\mathbf{q}\uparrow,\downarrow}(\mathbf{r}) = e^{-i k_1 x}\sum_{\bar{K}} U_{\mathbf{q}\uparrow,\downarrow\bar{K}}\, e^{i\mathbf{q}\cdot\mathbf{r}+i\bar{K}x}, \tag{62}$$

$$v_{\mathbf{q}\uparrow,\downarrow}(\mathbf{r}) = e^{i k_1 x}\sum_{\bar{K}} V_{\mathbf{q}\uparrow,\downarrow\bar{K}}\, e^{i\mathbf{q}\cdot\mathbf{r}-i\bar{K}x}, \tag{63}$$

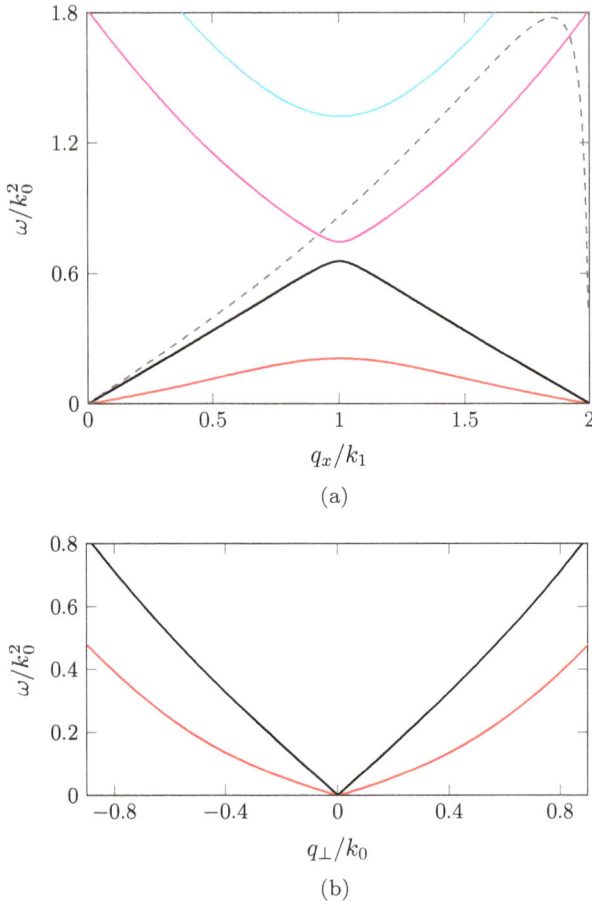

Fig. 12. (a). Lowest four excitation bands (solid lines) along the x direction ($q_\perp = 0$). The dashed line corresponds to the Feynman relation $\omega = q_x^2/2S(q_x)$. (b). Lowest two excitation bands in the transverse direction ($q_x = 0$). The parameters are the same as in Fig. 11.

where $\mathbf{q}$ is the wave vector of the excitation. This ansatz can also be used to calculate the density and spin-density dynamic response function, similarly to what we did in Sec. 3.1, by adding to the Hamiltonian a perturbation proportional to $e^{i(\mathbf{q}\cdot\mathbf{r}-\omega t)+\eta t}$ and $\sigma_z e^{i(\mathbf{q}\cdot\mathbf{r}-\omega t)+\eta t}$ with $\eta \to 0^+$, respectively.

The spectrum of the elementary excitations in the stripe phase is reported in Fig. 12 for the same parameters used in Fig. 11. We have considered both excitations propagating in the x direction orthogonal to the

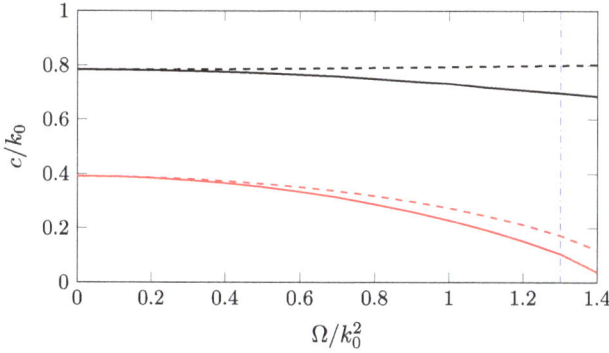

Fig. 13. Sound velocities in the first (red) and second (black) bands along the x (c_x, solid lines) and transverse ($c_\perp$, dashed lines) directions as a function of Ω. The blue dash-dotted line represents the transition from the stripe phase to the plane-wave phase. The values of the parameters G_1/k_0^2 and G_2/k_0^2 are the same as in Fig. 11.

stripes (labelled with the wave vector q_x) and in the transverse directions parallel to the stripes (identified by the wave vector $q_\perp$). A peculiar feature, distinguishing the stripe phase from the other uniform phases, is the occurrence of two gapless bands. The excitation energies along the x direction vanish at the Brillouin wave vector $q_B = 2k_1$, which is a usual situation in crystals. A similar double gapless band structure has been predicted recently in condensates with soft-core, finite-range interactions.[37,38,61]

In Fig. 13 we compare the sound velocities of the two gapless branches in the longitudinal (c_x) and transverse ($c_\perp$) directions. We find that c_x is always smaller than $c_\perp$, reflecting the inertia of the flow caused by the presence of the stripes. The value of $c_\perp$ in the second band (second sound) is well reproduced by the Bogoliubov expression $\sqrt{2G_1}$ (equal to $0.78\,k_0$ in our case) for the sound velocity. Notice that the sound velocity in the first band (first sound) becomes lower and lower as the Rabi frequency increases, approaching the transition to the plane-wave phase. The Bogoliubov solutions in the stripe phase exist also for values of Ω larger than the critical value $\Omega^{(I-II)} = 1.3\,k_0^2$, due to the first-order nature of the transition (effect of metastability).

5.2. *Static structure factor and static response function*

The nature of the excitation bands can be understood by calculating the static structure factors for the density and the spin-density operators, which

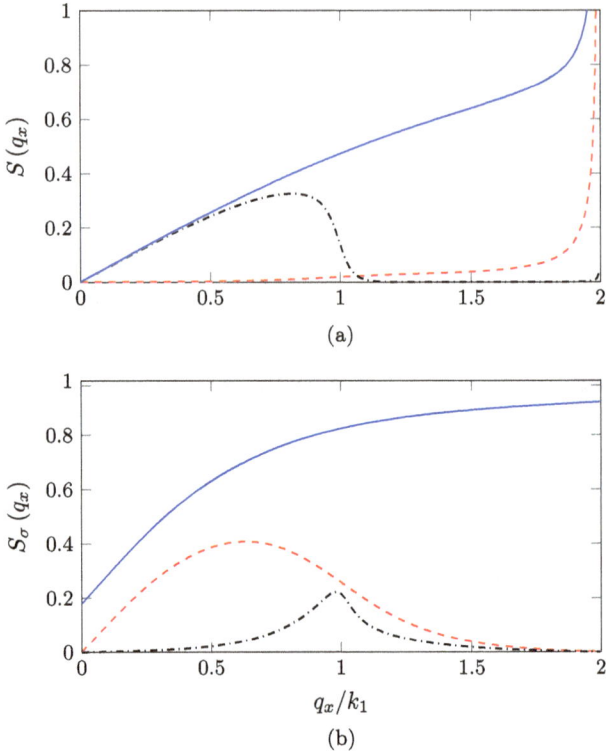

Fig. 14. Density (a) and spin-density (b) static structure factor as a function of q_x (blue solid line). The contributions of the first (red dashed line) and second (black dash-dotted line) bands are also shown. The parameters are the same as in Fig. 11.

can be written as

$$S(\mathbf{q}) = N^{-1} \sum_{\ell} |\langle 0|\rho_{\mathbf{q}}|\ell\rangle|^2 \tag{64}$$

and

$$S_\sigma(\mathbf{q}) = N^{-1} \sum_{\ell} |\langle 0|\sigma_{z,\mathbf{q}}|\ell\rangle|^2 \tag{65}$$

respectively. In these equations $\rho_{\mathbf{q}} = \sum_j e^{i\mathbf{q}\cdot\mathbf{r}_j}$ and $\sigma_{z,\mathbf{q}} = \sum_j \sigma_{z,j} e^{i\mathbf{q}\cdot\mathbf{r}_j}$ are the $\mathbf{q}$ components of the above-mentioned operators, while ℓ is the band index. In Fig. 14 we show the static structure factors for wave vectors along the x axis, as well as the contributions to the total sum coming from the two gapless branches ($\ell = 1, 2$). The figure clearly shows that, at small q_x, the lower branch is basically a spin excitation, while the upper branch is a density mode. The density nature of the upper branch, at small q_x, is further confirmed by the comparison with the Feynman relation $\omega = q_x^2/2S(q_x)$ (see Fig. 12(a)). A two-photon Bragg scattering experiment with laser frequencies far from resonance, being sensitive to the density response, will consequently excite only the upper branch at small q_x. Bragg scattering experiments actually measure the imaginary part of the response function, a quantity which, at enough low temperature, can be identified with the $T = 0$ value of the dynamic structure factor $S(q_x, \omega) = \sum_\ell |\langle 0|\rho_{q_x}|\ell\rangle|^2 \delta(\omega - \omega_{\ell 0})$, where $\omega_{\ell 0}$ is the excitation frequency of the ℓ-th state.[42] Notice that, differently from $S(q_x)$, the spin structure factor $S_\sigma(q_x)$ does not vanish as $q_x \to 0$, being affected by the higher energy bands as a consequence of the Raman term in Hamiltonian (2). As q_x increases, the lower branch actually reveals a hybrid character and, when approaching the Brillouin wave vector $q_B = 2k_1$, it is responsible for the divergent behavior of the density static structure factor (see Fig. 14(a)), which is again a typical feature exhibited by crystals.

It is worth pointing out that the occurrence of two gapless excitations is not by itself a signature of supersolidity and is exhibited also by uniform mixtures of BECs without spin-orbit and Raman couplings[62] as well as by the plane-wave phase of the Rashba Hamiltonian with $SU(2)$-invariant interactions ($G_2 = 0$).[63–65] Only the occurrence of a band structure, characterized by the vanishing of the excitation energy and by the divergent behavior of the structure factor at the Brillouin wave vector, can be considered an unambiguous evidence of the density modulations characterizing the stripe phase. The divergent behavior near the Brillouin zone is even more pronounced (see Fig. 15) if one investigates the static response function

$$\chi(q_x) = 2N^{-1} \sum_\ell \frac{|\langle 0|\rho_{q_x}|\ell\rangle|^2}{\omega_{\ell 0}}, \tag{66}$$

proportional to the inverse energy-weighted moment of the dynamic structure factor.

The divergent behaviors of $S(q_x)$ and $\chi(q_x)$ can be rigorously proven using the Bogoliubov[66] and the uncertainty principle[45,46] inequalities applied to systems with spontaneously broken continuous symmetries.

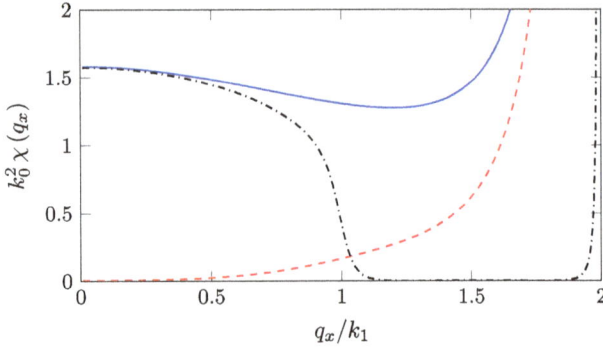

Fig. 15. Static response as a function of q_x (blue solid line). The contributions of the first (red dashed line) and second (black dash-dotted line) bands are also shown. The parameters are the same as in Fig. 11.

These inequalities are based, respectively, on the relations

$$m_{-1}(F)m_1(G) \geq |\langle [F, G] \rangle|^2 \tag{67}$$

and

$$m_0(F)m_0(G) \geq |\langle [F, G] \rangle|^2 \tag{68}$$

involving the k-th moments $m_k(\mathcal{O}) = \sum_{\ell}(|\langle 0|\mathcal{O}|\ell\rangle|^2 + |\langle 0|\mathcal{O}^{\dagger}|\ell\rangle|^2)\omega_{\ell 0}^k$ of the ℓ-th strengths of the operators $F = \sum_j e^{iq_x x_j}$ and $G = \sum_j (p_{x,j} e^{-i(q_x - q_B)x_j} + \text{H.c.})/2$, with $q_B = 2k_1$ the Brillouin wave vector defined above. The commutator $\langle [F, G] \rangle = q_x N \langle e^{iq_B x} \rangle$, entering the right-hand side of the inequalities, coincides with the relevant crystalline order parameter and is proportional to the density modulations of the stripes. The moments $m_{-1}(F)$ and $m_0(F)$ are instead proportional to the static response $\chi(q_x)$ and to the static structure factor $S(q_x)$, respectively. It is not difficult to show that the moments $m_1(G)$ and $m_0(G)$ are proportional, respectively, to $(q_x - q_B)^2$ and to $|q_x - q_B|$ as $q_x \to q_B$ due to the translational invariance of the Hamiltonian. This causes the divergent behaviors $S(q_x) \propto 1/|q_x - q_B|$ and $\chi(q_x) \propto 1/(q_x - q_B)^2$ with a weight factor proportional to the square of the order parameter. The value of the crystalline order parameter $\langle e^{iq_B x} \rangle$ is larger for larger values of Ω. For this reason it is useful to work with large values of the spin interaction parameter G_2, allowing for large values of the Raman coupling.[e] The experimental achievement of configurations with relatively large G_2 will be the subject of the next subsection.

5.3. *Experimental perspectives for the stripe phase*

As we have already anticipated, there is still no experimental evidence for the periodic modulations of the density profile in the stripe phase. The main reason is that, in the conditions of current experiments with spin-orbit-coupled ^{87}Rb BECs,[18,32] the contrast and the wavelength of the fringes are too small to be revealed. Another problem originates from the smallness of the difference $\Delta\mu$ between the chemical potentials in the plane-wave and the stripe phases, which, assuming $g_{\uparrow\uparrow} = g_{\downarrow\downarrow}$, is given by $\Delta\mu = 2G_2$ in the $\Omega = 0$ limit, and becomes even smaller at finite Ω. As a consequence, a tiny magnetic field (arising, for example, from external fluctuations) can easily bring the system into the spin-polarized plane-wave phase.

In Ref. 67 we have proposed a procedure to make the experimental detection of the fringes a realistic perspective, improving their contrast and their wavelength, and increasing the stability of the stripe phase against magnetic fluctuations. The idea is to trap the atomic gas in a 2D configuration, with tight confinement of the spin-up and spin-down components around two different positions, displaced by a distance d along the z direction. This configuration can be realized with a trapping potential of the form

$$V_{\text{ext}}(z) = \frac{\omega_z^2}{2}\left(z - \frac{d}{2}\sigma_z\right)^2 \tag{69}$$

with a sufficiently large value of ω_z. As a consequence of these trapping conditions, the overlap of the densities of the two spin components can be significantly quenched, and thus the effective interspecies coupling is reduced with respect to the intraspecies couplings. This yields a value of the parameter γ larger than in the $d = 0$ case, and consequently the critical Raman coupling $\Omega^{(\text{I--II})}$ can significantly increase (see Eq. (21)),

---

[e]For ^{87}Rb the value of G_2 is small and the divergency effect in $S(q_x)$ is weak. In this case, the sound velocity of the lowest band is small and the dispersion practically exhibits a q^2-like behavior at small q.

(a)

(b)

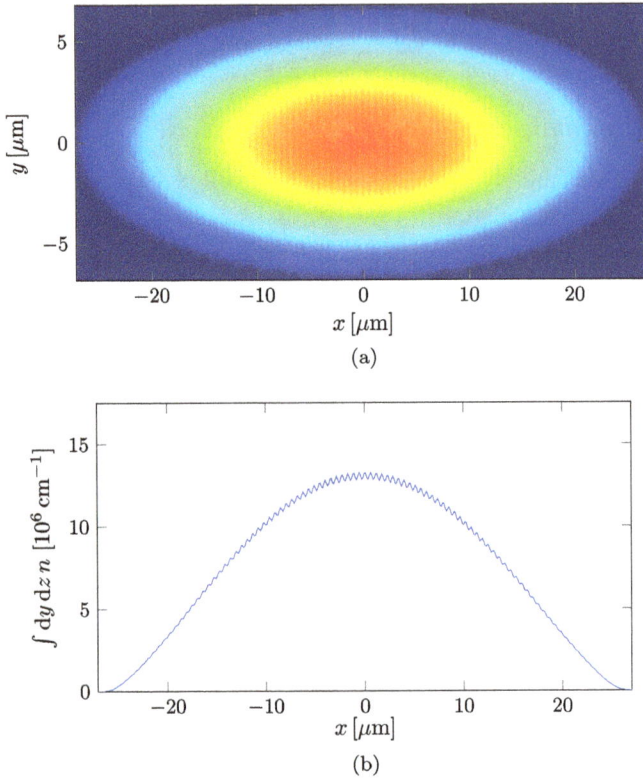

Fig. 16. Integrated density profiles $\int dz\, n$ (a) and $\int dy\, dz\, n$ (b) in the stripe phase, evaluated in the conditions described in the text, and without separation of the traps for the two spin components ($d = 0$).

allowing for the realization of the striped configuration with a high fringe contrast (15).[f]

Quantitative predictions for the novel configuration discussed above can be obtained by solving numerically the 3D Gross-Pitaevskii equation. In Figs. 16 and 17 we show the results for a gas of $N = 4 \times 10^4$ ^{87}Rb atoms confined by an harmonic potential with frequencies $\left(\omega_x, \omega_y, \omega_z\right) = 2\pi \times (25, 100, 2500)$ Hz, the scattering lengths equal to those reported in Sec. 2, $k_0 = 5.54\,\mu\mathrm{m}^{-1}$ and $E_r = h \times 1.77\,\mathrm{kHz}$ consistent with Ref. 18.

---

[f] Another important consequence is that, due to the increase of the value of γ, the critical density $n^{(c)}$ can be significantly lowered with respect to the value in the $d = 0$ case, becoming of more realistic achievement in future experiments.

(a)

(b)

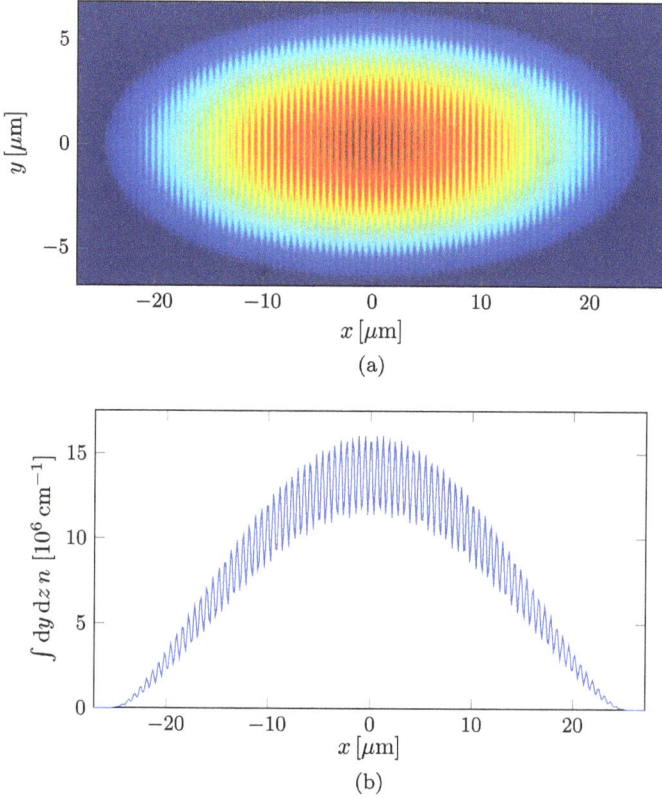

Fig. 17. Integrated density profiles $\int dz\, n$ (a) and $\int dy\, dz\, n$ (b) in the stripe phase, evaluated in the conditions described in the text, and with traps separated along z by a distance $d = a_z$, which helps increasing the visibility of the fringes with respect to Fig. 16.

Figure 16 corresponds to $d = 0$, while Fig. 17 corresponds to $d = a_z = 0.22\ \mu$m, a_z being the harmonic oscillator length along z. In both Figs. 16 and 17 we have chosen values of the Raman coupling equal to one half the critical value needed to enter the plane-wave phase, in order to ensure more stable conditions for the stripe phase. This corresponds to $\Omega = 0.095\ E_r$ in Fig. 16 and to $\Omega = 1.47\ E_r$ in Fig. 17. The density plotted in the top panels corresponds to the 2D density, obtained by integrating the full 3D density along the z direction; in the bottom panels we show the double integrated density $\int dy\, dz\, n$ as a function of the most relevant x variable. The figures clearly show that in the conditions of almost equal coupling constants (Fig. 16) the density modulations are very small, while their effect is strongly amplified in Fig. 17, where the interspecies coupling is reduced with respect to the intraspecies values.

The suggested procedure also has the positive effect of making the stripe phase more robust against fluctuations of external magnetic fields. Indeed, the reduction of the interspecies coupling and the increase of the local density, due to the tight axial confinement, yield a significant increase of the energy difference between the stripe and the plane-wave phases. For example, in the case considered above, for the configuration with a $d = a_z$ displacement of the two spin layers (Fig. 17) a magnetic detuning of about $0.35\ E_r$ is required to bring the system into the spin-polarized phase; in the absence of displacement (Fig. 16) the critical value for the magnetic detuning is instead much smaller ($\sim 0.001\ E_r$).

Let us finally address the problem of the small spatial separation of the fringes, given by π/k_1, which turns out to be of the order of a fraction of a micron in standard conditions. One possibility to increase the wavelength of the stripes is to lower the value of k_0 by using lasers with a smaller relative incident angle. In the following we discuss a more drastic procedure which consists of producing, after the realization of the stripe phase, a $\pi/2$ Bragg pulse with a short time duration (smaller than the time $1/E_r$ fixed by the recoil energy), followed by the sudden release of the trap. This pulse can transfer to the condensate a momentum k_B or $-k_B$ along the x direction, where k_B is chosen equal to $2k_1 - \epsilon$ with ϵ small compared to k_1. The $\pi/2$ pulse has the effect of splitting the condensate into various pieces, with different momenta. The situation is schematically shown in Fig. 18 for the

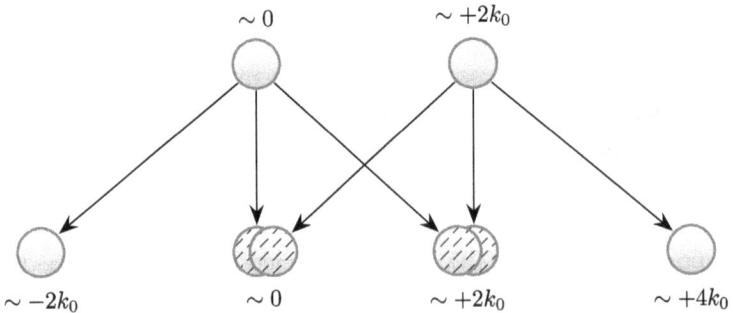

Fig. 18. Schematical description of the splitting of the spin-down component of the stripe wave function into different momentum components caused by a $\pi/2$ Bragg pulse transferring momentum $2k_1 - \epsilon$.

(a)

(b)

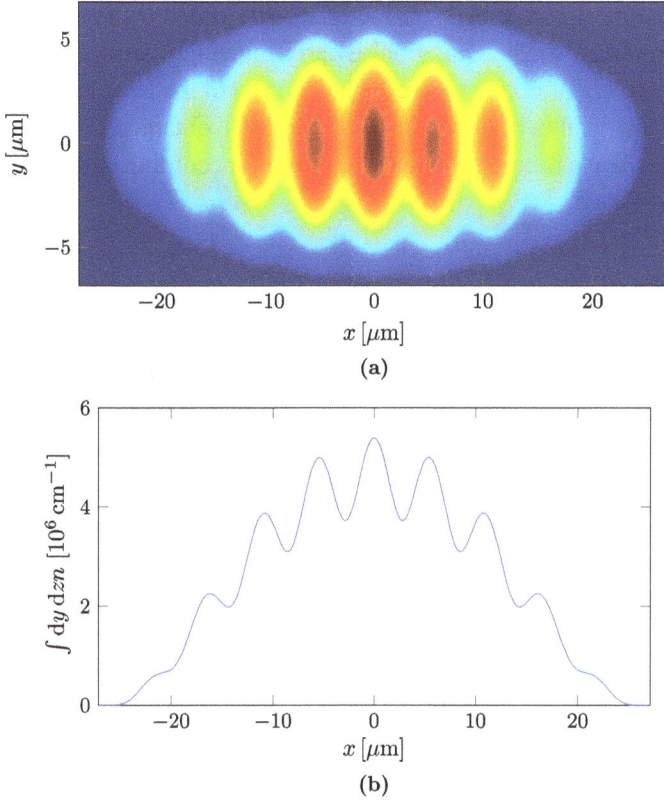

Fig. 19. Integrated density profiles $\int dz\, n$ (a) and $\int dy\, dz\, n$ (b) in the stripe phase, in the same conditions as Fig. 17, after the application of a $\pi/2$ Bragg pulse with transferred momentum $k_B = 1.8\, k_1$.

spin-down component, where the initial condensate wave function, which in the stripe phase is a linear combination with canonical momenta $\pm k_1$, corresponding to momenta $k_0 - k_1$ and $k_0 + k_1$ in the laboratory frame, after the Bragg pulse will be decomposed into six pieces. Two of them, those labeled in the lower part of the figure with momentum ~ 0, will be practically at rest after the pulse and are able to interfere with fringes of wavelength $2\pi/\epsilon$, which can easily become large and visible *in situ*. It is worth noticing that these two latter pieces originate from the two different momentum components of the order parameter (8) in the stripe phase and involve $1/3$ of the total number of atoms. The corresponding interference effect would be consequently absent in the plane-wave phase, where only one momentum component characterizes the order parameter. The other pieces produced by the Bragg pulse carry much higher momenta and will fly away rapidly after the release of the trap and of the laser fields. In Fig. 19 we show a typical behavior of the density profile obtained by modifying the condensate wave function in momentum space according to the prescription discussed above.

6. Conclusion

In this review we have illustrated some relevant static and dynamic properties of spin-orbit-coupled Bose-Einstein condensates in the simplest realization of a spin-$1/2$ configuration, characterized by equal Rashba and Dresselhaus couplings and vanishing or small magnetic detuning. The phase diagram of these Bose-Einstein condensates is characterized by the existence of three phases: the stripe, the plane-wave and the single-minimum phase. These phases merge in a characteristic tricritical point. The phase transition between the stripe and the plane-wave phase has a first-order nature, while the transition between the plane-wave and the single-minimum phase is of second order and is characterized by a divergent behavior of the magnetic polarizability. The stripe phase exhibits typical density modulations, which are the consequence of a mechanism of spontaneous breaking of translational invariance. The three phases discussed in the present paper exhibit interesting dynamical features, like the suppression of the dipole oscillation frequency in the presence of harmonic trapping and of the sound velocity close to the second-order phase transition, the appearance of a roton minimum in the plane-wave phase and the occurrence of a double gapless band structure in the excitation spectrum of the stripe phase. Some of these features have already been confirmed in recent experiments. Finally, we have discussed a procedure for the experimental exploration of the intriguing physics of the stripe phase, opening new perspectives for the identification of supersolid phenomena in ultracold atomic gases.

Acknowledgments

We wish to thank Lev P. Pitaevskii for many useful and stimulating discussions. This work has been supported by ERC through the QGBE grant and by Provincia Autonoma di Trento. The Centre for Quantum

Technologies is a Research Centre of Excellence funded by the Ministry of Education and National Research Foundation of Singapore.

A.1. Coefficients in the response function

The coefficients in the response function (33) can be expressed as follows. In phase II:

$$
a = -\frac{q^4}{4} + \left[\left(k_0^2 + 3k_1^2 \right) \cos^2 \alpha - 2 \left(k_0^2 - G_2 \right) + \frac{2G_2 k_1^2}{k_0^2} \right] q^2
$$

$$
+ 4 \left(k_0^2 - 2G_2 \right) \left[\left(k_0^2 - k_1^2 \right) \cos^2 \alpha - k_0^2 + \frac{2G_2 k_1^2}{k_0^2} \right],
$$

$$
b_0 = \frac{q^8}{16} - \left[\left(k_0^2 + k_1^2 \right) \cos^2 \alpha - k_0^2 - G_1 + G_2 \right] \frac{q^6}{2}
$$

$$
+ \left\{ \left(k_0^2 - k_1^2 \right)^2 \cos^4 \alpha \right.
$$

$$
- 2 \left[k_0^2 \left(k_0^2 - k_1^2 \right) + G_1 \left(k_0^2 + 3k_1^2 \right) - G_2 \left(k_0^2 - 5k_1^2 \right) \right] \cos^2 \alpha
$$

$$
\left. + k_0^2 \left(k_0^2 - 2G_2 \right) + 4G_1 \left(k_0^2 - G_2 \right) + 2 \left(k_0^2 - 2G_1 - 2G_2 \right) \frac{G_2 k_1^2}{k_0^2} \right\} q^4
$$

$$
- 8 \left(k_0^2 - 2G_2 \right) \left[\left(k_0^2 - k_1^2 \right) \left(G_1 + \frac{G_2 k_1^2}{k_0^2} \right) \cos^2 \alpha \right.
$$

$$
\left. - G_1 k_0^2 - \left(k_0^2 - 2G_1 - 2G_2 \right) \frac{G_2 k_1^2}{k_0^2} \right] q^2,
$$

$$
b_1 = q^4 + 4 \left[\left(k_0^2 - k_1^2 \right) \cos^2 \alpha + 2 \left(G_1 + G_2 \right) \right] q^2
$$

$$
+ 16 \left(k_0^2 - 2G_2 \right) \left(k_0^2 - k_1^2 \right) \frac{G_2}{k_0^2},
$$

$$
b_2 = -\frac{q^4}{2} - 2 \left[\left(k_0^2 - 3k_1^2 \right) \cos^2 \alpha + k_0^2 + G_1 - G_2 \right] q^2
$$

$$
- 4 \left(k_0^2 - 2G_2 \right) \left(k_0^2 - \frac{2G_2 k_1^2}{k_0^2} \right),
$$

with k_1 given by (17). In phase III:

$$a = -\frac{q^4}{4} - \left(\Omega - k_0^2 \cos^2 \alpha + 2G_2\right) q^2 - \Omega \left[\Omega - 2\left(k_0^2 \cos^2 \alpha - 2G_2\right)\right],$$

$$b_0 = \frac{q^8}{16} + \left[\Omega - 2\left(k_0^2 \cos^2 \alpha - G_1 - G_2\right)\right] \frac{q^6}{4}$$
$$+ \left[\Omega^2 - 4\left(k_0^2 \cos^2 \alpha - 2G_1 - G_2\right)\Omega \right.$$
$$\left. + 4\left(k_0^2 \cos^2 \alpha - 2G_1\right)\left(k_0^2 \cos^2 \alpha - 2G_2\right)\right] \frac{q^4}{4}$$
$$+ 2G_1\Omega \left[\Omega - 2\left(k_0^2 \cos^2 \alpha - 2G_2\right)\right] q^2,$$

$$b_1 = 0,$$

$$b_2 = -\frac{q^4}{2} - \left[\Omega + 2\left(k_0^2 \cos^2 \alpha + G_1 + G_2\right)\right] q^2 - \Omega \left(\Omega + 4G_2\right).$$

References

1. K. von Klitzing, The quantized Hall effect, *Rev. Mod. Phys.* **58**, 519 (Jul, 1986). doi: 10.1103/RevModPhys.58.519.
2. M. Z. Hasan and C. L. Kane, Colloquium, *Rev. Mod. Phys.* **82**, 3045 (Nov, 2010). doi: 10.1103/RevModPhys.82.3045.
3. X.-L. Qi and S.-C. Zhang, Topological insulators and superconductors, *Rev. Mod. Phys.* **83**, 1057 (Oct, 2011). doi: 10.1103/RevModPhys.83.1057.
4. F. Wilczek, Majorana returns, *Nat. Phys.* **5**, 614 (Sep, 2009). doi: 10.1038/nphys1380.
5. J. D. Koralek, C. P. Weber, J. Orenstein, B. A. Bernevig, S.-C. Zhang, S. Mack, and D. D. Awschalom, Emergence of the persistent spin helix in semiconductor quantum wells, *Nature*. **458**, 610 (Apr, 2009). doi: 10.1038/nature07871.
6. J. Dalibard, F. Gerbier, G. Juzeliūnas, and P. Öhberg, Colloquium, *Rev. Mod. Phys.* **83**, 1523 (Nov, 2011). doi: 10.1103/RevModPhys.83.1523.
7. M. V. Berry, Quantal Phase Factors Accompanying Adiabatic Changes, *Proc. R. Soc. A.* **392**(1802), 45 (1984). doi: 10.1098/rspa.1984.0023.
8. J. Ruseckas, G. Juzeliūnas, P. Öhberg, and M. Fleischhauer, Non-Abelian Gauge Potentials for Ultracold Atoms with Degenerate Dark States, *Phys. Rev. Lett.* **95**, 010404 (Jun, 2005). doi: 10.1103/PhysRevLett.95.010404.
9. S.-L. Zhu, H. Fu, C.-J. Wu, S.-C. Zhang, and L.-M. Duan, Spin Hall Effects for Cold Atoms in a Light-Induced Gauge Potential, *Phys. Rev. Lett.* **97**, 240401 (Dec, 2006). doi: 10.1103/PhysRevLett.97.240401.
10. K. J. Günter, M. Cheneau, T. Yefsah, S. P. Rath, and J. Dalibard, Practical scheme for a light-induced gauge field in an atomic Bose gas, *Phys. Rev. A.* **79**, 011604 (Jan, 2009). doi: 10.1103/PhysRevA.79.011604.

11. N. R. Cooper and Z. Hadzibabic, Measuring the Superfluid Fraction of an Ultracold Atomic Gas, *Phys. Rev. Lett.* **104**, 030401 (Jan, 2010). doi: 10.1103/PhysRevLett.104. 030401.

12. P. Hauke, O. Tieleman, A. Celi, C. Ölschläger, J. Simonet, J. Struck, M. Weinberg, P. Windpassinger, K. Sengstock, M. Lewenstein, and A. Eckardt, Non-Abelian Gauge Fields and Topological Insulators in Shaken Optical Lattices, *Phys. Rev. Lett.* **109**, 145301 (Oct, 2012). doi: 10.1103/PhysRevLett.109.145301.

13. I. B. Spielman, Raman processes and effective gauge potentials, *Phys. Rev. A.* **79**, 063613 (Jun, 2009). doi: 10.1103/PhysRevA.79.063613.

14. Y.-J. Lin, R. L. Compton, A. R. Perry, W. D. Phillips, J. V. Porto, and I. B. Spielman, Bose-Einstein Condensate in a Uniform Light-Induced Vector Potential, *Phys. Rev. Lett.* **102**, 130401 (Mar, 2009). doi: 10.1103/PhysRevLett.102.130401.

15. Y.-J. Lin, R. L. Compton, K. Jimenez-Garcia, J. V. Porto, and I. B. Spielman, Synthetic magnetic fields for ultracold neutral atoms, *Nature.* **462**, 628 (Dec, 2009). doi: 10.1038/ nature08609.

16. J. Struck, C. Ölschläger, R. Le Targat, P. Soltan-Panahi, A. Eckardt, M. Lewenstein, P. Windpassinger, and K. Sengstock, Quantum Simulation of Frustrated Classical Magnetism in Triangular Optical Lattices, *Science.* **333**(6045), 996 (2011). doi: 10.1126/science.1207239.

17. C. V. Parker, L.-C. Ha, and C. Chin, Majorana returns, *Nat. Phys.* **9**, 769 (Dec, 2013). doi: 10.1038/nphys2789.

18. Y.-J. Lin, K. Jimenez-Garcia, and I. B. Spielman, Spin-orbit-coupled Bose-Einstein condensates, *Nature.* **471**, 83 (Mar, 2011). doi: 10.1038/nature09887.

19. Y. A. Bychkov and E. I. Rashba, Oscillatory effects and the magnetic susceptibility of carriers in inversion layers, *J. Phys. C: Solid State Phys.* **17**(33), 6039 (1984). doi: 0.1088/0022-3719/17/33/015.

20. G. Dresselhaus, Spin-Orbit Coupling Effects in Zinc Blende Structures, *Phys. Rev.* **100**, 580 (Oct, 1955). doi: 10.1103/PhysRev.100.580.

21. P. Wang, Z.-Q. Yu, Z. Fu, J. Miao, L. Huang, S. Chai, H. Zhai, and J. Zhang, Spin-Orbit Coupled Degenerate Fermi Gases, *Phys. Rev. Lett.* **109**, 095301 (Aug, 2012). doi: 10.1103/PhysRevLett.109.095301.

22. L. W. Cheuk, A. T. Sommer, Z. Hadzibabic, T. Yefsah, W. S. Bakr, and M. W. Zwierlein, Spin-Injection Spectroscopy of a Spin-Orbit Coupled Fermi Gas, *Phys. Rev. Lett.* **109**, 095302 (Aug, 2012). doi: 10.1103/PhysRevLett.109.095302.

23. V. Galitski and I. B. Spielman, Spin-orbit coupling in quantum gases, *Nature.* **494**, 49 (Feb, 2013). doi: 10.1038/nature11841.

24. N. Goldman, G. Juzeliūnas, P. Öhberg, and I. B. Spielman, Light-induced gauge fields for ultracold atoms, *Rep. Progr. Phys.* **77**(12), 126401 (Nov, 2014). doi: 10.1088/0034-4885/77/12/126401.

25. H. Zhai, Degenerate quantum gases with spin-orbit coupling: a review, *Rep. Progr. Phys.* **78**(2), 026001 (Feb, 2015). doi: 10.1088/0034-4885/78/2/026001.

26. W. Zheng, B. Liu, J. Miao, C. Chin, and H. Zhai, Strong Interaction Effects and Criticality of Bosons in Shaken Optical Lattices, *Phys. Rev. Lett.* **113**, 155303 (Oct, 2014). doi:10.1103/PhysRevLett.113.155303.

27. Y. Li, L. P. Pitaevskii, and S. Stringari, Quantum Tricriticality and Phase Transitions in Spin-Orbit Coupled Bose-Einstein Condensates, *Phys. Rev. Lett.* **108**, 225301 (May, 2012). doi: 10.1103/PhysRevLett.108.225301.

28. M. Boninsegni and N. V. Prokof'ev, Colloquium, *Rev. Mod. Phys.* **84**, 759 (May, 2012). doi: 10.1103/RevModPhys.84.759.

29. T.-L. Ho and S. Zhang, Bose-Einstein Condensates with Spin-Orbit Interaction, *Phys. Rev. Lett.* **107**, 150403 (Oct, 2011). doi: 10.1103/PhysRevLett.107.150403.

30. Y. Li, G. I. Martone, and S. Stringari, Sum rules, dipole oscillation and spin polarizability of a spin-orbit coupled quantum gas, *EPL.* **99**(5), 56008 (2012).

31. L. D. Landau and E. M. Lifshitz, *Statistical Physics, Part 1*, 3rd edn. Oxford, Pergamon (1980).

32. J.-Y. Zhang, S.-C. Ji, Z. Chen, L. Zhang, Z.-D. Du, B. Yan, G.-S. Pan, B. Zhao, Y.-J. Deng, H. Zhai, S. Chen, and J.-W. Pan, Collective Dipole Oscillations of a Spin-Orbit Coupled Bose-Einstein Condensate, *Phys. Rev. Lett.* **109**, 115301 (Sep, 2012). doi: 10.1103/PhysRevLett.109.115301.

33. W. Zheng and Z. Li, Collective modes of a spin-orbit-coupled Bose-Einstein condensate: A hydrodynamic approach, *Phys. Rev. A.* **85**, 053607 (May, 2012). doi: 10.1103/Phys RevA. 85.053607.

34. G. I. Martone, Y. Li, L. P. Pitaevskii, and S. Stringari, Anisotropic dynamics of a spin-orbit-coupled Bose-Einstein condensate, *Phys. Rev. A.* **86**, 063621 (Dec, 2012). doi: 10.1103/PhysRevA.86.063621.

35. Y. Li, G. I. Martone, L. P. Pitaevskii, and S. Stringari, Superstripes and the Excitation Spectrum of a Spin-Orbit-Coupled Bose-Einstein Condensate, *Phys. Rev. Lett.* **110**, 235302 (Jun, 2013). doi: 10.1103/PhysRevLett.110.235302.

36. L. Santos, G. V. Shlyapnikov, and M. Lewenstein, Roton-Maxon Spectrum and Stability of Trapped Dipolar Bose-Einstein Condensates, *Phys. Rev. Lett.* **90**, 250403 (Jun, 2003). doi: 10.1103/PhysRevLett.90.250403.

37. T. Macrì, F. Maucher, F. Cinti, and T. Pohl, Elementary excitations of ultracold soft-core bosons across the superfluid-supersolid phase transition, *Phys. Rev. A.* **87**, 061602 (Jun, 2013). doi: 10.1103/PhysRevA.87.061602.

38. S. Saccani, S. Moroni, and M. Boninsegni, Excitation Spectrum of a Supersolid, *Phys. Rev. Lett.* **108**, 175301 (Apr, 2012). doi: 10.1103/PhysRevLett.108.175301.

39. S.-C. J. Ji, L. Zhang, X.-T. Xu, Z. Wu, Y. Deng, S. Chen, and J.-W. Pan, Softening of Roton and Phonon Modes in a Bose-Einstein Condensate with Spin-Orbit Coupling, *arXiv:1408.1755* (2014).

40. M. A. Khamehchi, Y. Zhang, C. Hamner, T. Busch, and P. Engels, Measurement of collective excitations in a spin-orbit-coupled Bose-Einstein condensate, *Phys. Rev. A* **90**, 063624 (Dec, 2014). doi: 10.1103/PhysRevA.90.063624.

41. L.-C. Ha, L. W. Clark, C. V. Parker, B. M. Anderson, and C. Chin, Roton-Maxon Excitation Spectrum of Bose Condensates in a Shaken Optical Lattice, *Phys. Rev. Lett.* **114**, 055301 (Feb, 2015). doi: 10.1103/PhysRevLett.114.055301.

42. L. P. Pitaevskii and S. Stringari, *Bose-Einstein Condensation*. Oxford University Press, New York (2003).

43. M. Klawunn, A. Recati, L. P. Pitaevskii, and S. Stringari, Local atom-number fluctuations in quantum gases at finite temperature, *Phys. Rev. A.* **84**, 033612 (Sep, 2011). doi: 10.1103/PhysRevA.84.033612.

44. S. Hoinka, M. Lingham, M. Delehaye, and C. J. Vale, Dynamic Spin Response of a Strongly Interacting Fermi Gas, *Phys. Rev. Lett.* **109**, 050403 (Aug, 2012). doi: 10.1103/PhysRevLett.109.050403.

45. L. P. Pitaevskii and S. Stringari, Uncertainty principle, quantum fluctuations, and broken symmetries, *J. Low Temp. Phys.* **85**(5–6), 377 (1991). ISSN 0022-2291. doi: 10.1007/BF00682193.

46. L. P. Pitaevskii and S. Stringari, Uncertainty principle and off-diagonal long-range order in the fractional quantum Hall effect, *Phys. Rev. B.* **47**, 10915 (Apr, 1993). doi: 10.1103/PhysRevB.47.10915.

47. D. M. Stamper-Kurn, H.-J. Miesner, S. Inouye, M. R. Andrews, and W. Ketterle, Collisionless and Hydrodynamic Excitations of a Bose-Einstein Condensate, *Phys. Rev. Lett.* **81**, 500 (Jul, 1998). doi: 10.1103/PhysRevLett.81.500.

48. T. Ozawa, L. P. Pitaevskii, and S. Stringari, Supercurrent and dynamical instability of spin-orbit-coupled ultracold Bose gases, *Phys. Rev. A.* **87**, 063610 (Jun, 2013). doi: 10.1103/PhysRevA.87.063610.

49. C. Wang, C. Gao, C.-M. Jian, and H. Zhai, Spin-Orbit Coupled Spinor Bose-Einstein Condensates, *Phys. Rev. Lett.* **105**, 160403 (Oct, 2010). doi: 10.1103/PhysRevLett.105.160403.

50. C.-J. Wu, I. Mondragon-Shem, and X.-F. Zhou, Unconventional Bose-Einstein Condensations from Spin-Orbit Coupling, *Chin. Phys. Lett.* **28**(9), 97102 (2011). doi: 10.1088/0256-307X/28/9/097102.

51. S. Sinha, R. Nath, and L. Santos, Trapped Two-Dimensional Condensates with Synthetic Spin-Orbit Coupling, *Phys. Rev. Lett.* **107**, 270401 (Dec, 2011). doi: 10.1103/PhysRevLett.107.270401.

52. T. Ozawa and G. Baym, Striped states in weakly trapped ultracold Bose gases with Rashba spin-orbit coupling, *Phys. Rev. A.* **85**, 063623 (Jun, 2012). doi: 10.1103/PhysRevA.85.063623.

53. D. A. Zezyulin, R. Driben, V. V. Konotop, and B. A. Malomed, Nonlinear modes in binary bosonic condensates with pseudo–spin-orbital coupling, *Phys. Rev. A.* **88**, 013607 (Jul, 2013). doi: 10.1103/PhysRevA.88.013607.

54. Z. Lan and P. Öhberg, Raman-dressed spin-1 spin-orbit-coupled quantum gas, *Phys. Rev. A.* **89**, 023630 (Feb, 2014). doi: 10.1103/PhysRevA.89.023630.

55. Q. Sun, L. Wen, W.-M. Liu, G. Juzeliūnas, and A.-C. Ji, Quantum Phase Transitions of Bilayer Spin-orbit Coupled Bose-Einstein Condensates, *arXiv:1403.4338* (2014).

56. W. Han, G. Juzeliūnas, W. Zhang, and W.-M. Liu, Supersolid with nontrivial topological spin textures in spin-orbit-coupled Bose gases, *Phys. Rev. A* **91**, 013607 (Jan, 2015). doi: 10.1103/PhysRevA.91.013607.

57. C. Hickey and A. Paramekanti, Thermal Phase Transitions of Strongly Correlated Bosons with Spin-Orbit Coupling, *Phys. Rev. Lett.* **113**, 265302 (Dec, 2014). doi: 10.1103/PhysRevLett.113.265302.

58. A. F. Andreev and I. M. Lifshitz, Quantum Theory of Defects in Crystals, *JETP.* **29**, 1107 (1969).

59. A. J. Leggett, Can a Solid Be "Superfluid"?, *Phys. Rev. Lett.* **25**, 1543 (Nov, 1970). doi: 10.1103/PhysRevLett.25.1543.

60. G. V. Chester, Speculations on Bose-Einstein Condensation and Quantum Crystals, *Phys. Rev. A.* **2**, 256 (Jul, 1970). doi: 10.1103/PhysRevA.2.256.

61. M. Kunimi and Y. Kato, Mean-field and stability analyses of two-dimensional flowing soft-core bosons modeling a supersolid, *Phys. Rev. B.* **86**, 060510 (Aug, 2012). doi: 10.1103/PhysRevB.86.060510.
62. C. J. Pethick and H. Smith, *Bose-Einstein Condensation in Dilute Gases*, 2 edn. Cambridge University Press, New York (2003).
63. R. Barnett, S. Powell, T. Graß, M. Lewenstein, and S. Das Sarma, Order by disorder in spin-orbit-coupled Bose-Einstein condensates, *Phys. Rev. A.* **85**, 023615 (Feb, 2012). doi: 10.1103/PhysRevA.85.023615.
64. X.-Q. Xu and J. H. Han, Emergence of Chiral Magnetism in Spinor Bose-Einstein Condensates with Rashba Coupling, *Phys. Rev. Lett.* **108**, 185301 (Apr, 2012). doi: 10.1103/PhysRevLett.108.185301.
65. R. Liao, Z.-G. Huang, X.-M. Lin, and W.-M. Liu, Ground-state properties of spin-orbit-coupled Bose gases for arbitrary interactions, *Phys. Rev. A.* **87**, 043605 (Apr, 2013). doi: 10.1103/PhysRevA.87.043605.
66. N. N. Bogoliubov, Quasimittelwerte in Problemen der statistischen Mechanik, *Phys. Abh. SU.* **6**, 1 (1962).
67. G. I. Martone, Y. Li, and S. Stringari, Approach for making visible and stable stripes in a spin-orbit-coupled Bose-Einstein superfluid, *Phys. Rev. A* **90**, 041604(R) (Oct, 2014). doi:10.1103/PhysRevA.90.041604.

INDEX

www.ingramcontent.com/pod-product-compliance
Lightning Source LLC
Chambersburg PA
CBHW050551190326
41458CB00007B/2003